Lecture Notes in Computer Science

Lecture Notes in Artificial Intelligence   13924

Founding Editor

Jörg Siekmann

Series Editors

Randy Goebel, *University of Alberta, Edmonton, Canada*
Wolfgang Wahlster, *DFKI, Berlin, Germany*
Zhi-Hua Zhou, *Nanjing University, Nanjing, China*

The series Lecture Notes in Artificial Intelligence (LNAI) was established in 1988 as a topical subseries of LNCS devoted to artificial intelligence.

The series publishes state-of-the-art research results at a high level. As with the LNCS mother series, the mission of the series is to serve the international R & D community by providing an invaluable service, mainly focused on the publication of conference and workshop proceedings and postproceedings.

Seifedine Kadry · Rajendra Prasath

Editors

Mining Intelligence and Knowledge Exploration

9th International Conference, MIKE 2023
Kristiansand, Norway, June 28–30, 2023
Proceedings

 Springer

Editors
Seifedine Kadry ⒾⒹ
Noroff University College
Kristiansand, Norway

Rajendra Prasath ⒾⒹ
Indian Institute of Information Technology
Sri City, India

ISSN 0302-9743   ISSN 1611-3349 (electronic)
Lecture Notes in Artificial Intelligence
ISBN 978-3-031-44083-0   ISBN 978-3-031-44084-7 (eBook)
https://doi.org/10.1007/978-3-031-44084-7

LNCS Sublibrary: SL7 – Artificial Intelligence

This Springer imprint is published by the registered company Springer Nature Switzerland AG
The registered company address is: Gewerbestrasse 11, 6330 Cham, Switzerland

Paper in this product is recyclable.

Preface

This volume contains the revised collection of papers presented at MIKE 2023: The 9th International Conference on Mining Intelligence and Knowledge Exploration, held during June 28–30, 2023, as a virtual event hosted by Noroff University College, Kristiansand, Norway. The main conference website is http://www.mike.org.in/2023/. MIKE 2023 received 87 qualified submissions from 12 countries and each qualified submission was reviewed by a minimum of two Program Committee members using the criteria of relevance, originality, technical quality, and presentation. A rigorous double-blind review process with the help of our distinguished program committee finally recommended 16 of those submissions for acceptance for oral presentation at the conference. Hence, the acceptance rate of full papers for this edition of MIKE was 25.29% and the acceptance rate of the promising contributions accepted as short papers was 18.39%.

The International Conference on Mining Intelligence and Knowledge Exploration (MIKE) is an initiative focusing on research and applications to various topics of human intelligence mining and knowledge discovery. Human intelligence has evolved steadily over many generations, and today human expertise excels in multiple domains and in knowledge-acquiring artifacts. The primary goal was to focus on the frontiers of human intelligence mining toward building a body of knowledge in this key domain. The focus was also to present state-of-the-art scientific results, to disseminate modern technologies, and to promote collaborative research in mining intelligence and knowledge exploration.

MIKE 2023 identified 10 tracks by topic, each led by 2-3 track coordinators to contribute and also to handle submissions falling in their areas of interest. The involvement of each of them along with the supervision of the Program Chairs ensured selection of quality papers for the conference. Each track coordinator took responsibility to fulfil the tasks assigned to him since we started circulating the first call for papers. This was reflected in every paper appearing in the proceedings with an impact in terms of the quality of the submissions.

The accepted papers were chosen on the basis of research excellence, which provides a body of literature for researchers involved in exploring, developing, and validating learning algorithms and knowledge-discovery techniques. Accepted papers were grouped into various subtopics including Knowledge Exploration in IoT, Medical Informatics, Machine Learning, Text Mining, Natural Language Processing, Cryptocurrency & Blockchain, and Application of Artificial Intelligence, and other areas not included in the above list. Researchers presented their work and had an excellent opportunity to interact with eminent professors and scholars in their area of research. All participants benefited from discussions that facilitated the emergence of new ideas and approaches.

We sincerely express our gratitude to Chaman Lal Sabharwal, Missouri University of Science and Technology, USA for his continued support of MIKE 2023. Several eminent scholars, including N. Subba Reddy, Gyeongsang National University, South

Korea, also extended their kind support in guiding us to organise the MIKE conference even better than the previous edition.

We express our sincere thanks to our Program Chairs Mazin Abed Mohammed from University of Anbar, Iraq; Venkatesan Rajinikanth from Saveetha School of Engineering, India; Rubén González Crespo from Universidad Internacional de La Rioja, Spain; and Muhammad Sharif from COMSATS University Islamabad, Pakistan for their guidance, suggestions and constant support, which were crucial in planning various activities of MIKE. We thank our workshop chair Sahar Yassine from Noroff University College, Norway for her support. Appreciations are due to Ricardo S. Alonso Rincón from Universidad Internacional de la Rioja, Spain for his support as the publicity chair. The unconditional support extended by all local organising committee members from Noroff University College, Norway, including Isah Lawal; Sahar Yassine; and Fabricio Bortoluzzi, is gratefully appreciated.

The first day of the conference started with the NUC Rector talk and then the opening remarks Seifedine Kadry from NUC, Norway and Rajendra Prasath from Indian Institute of Information Technology Sri City. Then three technical sessions took place in succession. The second-day events started with the keynote speech of Kiran Raja from the Norwegian University of Science and Technology (NTNU), Norway on Wireless Capsule Endoscopy for Early Pathology Detection, and thee more sessions took place after this keynote speech; and finally the third-day events started with the invited talk of Pierre Lison, from the Norwegian Computing Center, Norway on Privacy-Enhancing Natural Language Processing. Regular sessions and a special Hands-On Session on Unveiling the Black-Box: An Introduction to Explainable AI in Computer Vision, delivered by Julia El Zini.

A large number of eminent professors, well-known scholars, industry leaders, and young researchers participated in making MIKE 2023 a great success. We recognize and appreciate the hard work put in by each individual author of the articles published in these proceedings. We also express our sincere thanks to Noroff University College, Kristiansand, Norway for hosting MIKE 2023.

We thank the Technical Program Committee members and all reviewers for their timely and thorough participation in the reviewing process. We appreciate the time and effort put in by the members of the local organizing team at Noroff University College, Kristiansand, Norway and IIIT Sri City, India. We are very grateful to all our sponsors for their generous support to MIKE 2023.

Finally, we acknowledge the support of EasyChair in the submission and review processes.

We are very much pleased to express our sincere thanks to the Springer staff, for their support in publishing the proceedings of MIKE 2023.

June 2023

Seifedine Kadry
Rajendra Prasath

Organization

Advisory Committee

Adrian Groza	Technical University of Cluj-Napoca, Romania
Agnar Aamodt	Norwegian University of Science and Technology, Norway
Aidan Duane	Waterford Institute of Technology, Ireland
Alexander Gelbukh	Instituto Politécnico Nacional, Mexico
Amit A. Nanavati	IBM India Research Labs, India
Anil Vuppala	IIIT Hyderabad, India
Ashish Ghosh	Indian Statistical Institute, Kolkata, India
B. Yegnanarayana	IIIT Hyderabad, India
Bjorn Gamback	NTNU, Norway
Chaman Lal Sabharwal	Missouri University of Science and Technology, USA
Debi Prosad Dogra	Indian Institute of Technology, Bhubaneswar, India
Genoveva Vargas-Solar	CNRS, France
Grigori Sidorov	CIC - IPN, Mexico
Hrishikesh Venkataraman	IIIT Sri City, India
Iain Sutherland	Noroff University College, Norway
Ildar Batyrshin	Instituto Politécnico Nacional, Mexico
Kazi Shah Nawaz Ripon	Østfold University College, Norway
Krishnaiyya Jallu	IIITDM Kurnool, India
Mandar Mitra	Indian Statistical Institute, Kolkata, India
Manish Shrivastava	IIIT Hyderabad, India
Maunendra S. Desarkar	Indian Institute of Technology, Hyderabad, India
N. Subba Reddy	Gyeongsang National University, South Korea
Niloy Ganguly	Indian Institute of Technology, Kharagpur, India
Nirmalie Wiratunga	Robert Gordon University, UK
P. V. Rajkumar	Texas Southern University, USA
Paolo Rosso	Universitat Politecnica de Valancia, Spain
Philip O'Reilly	University College Cork, Ireland
Pinar Ozturk	NTNU, Norway
Radu Grosu	Vienna University of Technology, Austria
Rajarshi Pal	IDRBT, India
Ramon Lopez de Mantaras	IIIA - CSIC, Spain
Saurav Karmakar	GreyKarma Technologies, India

Sudeshna Sarkar	Indian Institute of Technology, Kharagpur, India
Sudip Misra	Indian Institute of Technology, Kharagpur, India
Susmita Ghosh	Jadavpur University, India
T. Kathirvalavakumar	VHNSN College (Autonomous), India
Tanmoy Chakraborty	IIIT Delhi, India
Tapio Saramäki	Tampere University of Technology, Finland
V. Ravi	IDRBT, India
Vasile Rus	University of Memphis, USA
Vasudeva Verma	IIIT Hyderabad, India
Yannis Stylianou	University of Crete, Greece

Technical Program Committee

Alexander Gelbukh	Instituto Politécnico Nacional, Mexico
Alexander Ryjov	Moscow State University, Russia
Ali Kalakech	Lebanese University, Lebanon
Anca Hangan	Technical University of Cluj-Napoca, Romania
Anca Marginean	Technical University of Cluj-Napoca, Romania
Aristidis Likas	University of Ioannina, Greece
Bernardete Ribeiro	University of Coimbra, Portugal
Birjodh Tiwana	LinkedIn Inc., USA
Bogdan Iancu	Technical University of Cluj-Napoca, Romania
Camelia Chira	Babeş-Bolyai University, Romania
Camelia Lemnaru	Technical University of Cluj-Napoca, Romania
Chaman Lal Sabharwal	Missouri University of Science and Technology, USA
Christos Georgiadis	University of Macedonia, Greece
Christos Schizas	University of Cyprus, Cyprus
Ciprian Oprisa	Technical University of Cluj-Napoca, Romania
Costin Badica	University of Craiova, Romania
Cristina Feier	University of Bremen, Germany
Debasis Ganguly	Dublin City University, Ireland
Denis Trcek	University of Ljubljana, Slovenia
Efthyvoulos Kyriacou	Frederick University, Cyprus
Eva Onaindia	Polytechnic University of Valencia, Spain
Farah Bouakrif	University of Jijel, Algeria
Florin Craciun	Babeş-Bolyai University, Cluj-Napoca, Romania
Florin Leon	Gheorghe Asachi Technical University of Iaşi, Romania
Frantisek Capkovic	Slovak Academy of Sciences, Slovakia
George Magoulas	Birkbeck, University of London, UK

George Tsekouras	University of the Aegean, Greece
Gheorghe Sebestyen	Technical University of Cluj-Napoca, Romania
Giorgio Gnecco	IMT School for Advanced Studies, Lucca, Italy
Gloria Inés Alvarez	Pontificia Universidad Javeriana Cali, Colombia
Goutham Reddy Alavalapati	Fontbonne University, USA
Hafiz Tayyab Rauf	University of Bradford, UK
Hakan Haberdar	University of Houston, USA
Hans Moen	University of Turku, Finland
Haralambos Mouratidis	University of Brighton, UK
Horia Cucu	University Politehnica of Bucharest, Romania
Ilias Sakellariou	University of Macedonia, Greece
Ioannis Anagnostopoulos	University of Thessaly, Greece
Ioannis Chamodrakas	University of Athens, Greece
Ioannis Hatzilygeroudis	University of Patras, Greece
Ioannis Karydis	Ionian University, Greece
Irina Dragoste	TU Dresden, Germany
Isidoros Perikos	University of Patras, Greece
Isis Bonet Cruz	EIA University, Colombia
Jacek Kabzinski	Technical University of Lodz, Poland
Jilin Hu	Aalborg University, Denmark
Jimmy Jose	National Institute of Technology, Calicut, India
Jose Maria Luna	University of Cordoba, Spain
Jose R. Villar	University of Oviedo, Spain
Juan Recio-Garcia	Universidad Complutense de Madrid, Spain
Katia Lida Kermanidis	Ionian University, Greece
Kazi Shah Nawaz Ripon	Østfold University College, Norway
Konstantinos Margaritis	University of Macedonia, Greece
Kostas Karpouzis	National Technical University of Athens, Greece
Kristína Machová	Technical University of Kosice, Slovakia
Lasker Ershad Ali	Peking University, China
Lasse Berntzen	University of South-Eastern Norway, Norway
Liviu P. Dinu	University of Bucharest, Romania
Maciej Ogrodniczuk	Institute of Computer Science, Polish Academy of Sciences, Poland
Manolis Maragoudakis	University of the Aegean, Greece
Marko Hölbl	University of Maribor, Slovenia
Martin Holena	Academy of Sciences, Czech Republic
Matteo Ceriotti	University of Duisburg-Essen, Germany
Mauro Gaggero	CNR, Italy
Michael Breza	Imperial College London, UK
Michel Aldanondo	Toulouse University, France
Mihaela Oprea	Petroleum-Gas University of Ploieşti, Romania

Contents

Multimodal Body Sensor for Recognizing the Human Activity Using DMOA Based FS with DL

M. Rudra Kumar[1], A. Likhitha[2(✉)], A. Komali[1,2], D. Keerthana[1,2], and G. Gowthami[1,2]

[1] Department of CSE, G. Pullaiah College of Engineering and Technology, Kurnool, India
[2] G. Pullaiah College of Engineering and Technology, Kurnool, India
ambalalikhitha11@gmail.com

Abstract. The relevance of automated recognition of human behaviors or actions stems from the breadth of its potential uses, which includes, but is not limited to, surveillance, robots, and personal health monitoring. Several computer vision-based approaches for identifying human activity in RGB and depth camera footage have emerged in recent years. Techniques including space-time trajectories, motion indoctrination, key pose extraction, tenancy patterns in 3D space, motion maps in depth, and skeleton joints are all part of the mix. These camera-based methods can only be used inside a constrained area and are vulnerable to changes in lighting and clutter in the backdrop. Although wearable inertial sensors offer a potential answer to these issues, they are not without drawbacks, including a reliance on the user's knowledge of their precise location and orientation. Several sensing modalities are being used for reliable human action detection due to the complimentary nature of the data acquired from the sensors. This research therefore introduces a two-tiered hierarchical approach to activity recognition by employing a variety of wearable sensors. Dwarf mongoose optimization process is used to extract the handmade features and pick the best features (DMOA). It predicts the composite's behavior by emulating how DMO searches for food. The DMO hive is divided into an alpha group, scouts, and babysitters. Every community has a different strategy to corner the food supply. In this study, we tested out a number of different methods for video categorization and action identification, including ConvLSTM, LRCN and C3D. The projected human action recognition (HAR) framework is evaluated using the UTD-MHAD dataset, which is a multimodal collection of 27 different human activities that is available to the public. The suggested feature selection model for HAR is trained and tested using a variety of classifiers. It has been shown experimentally that the suggested technique outperforms in terms of recognition accuracy.

Keywords: Human action recognition · Dwarf mongoose optimization algorithm · Camera-based approaches · Key poses extraction · Convolutional Neural Network

1   Introduction

Ubiquitous sensing, which uses data collected by sensors placed strategically about a building, has become increasingly popular in recent years [1]. Wearable sensor research for (HAR) has exploded in recent years, thanks in large part to its widespread potential use in fields as diverse as sports, interactive gaming, healthcare, and other monitoring schemes. Multimodal HAR is best understood as a variation on the long-honoured series classification problem [2], wherein a sliding time window is used to partition incoming sensor data into discrete time intervals from which discriminative features may be extracted. Techniques, such as may be used to further distinguish each time frame [3]. In addition, it can be challenging for shallow learning to capture the essential elements of complicated actions, and feature selection is often a laborious process [5]. Studies into automatic feature extraction with little human effort are of paramount importance as a means of addressing the aforementioned issues. The field of multimodal HAR is shifting its focus from surface learning to deep learning at the moment [6].

To improve system performance and do away with the requirement for hand-crafted features, recent research in sensor-based HAR has focused heavily on deep learning, in which many layers are layered to build (DNNs) [7, 8]. In particular, the rich representation power of (CNNs) has substantially advanced the performance of HAR. DNNs will improve in performance as their model capacities for rich representation grow, but this will unavoidably increase the need for highly labelled data. Annotated or "ground truth labelled" training data is a source of difficulty for deep HAR identification [9, 10]. Annotating the ground truth requires the annotator to sift through raw sensor data and physically identify all activity instances. This is a time-consuming and costly process. As compared to data captured by other sensor modalities, such cameras, the time series data recorded by multimodal embedded sensors like accelerometers and gyroscopes is far more challenging to comprehend [11]. To effectively segment and classify a specific activity from a lengthy needs significant human work. Thus, while these DNN models can automatically extract relevant features for categorization, they still need precise truth, which would necessitate significantly more human work to provide an ideal training dataset for HAR in a supervised learning situation [12].

Due to its independence on the kind, distance, and arithmetical scale of distinct features derived from numerous sensory modalities [13], decision-level fusion has been the primary focus of existing research for multimodal HAR. Furthermore, the final for classification has fewer dimensions after decision-level fusion, and no post-processing of the retrieved features is required. Independent and stand-alone categorization choices pertaining to each sense modality, which are subsequently fused using some soft rule to generate the final conclusion, is the main shortcoming of the decision-level fusion [14]. On the other hand, feature-level fusion is useful for gathering features simultaneously from several sensors and integrating them to produce enough information for a sound judgement [15].

2 Related Works

Using machine and deep learning (DL) models, Pradhan and Srivastava [16] categorized multi-modal physiological inputs. Dahou et al. [17] provide a methodology to increase the performance of several applications using a wide variety of data kinds by addressing the large dimensionality of data transported via the SIoT system.

Islam et al. [18] recommend a fusion procedure for activity recognition using a multi-head (CNN) equipped with a Convolution Block Attention Module (CBAM) to process the visual data and a (ConvLSTM) to handle the time-sensitive multi-source sensor information. In order to evaluate and recover channel and spatial dimension attributes, the three CNN sub-architectures and CBAM for visual data are implemented.

Novel system architecture presented by Zhang et al. [19] consists of three parts: feature selection using an oppositional and chaos particle swarm optimization (OCPSO) algorithm, a multi-input (MI-1D-CNN) that takes advantage of signals, and deep decision fusion (DDF) that combines D-S evidence theory and entropy. Using the UCI HAR and WIDSM datasets, the suggested architecture is tested.

Using the combination of EEG and face video clips, Muhammad et al. [20] describe a multimodal emotion identification approach based on deep canonical correlation analysis (DCCA). We use a two-stage framework in which the first stage uses features extracted from a single modality to recognize emotions, and the second stage combines the highly correlated features from the two modalities and classifies the data. After fusing highly correlated data using a DCCA-based method, the SoftMax classifier was then used to categorize faces into one of three fundamental human emotion categories: joyful, neutral, or sad. The suggested method was explored using the MAHNOB-HCI and DEAP public datasets. The average accuracy of the experimental findings was 93.86% on the MAHNOB-HCI dataset and 91.54% on the DEAP dataset. By contrasting the proposed framework with other efforts, we were able to assess its competitiveness and provide justification for its exclusivity in the pursuit of this level of precision.

Human gait identification was the focus of Jahangir et al. [21].'s novel two-stream deep learning approach. In the first stage, we discussed a method for improving contrast by combining data from local and global filters. In the second stage, data augmentation is carried out to expand the dimensionality of the raw dataset (CASIA-B). Third, we use deep transfer learning using the supplemented dataset. Fourth, a serial-based method is utilized to combine the extracted features of the two streams; and fifth, an enhanced method is employed to further optimize the fusion. Eight different angles from the CASIA-B dataset were used in the experimentation procedure, with results of 97.3, 98.6, 97.7, 96.5, 92.9, 93.7, 94.7, and 91.2% accuracy. The results of head-to-head comparisons with SOTA methods revealed increased precision and decreased processing time.

3 Proposed System

We begin with a brief description of the experimental dataset, followed by a discussion of the methodology and metrics utilized in the experiments. We then detail how our suggested framework may be put into action. We conclude with a discussion of the qualitative results, which should give you some good ideas about the recommended approach.

3.1 Dataset and Implementation Details

The suggested technique was tested using the UTD-MHAD dataset, which is a publicly available multimodal HAR dataset consisting of 27 human activities performed by 8 people. Figure 1 shows a list of these activities along with several visual representations.

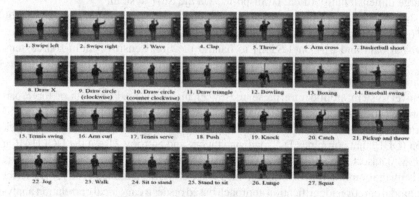

Fig. 1. Sample Images of the dataset [22]

Each participant performed each task four times. As a result, we have 8 participants × 4 trials per action × 27 actions totalling 864 trimmed data sequences. Three data sequences were corrupted during data recording, therefore after cleaning up the dataset, only 861 sequences remained. Both a Microsoft Kinect sensor (30 frames and a wearable inertial sensor (50 samples per second) were used to acquire the information in a controlled indoor environment. In order to record triaxial acceleration triaxial angular velocity, a Bluetooth-enabled hardware module was employed as a wearable inertial sensor (using a gyroscope). During activities 1–21, the participant wore the sensor on their right wrist; for actions 22–27, the sensor was attached to their right thigh. Each segmented action trial in the dataset is represented by four files, one for each of the four sensory modalities included in the dataset.

3.2 Feature Extraction

In particular, we made use of handmade features; the following sections outline each method in depth.

3.2.1 Handcrafted Features

Techniques for extracting features by hand are easy to implement and need less computing power. Simple statistical procedures more intricate frequency domain-based features, can be used to compute them on time series data. Table 1 summarizes the calculated characteristics and is followed by a detailed explanation of each. Each dimension of features received its own set of statistical calculations.

a) Extreme: Let X is the feature course. The $Max(X)$ function finds and revenues the largest feature value $x_i \in X$.

Table 1. List of Handcrafted Topographies

Skewness	Extreme
Norm of SOM	Percentile 50
Spectral energy	Percentile 80
Kurtosis	Minimum
Auto-correlation	Average
First-Order Mean (FOM)	Standard-deviation
Norm of FOM	Zero crossing
Second-order mean (SOM)	Percentile 20
Spectral entropy	Interquartile

b) Least: With X as input, the Min(X) function will locate the minimum story value (xi X) and return it.

c) Average: When there are N possible tale values, the average earnings are equal to the value in the middle of feature vector X. As in,

$$Average(X) = \mu = \frac{\sum_{i=1}^{n} x_i}{N} \tag{1}$$

d) Standard Deviation: It defines the amount of difference in feature vector $X = \{x1, x2 \ldots xN\}$ and can be calculated using the following preparation:

$$Stdev(X) = \sigma = \sqrt{\frac{1}{N} \sum_{i=1}^{N} (x_i - \mu)^2} \tag{2}$$

e) The frequency with which the signal value passes zero in either direction is an indicator of how quickly or slowly the activity is changing.

f) The term "percentile" is used to describe a score where a specified fraction of all possible responses fall below that value. The pth percentile is defined as the number at which no more than (100 p) % of the capacities are lower than this value and no more than 100(1 p) % are higher than this value. The 25th percentile, for instance, indicates that the value is larger than 25 other values but lower than 75 other feature values.

g) To calculate the interquartile range, use the difference among the first and third quartiles.

h) The skewness of a distribution is a measure of how far off centre the data of relative to the mean:

$$Sk = \frac{1}{N\sigma^3} \sum_{i=1}^{n} (x_i - \mu)^3 \tag{3}$$

i) Kurtosis: To what extent the distribution's tails deviate from the normal distribution's tails is measured by the statistic. A larger kurtosis number indicates that there are more extreme deviations, or outliers, in the data. It may be calculated mathematically as:

$$Kr = \frac{1}{N\sigma^4} \sum_{i=1}^{n} (x_i - \mu)^4 \qquad (4)$$

j) Auto-correlation: It is a statistical method for determining how closely one set of the time series data is related to its own lagged version across a range of time periods and it may be calculated as:

$$r_k = \frac{\sum_{i=1}^{N-k} (x_i - \mu)(x_{i+k} - \mu)}{\sum_{i=1}^{N} (x_i - \mu)^2} \qquad (5)$$

k) Order Mean Values: They are derived from the sorted list of numbers (in ascending order). That is, in an ordered collection of features X, the first ordered mean is just the smallest sample value X1, the second ordered mean is the value X2, and so on.

l) Norm Values: They help determine how far away from zero a feature vector actually is. There were two metrics we used: L1-norm.

m) Spectral Energy: We remember that in the recorded data numerous sensors are utilized to access the human actions; these sensors may be thought of as a function whose amplitude varies with time. The signal was changed from a time series to a frequency range using the Fourier transform, and the Spectral energy formulation was used to determine the energy levels at each frequency. The z value is the total amplitude squared of the frequencies present (n). As in,

$$S_E = \sum_{i=1}^{N} F(n)^2 \qquad (6)$$

n) It is also calculable using normalized frequency spectra. As in,

$$\hat{F}(n) = \frac{F(n)}{\sum_{i=1}^{N} F(n)} \qquad (7)$$

o) The normalized form of Eq. (6) is as follows:

$$NS_E = \sum_{i=1}^{N} \hat{F}(n)^2 \qquad (8)$$

p) Spectral Entropy: It is a way to quantify the spectral distribution of a signal in terms of frequency, and it is entropy. One possible mathematical description of spectral entropy is as follows:

$$S_{EN} = -\sum_{i=1}^{N} \hat{F}(n) \times log\hat{F}(n) \qquad (9)$$

These 18 features are the product of computations performed on each column in the features set for a single action and are joined together in a single row. Given that the input data is 61-dimensional, the resulting handmade feature will have a dimension of 1 (18 61) (or 1 1098). We tested the recognition accuracy of these generated features using a DL model.

3.3   Classification Using DL Models

3.3.1   Long-term Recurrent Convolutional Network (LRCN)

The goal of LRCN [23, 24] is to use convolution neural networks to extract spatial data from each frame. In order to categorise the data, the results from the convolutional networks are fed into a Bi-LSTM network, which combines the retrieved spatial characteristics with the temporal features. These models require an input size of 90 by 90 pixels. Convolutional filters are modelled as a matrix (in our example, 3×3 in size) with a random set of values that convolve across the picture and calculate the dot operation, and the output is then sent on to the next layer in the custom CNN model. Using convolution over k channels, the following Eqs. (10)–(13) summarise an input frame and provide a matrix as a result.

$$A_o^{(m)} = g_m(w_{ok}^{(m)} * A_k^{(m-1)} + b_o^{(m)}) \tag{10}$$

$$W - ok * A_k[s, t] = a_{p,q} * b \tag{11}$$

$$a = A_k[s + p, t + q] \tag{12}$$

$$b = w_{ok}[P - 1 - p, Q - 1 - q] \tag{13}$$

As per the above equations, max pooling is used to minimise the number of parameters after each convolutional layer in the network that lightens the convolutional burden. The rate of output is stable at this point. Our model used Rectified Linear Unit (ReLU), as seen in Eq. (16), and SoftMax, which converts a system's output into a probability distribution across projected classes. We imported an ImageNet-trained VGG-16 network and deleted the top layer to use its features in the VGG-LSTM model. Time-distributed layering was followed by a 256-filter bidirectional lstm, 256-filter ReLU-activated dense layering, and a final 2-neuron output layer.

$$y = A.x + b \tag{14}$$

$$y_i = \sum_{j=1}^{i} (A_{ij}, x_j) + b_i \tag{15}$$

$$y = \max(0, x) \tag{16}$$

$$y = A.x + b \tag{17}$$

$$y_i = \sum_{j=1}^{i} (A_{ij}x_j) + b_i \tag{18}$$

3.3.2 Convolutional Long Short-Term Memory (CLSTM) [25, 26]

While LSTM is limited to the temporal domain, we have also utilised ConvLSTM, which can be applied to the spatial domain. To do this, we employ ConvLSTM with spatially-oriented tensor inputs, cell outputs, hidden states, and gates. While both ConvLSTM and LSTM have a similar architecture, the two models diverge in how they handle transitions from input to state and from state to state. The activation function, convolution operator, and Hadamard product are respectively denoted by the symbols "", "," and "0" in the following Eqs. (19)–(24).

$$i_t = \sigma\left(w_{xi}x_t + w_{hiht-1} + w_{ci}{}^\circ c_{t-1} + b_i\right) \tag{19}$$

$$f_t = \sigma\left(w_{xf}x_t + w_{hfh}t - 1 + w_{cf}{}^\circ c_{t-1} + b_f\right) \tag{20}$$

$$\tilde{c}_t = \tanh(w_{xc}x_t + w_{hc}h_{t-1} + b_c) \tag{21}$$

$$c_t = f_t{}^\circ c_{t-1} + i_t{}^\circ c_t \tag{22}$$

$$o_t = \sigma(w_{xo}x_t + w_{h0}h_{t-1} + w_{co}\sigma c_t + b_o) \tag{23}$$

$$h_t = o_t \tanh(c_t) \tag{24}$$

The aforementioned formulas allow us to translate between the input value X_t and the output value Ht1 of the last neuron, where Ct1 is the current location. The convolution filter has a kernel size of k by k, where k is the dimension of kernel. In order to extract features from a movie, ConvLSTM reads in frames as input and performs a multidimensional convolution operation on each frame. To extract features more efficiently than the CNN model, ConvLSTM may transport and process input in both the inter-layer and the intro-layer.

3.3.3 3D Convolutional Neural Networks (C3D)

In contrast to 2D-CNNs, C3D [27–29] can extract both temporal and spatial data from videos. This is due to the fact that 2D convolution applied to a video section compresses the temporal features after convolving, leading to an overall feature map that fails to accurately reflect any motion. A 3D filter kernel is created by stacking many frames together to create the 3D cube needed for the 3D convolution. Frames × Height × Width × Channels in the following format: 25 × 90 × 90 × 3. A ReLU activation function follows the 64 filters in the first 3D convolutional layer. The next step is a max pooling, which takes the most notable features from each feature map patch and calculates their maximum value.

4 Results and Discussion

Evaluation metrics including false positive rate (FPR), error rate (ER), accuracy (AUC), true positive rate (TPR), and precision (P) are used to make predictions about HAR detection (see Tables 2 and 3).

$$TPR = \frac{True\ Positive}{False\ Negative + True\ Positive} \tag{25}$$

$$FPR = \frac{False\ Positive}{True\ Negative + False\ Positive} \tag{26}$$

$$Precision = \frac{True\ Positive}{True\ Positive + False\ Positive} \tag{27}$$

$$Error_{Rate} = \frac{False\ Positive + False\ Negative}{False\ Negative} \tag{28}$$

Table 2. Analysis of Various DL Classifiers without DMOA

Algorithms	TPR (%)	FPR (%)	Accuracy (%)	Error Rate
MLP	83.0	9.6	89.2	0.30
DBN	89.7	8.1	90.5	0.23
C3D	93.8	5.3	92.6	0.15
CLSTM	95.6	4.5	94.3	0.18
LRCN	96.8	2.5	96.4	0.16

In the analysis of TPR, the three proposed models achieved nearly 93% to 96%, DBN achieved 89.7% and MLP achieved 83%. When the models are tested with FPR, DBN and MLP achieved 9.6% and 8.1%, where the three models of proposed approach achieved 2.5% to 5.3%. The error rate is very low in C3D, CLSTM and LRCN, where DBN and MLP has high error rate. i.e., 0.30 and 0.23 (see Figs. 2 and 3).

Table 3. Analysis of Various DL Classifiers with DMOA

Algorithms	TPR (%)	FPR (%)	Accuracy (%)	Error Rate
MLP	89.7	8.9	91.2	0.28
DBN	91.8	7.2	92.1	0.20
C3D	94.4	4.6	95.2	0.13
CLSTM	96.1	3.5	96.4	0.15
LRCN	98.6	1.4	98.4	0.14

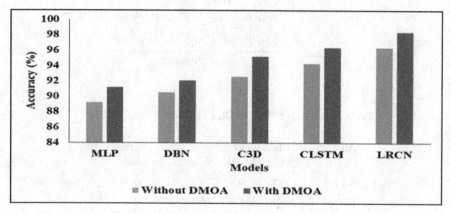

Fig. 2. Accuracy Validation

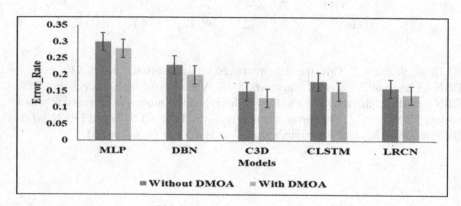

Fig. 3. Error_Rate Presentation

5 Conclusion

Dwarf Mongoose Optimization is a suggested optimization-based Feature selection method in this study for human action recognition. In order to detect an action, the proposed system combines the data derived from several sense modalities utilising a supervised trifecta of deep learning methods. The extensive experimental findings validate the validity of our proposed strategy for human action classification in comparison to standalone sensor modalities. Furthermore, as compared to state-of-the-art deep CNN approaches, the system's recognition accuracy is enhanced while computational cost is decreased by fusing time domain information calculated from inertial sensors with those from depth/RGB movies. In addition, it does not utilize Multi-view HAR, and the subject whose actions are being identified maintains their current orientation with relation to the camera. Further work will involve expanding the suggested HAR technique to compensate for these deficiencies. Moreover, we hope to explore the many uses for the suggested fusion architecture by utilizing an RGB-D camera and a set of wearable inertial sensors.

References

1. Yadav, S.K., Tiwari, K., Pandey, H.M., Akbar, S.A.: A review of multimodal human activity recognition with special emphasis on classification, applications, challenges, and future directions. Knowl.-Based Syst. **223**, 106970 (2021)
2. Zhao, H., Miao, X., Liu, R., Fortin, G.: Multi-sensor information fusion based on machine learning for real applications in human activity recognition: state-of-the-art and research challenges. Inf. Fusion **80**, 241–265 (2022)
3. Ferrari, A., Mocci, D., Mobile, M., Napolitano, P.: Trends in human activity recognition using smartphones. J. Reliable Intell. Environ. **7**(3), 189–213 (2021)
4. Islam, M.M., Iqbal, T.: Multi-gat: a graphical attention-based hierarchical multimodal representation learning approach for human activity recognition. IEEE Robot. Autom. Lett. **6**(2), 1729–1736 (2021)
5. Rani, S., Babar, H., Coleman, S., Singh, A., Allandale, H.M.: An efficient and lightweight deep learning model for human activity recognition using smartphones. Sensors **21**(11), 3845 (2021)
6. Khan, I.U., Afzal, S., Lee, J.W.: Human activity recognition via hybrid deep learning based model. Sensors **22**(1), 323 (2022)
7. Challa, S.K., Kumar, A., Samwell, V.B.: A multibranch CNN-BiLSTM model for human activity recognition using wearable sensor data. Vis. Comput. **38**(12), 4095–4109 (2022)
8. Xiao, Z., Xu, X., Xing, H., Song, F., Wang, X., Zhao, B.: A federated learning system with enhanced feature extraction for human activity recognition. Knowl.-Based Syst. **229**, 107338 (2021)
9. Zhang, S., et al.: Deep learning in human activity recognition with wearable sensors: a review on advances. Sensors **22**(4), 1476 (2022)
10. Ramanujan, E., Perumal, T., Padmavathi, S.: Human activity recognition with smartphone and wearable sensors using deep learning techniques: a review. IEEE Sens. J. **21**(12), 13029–13040 (2021)
11. Wang, D., Yang, J., Cui, W., Xie, L., Sun, S.: Multimodal CSI-based human activity recognition using GANs. IEEE Internet Things J. **8**(24), 17345–17355 (2021)

12. Hamad, R.A., Kimura, M., Yang, L., Woo, W.L., Wei, B.: Dilated causal convolution with multi-head self-attention for sensor human activity recognition. Neural Comput. Appl. **33**, 13705–13722 (2021)
13. Gu, F., Chung, M.H., Chignell, M., Valaee, S., Zhou, B., Liu, X.: A survey on deep learning for human activity recognition. ACM Comput. Surv. (CSUR) **54**(8), 1–34 (2021)
14. Garcia, K.D., et al.: An ensemble of autonomous auto-encoders for human activity recognition. Neurocomputing **439**, 271–280 (2021)
15. Tasnim, N., Islam, M.K., Baek, J.H.: Deep learning based human activity recognition using spatio-temporal image formation of skeleton joints. Appl. Sci. **11**(6), 2675 (2021)
16. Pradhan, A., Srivastava, S.: Hierarchical extreme puzzle learning machine-based emotion recognition using multimodal physiological signals. Biomed. Signal Process. Control **83**, 104624 (2023)
17. Dahou, A., Chelloug, S.A., Alduailij, M., Elaziz, M.A.: Improved feature selection based on chaos game optimization for social internet of things with a novel deep learning model. Mathematics **11**(4), 1032 (2023)
18. Islam, M.M., Nooruddin, S., Karray, F., Muhammad, G.: Multi-level feature fusion for multimodal human activity recognition in internet of healthcare things. Inf. Fusion **94**, 17–31 (2023)
19. Zhang, Y., Yao, X., Fei, Q., Chen, Z.: Smartphone sensors-based human activity recognition using feature selection and deep decision fusion. IET Cyber-Phys. Syst.: Theory Appl. **8**, 76–90 (2023)
20. Muhammad, F., Hussain, M., Aboalsamh, H.: A bimodal emotion recognition approach through the fusion of electroencephalography and facial sequences. Diagnostics **13**(5), 977 (2023)

Detection of Chicken Disease Based on Day-Age Using Pre Trained Model of CNN

K. Sreenivasulu[1], H. Aini Sosan Khan[2(✉)], K. Damini[2], M. Akhila[2], and G. Bharathi[2]

[1] Department of CSE, G. Pullaiah College of Engineering and Technology, Kurnool, India
[2] G. Pullaiah College of Engineering and Technology, Kurnool, India
ainisosankhan@gmail.com

Abstract. The rapid advancement of computer vision procedures has made the artificial raising of animals a more viable option for actual production settings. One such example is the critical need of enhancing the accuracy of day-age identification of birds in the poultry breeding industry. Within a time frame of 100 days, this article addresses the challenge of accurately categorising the age of hens. In actual application settings, when data volumes are large and device computing capabilities vary, it is crucial to make the most of the processing capabilities of edge computing devices without compromising data accuracy. An accurate deep learning-based model is proposed in this study for use in edge computing. This article takes a pre-trained DarkNet-53 model into account, performs model to make it suitable for low computing power circumstances, and runs a series of focused tests to ensure the model's efficacy. Classification accuracy is enhanced by 3% associated to the baseline model and 2% compared to AlexNet, VGG, and ResNet. It is possible to improve detection accuracy while still fulfilling the needs of the real detection situation. Eventually, this article created a mobile-optimized detection software and a whole image-acquisition system for aviaries.

Keywords: Deep learning-based model · Pre-trained DarkNet-53 · Day-age detection · Chicken Disease · Edge computing devices

1 Introduction

With the shift towards industrial-scale operations and meticulous administration in the livestock industry, farm intelligence has emerged as a necessary tool. To do this, it is necessary to accurately identify each livestock individual throughout the breeding process. Several subsequent breeding and conservation activities, including as monitoring development phases, (BCS) [1, 2], and altering breeding strategies, rely on the ability to identify and manage every individual fowl. Poor medical care, incorrect heat detection, and delayed reproduction may occur from not keeping track of animals as they age, leading to decreased output, poor animal death. In this study, we investigate hens' abilities to recognise human faces. Table 1 displays the breakdown of a chicken's developmental stages as a function of its age in days [3].

© The Author(s), under exclusive license to Springer Nature Switzerland AG 2023
S. Kadry and R. Prasath (Eds.): MIKE 2023, LNAI 13924, pp. 13–24, 2023.
https://doi.org/10.1007/978-3-031-44084-7_2

Industrial-scale operations and meticulous administration in the livestock industry, farm intelligence has emerged as a necessary tool. To do this, it is necessary to accurately identify each livestock individual throughout the breeding process. Several subsequent breeding and conservation activities, including as monitoring development phases, (BCS) [1, 2], and altering breeding strategies, rely on the ability to identify and manage every individual fowl. Poor medical care, incorrect heat detection, and delayed reproduction may occur from not keeping track of animals as they age, leading to decreased output, poor animal death. In this study, we investigate hens' abilities to recognise human faces. Table 1 displays the breakdown of a chicken's developmental stages as a function of its age in days [3].

Table 1. Development phases of chicken branded by day-old eternities.

Living circumstances	Growth Phases
Newborn-60 days of stage	Chicks
61–150 days of age.upbringing chickens	Breeding stage1
151 days of age and overhead	Adult stage
Hens that have not started laying eggs; Breeding roosters that have not yet been bred	Reserved chickens

The phases of life are often measured by certain chicken characteristics [4]. To be more specific: (1) Beak. A chick's beak is pointed, slender, and sharp. During eight months of outdoor foraging, adult hens have robust, short beaks with a firm, smooth tip and broad, rough corners of the mouth. Tumor in the nasal cavity, case 2. Chickens with nasal tumours have a light red colour when they are young, and a light pink colour when they are two years old or older [5]. Chickens with nasal tumours are pink and tough when they are four or five years old. Toe and feather traits are also included in an empirical evaluation. (3) Toes. Baby birds' feet are brilliant red and covered in small, silky scales. Adult chickens have stocky bodies and dark red feet that are covered in thick, hard scales. They have tough, curved toenails. Feathers, four. The size of a chick's wing is correlated with its age in months [6]. However these visual approaches can only loosely categorise hens, not their age or developmental stage.

Nowadays, most conventional farms employ manual recordkeeping and subjective evaluations to distinguish between chick developmental phases; this practise might entail considerable effort. The optical density test technique measures [7] to infer the day-age of chickens, thanks to the advancement of computer vision and its use in the sector. Nevertheless, this bone density measurement technique cannot be used on living chickens. This means that eliminating guesswork and precisely pinpointing a bird's age at hatch is an absolute need for contemporary chicken rearing. Hence, using AI to produce a set of fewer personal [8]. Laying chickens of varying ages need vastly varied diets and care strategies. It is important to have a thorough understanding of the reserve chicken period's fluctuating amino acid demands [9]. Lastly, the economic advantages may be increased by precisely determining the broiler thanks to the exact identification of chicken day-age. The broilers' day-age at slaughter, division, and sale all affect the quality of their meat [10]. Accurate day-age judgement is used to determine when chickens are killed, allowing producers to respond flexibly to market demands for a variety of chicken products with meat of optimal texture [11].

Recent years have seen the development of several computerised pattern recognition systems using camera traps for classification, allowing for the rapid and precise identification of species for agricultural purposes [12]. Nevertheless, as the number of recorded cameras grows rapidly, a solution -point training would take a long time, and the key will be figuring out how to make use of the vast amount of computational power available at the edge. These investigations pave the path for future work in computer-vision-based chicken categorization. In light of the above, this work offers an edge-computing-compatible, high-precision DL-based chicken classification model aged 100 days. Part 2 delivers a swift of relevant literature, while Sect. 3 details the suggested model and its rationale. In Sects. 4 and 5, we see the validation analysis using scientific input.

2 Related Works

To enhance the performance of caged hens, Yang et al. [13] presented CCD pix2pixHD. The suggested defencing algorithm successfully recognises the wire mesh of the coop and restores the full shapes of the chickens. The cage wire mesh was detected with a 94.71% success rate in the test set, while the image recovery process yielded a peak PSNR of 25.24 dB and an SSIM of 90.04%. This research standpoint of the most fundamental separate detection in the caged chicken detection challenge to validate the viability of the approach described here. Hence, we tested the defencing algorithm with YOLOv5s, YOLOv5m, YOLOv5l, and YOLOv5x to ensure its efficacy. According to the findings of the experiments, the detection accuracy of the caged hens increased by 16.1%, 12.1%, 7.3%, and 5.4% when the defencing algorithm was used. A 29.1%, 16.4%, 8.5%, and 6.8% increase in recall accuracy was recorded. This is the first time a deep learning-based defencing technique has been used on confined hens, and the results are promising. This paper's approach of removing cage wire mesh has the potential to be very effective, and it also serves as a useful technical reference for future poultry researchers.

To conduct thorough mining and analysis of Raman spectrum data, Sun et al. [14] choose to use a convolutional neural network (CNN), which excels at addressing multi-classification tasks. The instrument settings were optimised by evaluating the effects of three laser wavelengths (532 nm, 638 nm, and 785 nm). In the end, it was determined that the best wavelength for detecting Salmonella was 532 nm. The Raman spectrum must be removed in a pre-processing phase. This research assessed the accuracy of CNN models trained with and without five different spectral pre-processing techniques: The findings showed that SG paired with SNV was the most effective spectrum pre-processing strategy for predicting Salmonella serotypes using Raman spectroscopy, with a training set accuracy of 98.7% and a test set accuracy of over 98.5% using a Convolutional Neural Network (CNN) model. This approach of preprocessing spectral data is more precise than others. In conclusion, this study shows that three different serotypes of Salmonella can be quickly identified and closely related bacterial species. This is crucial in stopping the spread of food-borne diseases and illnesses.

To train models for object recognition division, where the objects are chicks in photos collected on the farm, Cakic et al. [15] proposes using an existing IoT farming infrastructure and running deep learning on HPC offline. To improve upon the current digital poultry farm platform, the models may be transferred from HPC to edge AI devices, giving rise to a novel computer vision toolkit. Such cutting-edge sensors allow for the implementation of features like chicken counting, dead chicken identification, and even chicken weight and growth assessment. Early illness identification and better decision-making might be made possible by combining these features with ambient parameter monitoring. AutoML was used to determine the best Faster R-CNN construction for chicken identification and segmentation on the available dataset. Hyper parameter optimisation was performed further on the chosen architectures, resulting in an improvement in accuracy from 85% to 98% for object identification and from 90% to 98% segmentation. These representations were deployed to edge AI devices and tested in real-world chicken farms in an online environment. Although these early findings are encouraging, there is still room for improvement in both the dataset and the prediction models.

Tao et al. [16] combine hyperspectral microscopic imaging with deep learning to create a cutting-edge optical approach for the quick identification of diseases. In order to learn more about the microbes, researchers have created a further extract the species-dependent properties recorded discrimination, a fully-contained deep learning network based on feature fusion was developed (named BI-Net). We trained BI-Net using a massive dataset we created to distinguish between common infections, and then used it to categorise rare pathogens via transfer learning. In-depth research showed that BI-Net picked up on species-specific traits, with 92% classification accuracy and Kappa coefficients for both common and rare species. This approach significantly exceeded the state-of-the-art, indicating that it has great potential for use in clinical settings.

To better identify pecking habits and possible damages in cage-free settings, Subedi et al. [17] and [18] set out to build a machine vision technique by putting new models to the test. Cage-free laying hens' FP behaviour was tracked using two different to YOLOv5s-pecking, YOLOv5x-pecking performed 3.1%, 5.6%, and 5.2% better in precision, recall, and Map using a dataset of 1924 pictures. As compared to the YOLOv5x-pecking model, the YOLOv5s-pecking model is 80% smaller in size, using 75% less GPU RAM and 80% less time to train on the same dataset. As a result, the YOLOv5s-pecking model was deemed to be the most effective. As far as we know, this is the first research to use YOLOv5 models to observe negative behaviours in cage-free chickens. To ensure the safety of hundreds of millions of laying hens, this model may be used as a starting point for creating an autonomous model that can monitor pecking damages in commercial cage-free homes in real time.

To boost detection accuracy, Ren et al. [19] presented an attention encoder structure for feature extraction from chicken images. To test the efficacy of the suggested attention encoder in light of the dataset's inconsistencies, several data improvement strategies were presented, including Cutout, CutMix, and MixUp. For the purposes of this article, the structure was implemented in many popular CNNs for the purposes of evaluation and repeated ablative experiments. Using the ResNet-50 was able to increase the accuracy of chicken age identification to 95.2%, as shown by the final testing results. Eventually, a mobile-optimized detection application and a whole image-acquisition system for aviaries are developed in this article.

3 Proposed System

3.1 Dataset Analysis

The data utilised in this research was gathered with the endorsement of the Animal Ethics Committee of China Agricultural University Province. Table 2 displays the data set's daily distribution. The dataset used in this work was manually gathered by researchers at China's Guangdong Academy of Agricultural Sciences. From January 2021 to October 2022, these photographs were taken by researchers using a Canon 5D digital camera. Information about this dataset is shown in Fig. 1; the dimensions of each picture are 6720 by 4480 pixels.

The properties of the dataset utilised in this work are as shown in the table above.

One, training and learning on a single client is challenging and time-consuming due to the size of our collection.

The large number of classes in our dataset makes accurate categorization with a single model challenging.

Third, our dataset is not uniformly distributed; some sections have many recent photographs while others have few or none at all.

This work presents a distributed training approach based on DL since it is challenging to train a single client and acquire a high accuracy perfect due to the aforementioned properties of the dataset and the use situation.

3.2 Dataset Rebalancing and Recovery

The dataset used in this article is unbalanced, as previously mentioned. This work prioritises the dataset for re-balancing prior to training in order to avoid the model from ignoring the categories with few photos. The following three techniques for improving data are employed:

One) Alterations to geometry. Flipping, rotating, cropping, distorting, and scaling are all examples of geometric changes of a picture.

Changes in colour, number two. Although the aforementioned geometric operations may pick portions of a picture or rearrange its pixels, they do not alter the image's actual content. Data improvement techniques like noise, blur, colour transformation, erase, fill, etc., fall under the category of colour transformation since they alter the image's content.

Thirdly, improving data from several samples. In contrast to SMOTE [20] and [21], SamplePairing [22], multisampling data augmentation techniques leverage existing samples to create new ones.

This work employs the training procedure suggested by [23–25] which splits all pictures into training and test sets with a 7:3 split, with the test set yielding the findings discussed in Sect. 4.

Table 2. Delivery of datasets among classes.

Day-Age	Images-1	Images-2	Images-3	Images-4	Day-Age	Day-Age	Day-Age
1	286	268	249	273	2	3	4
5	251	275	269	263	6	7	8
9	278	266	266	250	10	11	12
13	260	3176	3164	3150	14	15	16
17	3171	3140	3133	3140	18	19	20
21	3120	3117	3080	3050	22	23	24
25	3070	3066	3052	3040	26	27	28
29	3050	3024	2928	-	30	31	32
33	-	-	2916	2903	34	35	36
37	2933	2926	2925	2913	46	47	48
41	2922	2966	2928	2923	42	43	44
45	2961	2926	2925	2913	46	47	48
49	2922	3020	-	-	50	51	52
53	-	-	-	-	54	55	56
57	-	-	-	-	58	59	60
61	-	-	-	-	62	63	64
65	-	-	-	765	66	67	68
69	741	630	622	619	70	71	72
73	646	629	628	-	74	75	76
77	612	648	644	612	78	79	80
81	635	648	647	639	82	83	84
85	644	648	647	675	86	87	88
89	633	649	-	648	90	91	92
93	647	638	622	652	94	95	96
97	653	650	641	652	98	99	100

3.3 Classification

DarkNet53 is used here for didactic reasons. During training, TL is used to pull features from the world's average pool layer.

3.3.1 Modified DarkNet-35 Model

The convolutional neural network DarkNet-53 has 53 layers. The YOLOv3 object identification mechanism relies on this. By merging Resnet's features, it can guarantee super issue caused by a too-deep network. Model architecture of the DarkNet-53 network.

Fig. 1. Display of the chicken dataset from diverse day-ages

Residual network and deep residual network are brought together. Layers of 1 1 convolution and 3 3 convolution, followed by residual blocks, are included. This is how we characterise the convolutional layer:

$$a_m^n = \sum_{j \in X_i} a_j^{n-1} * y_{jm}^n + z_m^n \tag{1}$$

In Eq. (1), the input picture is warped by multiple convolution kernels to generate m distinct feature maps a mn, with the m feature map serving as the representation for layer n. The convolution operation is represented by the symbol *. X i is the image's feature vector, and y jn is the jth element of the mth layer's convolution kernel.

After the first layer, the batch normalisation (BN) layer is the next crucial one.

$$a_{out} = \frac{\propto \left(a_m^n - \partial\right)}{\sqrt{\omega^2 + \varphi}} + \gamma \tag{2}$$

In Eq. (2), a stands for the scaling factor, m for the average of the outputs, w for the input variance, j for a constant offset, and a out for the outcome of the convolution computation. a out stands for the output of BN. Batch Normalization is used to standardise the output so that it conforms to the distribution of the batch of eigenvalues' coefficients. The subsequent convolutional layer is designed to hasten network convergence and prevent overfitting. The subsequent layer is an activation layer as well. A leaky is raised by this function.

$$x_j = \begin{cases} y_i, & if \quad a_{out} \geq 0 \\ \frac{y_i}{b_j} & if \quad a_{out} < 0 \end{cases} \tag{3}$$

For the purpose of Eq. (3), we will refer to the input value as y i, the activation value as xj, and the fixed parameter in the range $[-1, +1]$ as b j. The network's pooling layer is also crucial. This layer is used to reduce the network's weight sample size. This network employs a max-pooling layer. The last illustration shows how features (weights) may be integrated into a single layer. In the output layer, the retrieved features are subsequently labelled. This model has a depth of 53, a size of 155 MB, a sum of parameters of 41.6 million, and an input picture size of 256×256 pixels.

3.3.2 Transfer Learning

In (TL) is the process of applying information from one task to another. One practical way to increase the learning agent is to reuse or transfer data from previously learned tasks for the newly learnt tasks. In this case, deep feature extraction is performed using TL. To achieve this goal, we first use TL to hone in on a model that has been pre-trained. TL has the following mathematical definition:

Two f(y), where y = y 1, 2, y 3,…y n Y, characterise a domain d = Y,p(y). Either the marginal probabilities (p(p)p(q)) of the two domains are different.

Given a domain d, there are two parts to the task tX,g(.): a label space X and a prediction function g (.); the latter is hidden but may be inferred from the training data m j,n j j1,2,3,…N, where m j Y and n j X. As p(n j |m j) is equivalent to f(m j) from a probabilistic perspective, we may rewrite the task t as t = X,P(x|Y). Dissimilar tasks may have different label spaces (_p q) or conditional probability distributions (p(p q)p(p q)). This improved deep model benefits from the information learned from the original model's domain of origin (target domain). Stochastic gradient descent is then used as the learning technique, with a learning rate of 0.001, a mini batch size of 16, 200 epochs, and a mini batch size of 16. The improved deep model's Global Average Pooling (GAP) layer serves as the source for the features. Two redesigned optimisation techniques are then used to fine-tune the retrieved characteristics.

4 Results and Discussion

4.1 Experimental Platform

This study built a PC with an NVIDIA graphics card using the suggested technique based on PyTorch. The configurations of these various mobile platforms are illustrated in Table 3; they range from smartphones to development boards.

Table 3. Swift of heterogeneous strategies

Device	Total	Network	DRAM	Processor
NVIDIA Jetson Nano	2	Ethernet	8 GB	Cortex A57
HUAWEI Mate 20	2	Wifi	6 GB	Kirin 980
Samsung Galaxy Fold 2	1	Wifi	8 GB	Snapdragon 865

Different CPUs, memory, and network configurations are used by each of these gadgets, as indicated in the table above.

4.2 Performance Measure

The diagnosis are intended as:

$$Accuracy = \frac{TP + TN}{TP + FP + FN + TN} \tag{4}$$

$$Precision = \frac{TP}{TP + FP} \tag{5}$$

$$Recall = \frac{TP}{TP + FN} \tag{6}$$

$$F1Score = \frac{2 * Recall * Precision}{Recall + Precision} \tag{7}$$

where True Positives (TP) – designate positive belongings that are properly recognized as positive cases. False Positives (FP) – specify negative cases that are incorrectly acknowledged as positive cases. True Negatives (TN) – signpost undesirable cases that are properly identified as negative cases. False negatives (FN) – signpost positive cases that are incorrectly identified as negative cases. Based on the confusion matrix, other related performance can be resulting (see Table 4).

Table 4. Comparitive analysis of Proposed Pre-trained model with existing models

DL Classifiers	Class	Precision	recall	f1-score	Average Accuracy
ResNet	Disease	0.94	0.94	0.94	
	Moderate	0.90	0.90	0.90	
	Non-Disease	0.92	0.92	0.92	0.929825
VGGNet	Disease	0.94	0.94	0.94	
	Moderate	0.90	0.90	0.90	
	Non-Disease	0.92	0.92	0.92	0.929825
AlexNet	Disease	0.94	0.96	0.96	
	Moderate	0.97	0.89	0.93	
	Non-Disease	0.96	0.94	0.95	0.9508577
Proposed DarkNet	Disease	0.97	0.98	0.98	
	Moderate	0.98	0.94	0.99	
	Non-Disease	0.98	0.96	0.98	0.962912

In the analysis of average accuracy, ResNet achieved 0.92, VGGNet achieved 92%, AlexNet achieved 95% and proposed model achieved 96%. For non-disease analysis, the proposed model attained 98% of precision, 96% of recall and 98% of F1-score, where the existing models achieved 92% to 96% of precision, 92% to 94% of recall and 92% to 95% of F1-score. From this analysis, it is clearly proves that the proposed model achieved better performance. Figure 2 presents the graphical analysis of proposed pre-trained model for various metrics.

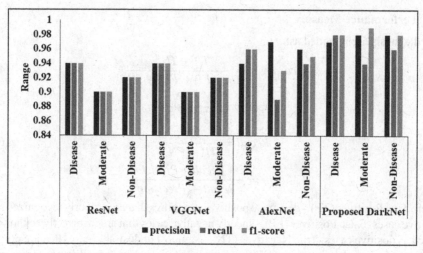

Fig. 2. Comparison of proposed model with existing techniques

5   Conclusion

Using a network and a deep learning architecture, the authors of this research present a solution to the issue of poor accuracy in chicken day-age detection in real-world settings. The following are the most significant results from this study:

1. In order to rapidly execute model training and parameter updating across several terminals, the authors of this study suggest a pre-trained model of CNN mechanism.
2. This study builds a lightweight archetypal based on the popular CNN net and implements it on the edge devices to deal with the widespread problem of insufficient processing power at the end user's disposal.
3. Third, several tests are carried out in this study to prove that this method works. This approach is 92.4% to 96.1% more accurate than the reference models.
4. The detection of day-old chickens provides a test case for the approach suggested in this research. Our long-term goal is to apply this method to a wide variety of farming settings, particularly large-scale smart agriculture applications like detecting and identifying individual animals in industrial-scale livestock operations.

References

1. Ahmed, G., Malick, R.A.S., Akhunzada, A., Zahid, S., Sagri, M.R., Gani, A.: An approach towards IoT-based predictive service for early detection of diseases in poultry chickens. Sustainability **13**(23), 13396 (2021)
2. Cuan, K., Zhang, T., Li, Z., Huang, J., Ding, Y., Fang, C.: Automatic newcastle disease detection using sound technology and deep learning method. Comput. Electron. Agric. **194**, 106740 (2022)
3. Neethirajan, S.: ChickTrack–a quantitative tracking tool for measuring chicken activity. Measurement **191**, 110819 (2022)

4. Adebiyi, A.I., Mcilwaine, K., Oluwayelu, D.O., Smyth, V.J.: Detection and characterization of chicken astrovirus associated with hatchery disease in commercial day-old turkeys in southwestern Nigeria. Adv. Virol. **166**, 1607–1614 (2021)

5. Bakar, M.A.A., Ker, P.J., Tang, S.G., Lee, H.J., Zainal, B.S.: Classification of unhealthy chicken based on chromaticity of the comb. In: 2022 IEEE International Conference on Computing (ICOCO), pp. 1–5. IEEE (2022)

6. Thavamani, S., Vijayakumar, J., Sruthi, K.: GLCM and K-means based chicken gender classification. In: 2021 Smart Technologies, Communication and Robotics (STCR), pp. 1–5. IEEE (2021)

7. Almashhadany, D.A.: Detection of antimicrobial residues among chicken meat by simple, reliable, and highly specific techniques. SVU-Int. J. Vet. Sci. **4**(1), 1–9 (2021)

8. Gu, K.: Development of nanobody-horseradish peroxidase-based sandwich ELISA to detect Salmonella Enteritidis in milk and in vivo colonization in chicken. J. Nanobiotechnol. **20**(1), 167 (2022)

9. Syahruni, S.: Development of lateral flow assay based on anti-IBDV IgY for the rapid detection of Gumboro disease in poultry. J. Virol. Methods **291**, 114065 (2021)

10. Basit, M.S.I., Mamun, M.A., Rahman, M.M., Noor, M.: Isolation and molecular detection of mycoplasma gallisepticum in commercial layer chickens in Sylhet, Bangladesh. World's Vet. J. **11**(4), 614–620 (2021)

11. Vizzini, P., et al.: Highly sensitive detection of campylobacter spp In chicken meat using a silica nanoparticle enhanced dot blot DNA biosensor. Biosens. Bioelectron. **171**, 112689 (2021)

12. Saikia, M., Bhattacharjee, K., Sarmah, P.C., Deka, D.K.: Comparative evaluation of direct smear and culture methods for detection of trichomonas gallinae infection in pigeon and chicken of Assam. Int. J. Curr. Sci. Res. Rev. **5**(11), 4331–4335 (2022)

13. Yang, J., Zhang, T., Fang, C., Zheng, H.: A defencing algorithm based on deep learning improves the detection accuracy of caged chickens. Comput. Electron. Agric. **204**, 107501 (2023)

14. Sun, J.: Rapid identification of salmonella serovars by using Raman spectroscopy and machine learning algorithm. Talanta **253**, 123807 (2023)

15. Cakic, S., Popovic, T., Krco, S., Nedic, D., Babic, D., Jovovic, I.: Developing edge AI computer vision for smart poultry farms using deep learning and HPC. Sensors **23**(6), 3002 (2023)

16. Tao, C., Du, J., Wang, J., Hu, B., Zhang, Z.: Rapid identification of infectious pathogens at the single-cell level via combining hyperspectral microscopic images and deep learning. Cells **12**(3), 379 (2023)

17. Subedi, S., Bist, R., Yang, X., Chai, L.: Tracking pecking behaviors and damages of cage-free laying hens with machine vision technologies. Comput. Electron. Agric. **204**, 107545 (2023)

18. Ren, Y., et al.: A high-performance day-age classification and detection model for chick based on attention encoder and convolutional neural network. Animals **12**(18), 2425 (2022)

19. Chawla, N.V., Bowyer, K.W., Hall, L.O., Kegelmeyer, W.P.: SMOTE: synthetic minority over-sampling technique. J. Artif. Intell. Res. **16**, 321–357 (2002)

20. Dwaram, J.R., Madapuri, R.K.: Crop yield forecasting by long short-term memory network with Adam optimizer and Huber loss function in Andhra Pradesh, India. Concurrency Comput.: Pract. Experience **34**(27), e7310 (2022)

21. Venkata Chalapathi, M.M., Rudra Kumar, M., Sharma, N., Shitharth, S.: Ensemble learning by high-dimensional acoustic features for emotion recognition from speech audio signal. Secur. Commun. Netw. **2022**, 1–10 (2022). https://doi.org/10.1155/2022/8777026

22. Inoue, H.: Data augmentation by pairing samples for images classification. arXiv 2018, arXiv: 1801.02929

23. Rudra Kumar, M., Pathak, R., Gunjan, V.K.: Diagnosis and medicine prediction for covid-19 using machine learning approach. In: Kumar, A., Zurada, J.M., Gunjan, V.K., Balasubramanian, R. (eds.) Computational Intelligence in Machine Learning: Select Proceedings of ICCIML 2021, pp. 123–133. Springer Nature Singapore, Singapore (2022). https://doi.org/10.1007/978-981-16-8484-5_10
24. Rudra Kumar, M., Pathak, R., Gunjan, V.K.: Machine learning-based project resource allocation fitment analysis system (ML-PRAFS). In: Kumar, A., Zurada, J.M., Gunjan, V.K., Balasubramanian, R. (eds.) Computational Intelligence in Machine Learning. LNEE, vol. 834, pp. 1–14. Springer, Singapore (2022). https://doi.org/10.1007/978-981-16-8484-5_1
25. Krizhevsky, A., Sutskever, I., Hinton, G.: Imagenet classification with deep convolutional neural networks. In: Proceedings of the Advances in Neural Information Processing Systems 25 (NIPS 2012), Lake Tahoe, NV, USA

Image Captioning Using Xception-Long Short-Term Memory

Nisha Panchal[1,2(✉)] and Dweepna Garg[1]

[1] Department of Computer Engineering, Devang Patel Institute of Advance Technology and Research (DEPSTAR), Faculty of Technology and Engineering (FTE), Charotar University of Science and Technology (CHARUSAT), Changa, Anand, Gujarat, India
nishpanchal132@gmail.com
[2] U & P U Patel Department of Computer Engineering, Chandubhai S. Patel Institute of Technology (CSPIT), Faculty of Technology and Engineering (FTE), Charotar University of Science and Technology (CHARUSAT), Changa, Anand, Gujarat, India

Abstract. Deep Learning has shown great potential in developing applications capable of automatically generating captions or descriptions for images and video frames. The critical components of this process are image processing and natural language processing, which play a crucial role in captioning images and videos. These applications can be used in several areas, including robotic vision systems, assisting people with visual impairments, generating metadata for search engines, answering visual questions, visual grounding, and more. This paper discusses various working algorithms such as a combination of CNN-RNN, encoder-decoder, attention mechanisms, and transformation models with evaluation matrices, datasets, and limitations of existing models. Xception-LSTM shows great potential compared to the traditional encoding-decoding model using BLEU and METEOR evaluation matrics.

Keywords: Image captioning · Xception · Long-short term memory (LSTM) · CNN (convolution neural networks) · RNN (Recurrent neural networks)

1 Introduction

Image processing is an essential aspect of computer science and has substantial relevance across various fields, including object detection, scene interpretation, and visual recognition. Dedicated hardware was used by researchers for executing imaging techniques to get appropriate results, especially for rigid objects before the emergence of deep learning. However, CNN and RNN driven by deep learning have important influences on visual-to-text generation, demonstrating remarkable progress recently.

The task of describing a scenario depicted in an image or video clip comes naturally to humans, but it poses significant challenges for machines. To tackle this issue, computer scientists are exploring methods to integrate the ability to comprehend human language with the capability to automatically extract and analyze visual data, thereby enabling machines to perform similar tasks. Although, extracting objects with their actions from

S. Kadry and R. Prasath (Eds.): MIKE 2023, LNAI 13924, pp. 25–33, 2023.
https://doi.org/10.1007/978-3-031-44084-7_3

the image and producing crisp as well as relevant sentences needs much substantial work in comparison to a simple image recognition task.

Image and video caption generation primarily involve analyzing an image's features and generating a corresponding textual description. As this field demands visual and textual mastery proficiency, it utilizes a blend of CV and NLP techniques to translate image comprehension from feature vectors into words arranged in the proper sequence. The captioning method must capture the objects in the given scenario as well as their traits, actions, and interrelationships.

Therefore, the most common method for image captions is the encoder-decoder architecture, which combines a Convolutional Neural Network (CNN) to encode image features and a Recurrent Neural Network (RNN) to generate a caption.

It has a clear separation of tasks – The CNN is responsible for encoding the image features, while the RNN is responsible for generating the caption. This separation of tasks makes it easier to debug and analyze the model.

Overall, the Encoder-Decoder architecture is a popular choice for image captioning due to its effectiveness, flexibility, simplicity, and clear separation of tasks.

This paper mentions the following details in upcoming sections, which are related work of image captioning, the Proposed Methodology, Results, and Discussion, and at the end conclusion and future work.

2   Related Work

Our research has involved an in-depth exploration of numerous studies about image captioning, encompassing a range of techniques, datasets, and evaluation methodologies. CNN is often used to extract features from an image. These features are then used as input to a language model that creates the image caption. CNNs are trained on large image datasets and can learn to recognize patterns and features in images [1, 3–5, 7, 8, 12, 15, 17]. RNN takes as input the output of the previous step (which is a word embedding) and the visual features that the CNN extracted from the image. And then generates the next word in the caption [3, 8].

Encoder-decoder models are a type of neural network architecture that leverages an encoder component to extract features from an input and a decoder component to generate an output. In the context of image captioning, the encoder is typically implemented using a convolutional neural network (CNN), which extracts salient features from the input image. On the other hand, the decoder is usually implemented using a recurrent neural network (RNN), which generates the caption based on the features extracted by the encoder. [2, 17]. This innovative approach has served as a starting point for subsequent research in the area of image captioning in 2015 [18].

Subsequently, the author of [19] introduced a novel approach to simultaneously train a CNN and an RNN for generating captions by aligning image regions with their corresponding linguistic units. To facilitate their experimentation, the authors employed the COCO dataset, which has since emerged as a widely accepted benchmark for assessing the effectiveness of image captioning models. Of significance, this paper also introduced the CIDEr score, a widely used evaluation metric for image captioning models [19].

After that, an attention-based approach to image captioning was introduced [20], where the model learns to selectively attend to different image regions when generating captions. The authors showed that attention mechanisms improve caption quality and reduce ambiguity, and proposed a novel "hard" attention mechanism that can be trained using backpropagation [20]. Attention mechanisms play a vital role in enabling image captioning models to focus on the most pertinent aspects of an image during caption generation. Rather than solely depending on the global image features, attention mechanisms allow these models to selectively concentrate on specific regions of the image that are most relevant to the current context of the caption being generated [2, 5, 13].

Thereafter bottom-up and top-down attention mechanism combines object-level features with region-level features to generate captions introduced [21]. This paper introduced a new dataset called Visual Genome, which contains more detailed object and attribute annotations than other datasets used for image captioning. This paper also introduced a new evaluation metric called SPICE, designed to measure the semantic similarity between generated and human captions [21].

Afterward, a new pre-training approach for image captioning that combines vision and language tasks to learn joint representations of images and captions was introduced [22]. The authors of this paper use a Transformer-based architecture that is pre-trained on a large corpus of image-caption pairs and shows that their method achieves state-of-the-art performance on several benchmark datasets.

Following that, a new Transformer-based architecture for image captioning that uses a meshed-memory mechanism to selectively attend to different regions of the image and the caption was introduced [23]. The authors show that their method outperforms other Transformer-based models and achieves state-of-the-art performance on the COCO dataset. Transformer Models Transformers are a relatively new development in the field of natural language processing and have proven to be highly successful in tasks such as text generation and machine translation. This is mainly because they use a self-attention mechanism that allows them to process input sequences simultaneously, making them well-suited for processing long input sequences such as captions. When creating captions, Transformer models are usually equipped with an encoder for extracting image features and a decoder for generating captions [11].

Visual Question answering is one of the major applications of image captioning that is mentioned in [24]. This paper proposes a new pre-training approach for image captioning using a single encoder to encode images and captions jointly.

A new approach was brought up in [25] for generating image captions by parallelizing the decoding process to improve efficiency. The authors propose a hierarchical structure for the caption that allows the model to generate the words in a parallel and efficient manner. The authors show that their method achieves state-of-the-art performance on the COCO dataset and is significantly faster than other models.

Throughout the years, numerous encoder-decoder techniques have emerged, employing different variations of Convolutional Neural Networks (CNN) and Recurrent Neural Networks (RNN).

3 Methodology

A combination of CNN-LSTM is a commonly used Neural network in image captioning [3, 5–9, 12, 13, 15, 17]. In the proposed model Xception model is used for feature extraction. The Xception takes an input image and outputs a vector of visual features as the following equation.

$$V = Xception(I) \tag{1}$$

In the given context, before utilizing the Xception model, a series of image preprocessing techniques were employed to adequately prepare the image. Consequently, the Xception model was applied to extract the feature vector associated with the image. In the current context, In Eq. (1) the variable "I" represents the input image, while "V" refers to the vector of visual features extracted from it.

The Xception model's utilization of depthwise separable convolutions, in contrast to traditional CNN models, delivers notable improvements in computational efficiency and speed. This architectural choice enables the model to analyze spatial relationships and feature interactions more effectively while minimizing redundant computations. Therefore, the Xception model is used for encoding features from the image compared to the traditional one.

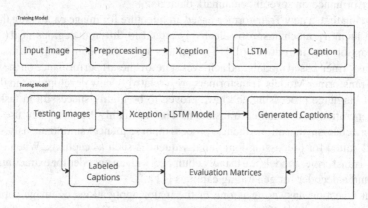

Fig. 1. Proposed Model

In the case of the decoding side, LSTM is a type of RNN that is frequently used for image captioning tasks because it is better suited for capturing long-term dependencies in sequential data such as natural language [3, 5–9, 12–13, 15, 17]. Figure 1 depicts the flow of the proposed model with training and testing bifurcations.

In image captioning, a critical task involves the model's ability to comprehend the context of the image and produce a fitting caption accordingly. To this end, the LSTM network is employed to process the image features extracted by an Xception. The LSTM network generates a sequential set of words, which are then combined to form a grammatically sound and semantically coherent sentence.

Compared to conventional RNNs, LSTM networks possess an added memory cell capable of preserving and retrieving information over extended periods. This feature enables LSTMs to more effectively manage long-term dependencies, a challenge for traditional RNNs. Additionally; LSTMs overcome the issue of vanishing gradients that are commonly encountered in traditional RNNs, thereby increasing the effectiveness and efficiency of the network [5, 6, 9].

Moreover, LSTM networks can also selectively forget or remember information from the previous time step, making them well-suited for tasks where the model needs to maintain a context for a long period.

Therefore, we have used LSTM networks over traditional RNNs for image captioning tasks because of their ability to better capture the complex dependencies and long-term context of natural language data.

The LSTM takes the vector of visual features from the Xception and generates a sequence of words that form the image caption. The LSTM does this by processing each word in the sequence one at a time and updating its internal state based on the previous words in the sequence. This can be represented in Eq. (2)

$$h_t = LSTM\left(V, h_{\{t-1\}}\right) \tag{2}$$

$$y_t = Softmax\left(W_{\{hy\}h_t} + b_y\right) \tag{3}$$

where V is the visual features at time t, h_t is the internal state of the LSTM at time t, y_t is the output probability distribution over the vocabulary at time t, and $W_{\{hy\}}$ is the weight matrix connecting the LSTM output to the vocabulary, and b_y is the bias term.

The Softmax function is used to convert the output of the LSTM to a probability distribution over the vocabulary so that the network can predict the next word in the sequence based on the probability of each possible word as per Eq. (3).

In this paper, we have used the Flickr8k dataset. It contains 8000 images [26], each with five different captions provided by human annotators. The dataset is divided into the train, validation, and test sets, and is often used for evaluating image-captioning models.

Figure 2 depicts the outcome of our proposed model where we have used the start and end keywords to indicate the starting and ending of the captions.

4   Results and Discussion

Several evaluation methods are used for image captioning, including:

1. BLEU:
     It is an evaluation metric used in natural language processing to measure the quality of machine-generated translations. It compares the n-gram overlap between the sentences. BLEU scores range from 0 to 1, with higher scores indicating a better-quality translation. [2, 4, 5, 8, 11–17].
2. ROUGE:
     The ROUGE evaluation matrix consists of several metrics, including ROUGE-1, ROUGE-2, and ROUGE-L. ROUGE-1 calculates the overlap of unigrams (single words) between the generated and reference summaries. ROUGE-2 calculates the overlap of bigrams (pairs of adjacent words), while ROUGE-L measures the longest common subsequence between the generated and reference summaries [8, 10, 12, 15, 16].

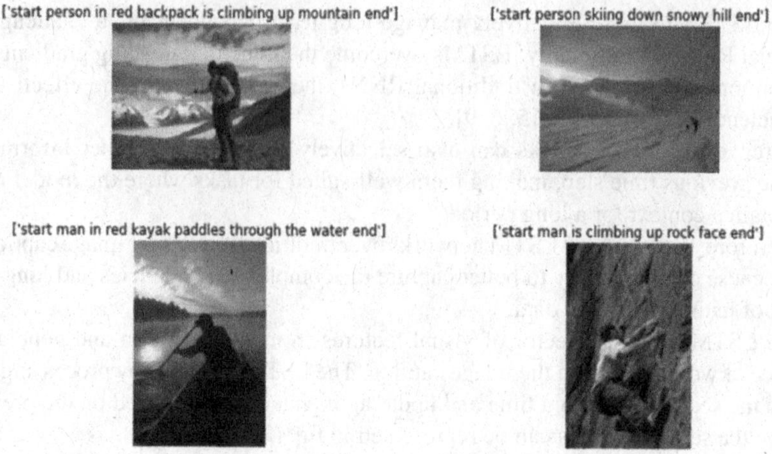

['start person in red backpack is climbing up mountain end']   ['start person skiing down snowy hill end']

['start man in red kayak paddles through the water end']   ['start man is climbing up rock face end']

Fig. 2. Results of Xception-LSTM model on Flicker30K dataset

3. METEOR:

It uses a combination of unigram precision, recall, and alignment-based metrics to evaluate the similarity between the sentences. METEOR is designed to handle nuances of natural language such as synonyms, paraphrases, and word order variations, [5, 6, 8, 10, 12, 13, 15–17].

It should be emphasized that a holistic assessment of image captioning systems cannot rely solely on a single metric. Instead, a blend of multiple evaluation techniques is usually employed to achieve a more comprehensive and precise evaluation of image captioning system performance.

The Xception-LSTM model evaluates the quality of the captions generated using BLEU and METEOR matrices shown in Fig. 3. BLEU evaluation works on n-gram overlapping words where the n value changes from 1 to 4. Based on the value of n results decrease. With that METEOR works on the ordering of the generated caption compare to the labeled caption.

5 Limitation

Despite significant advancements in the field of image captioning in recent years, there remain several challenges and issues that require attention and resolution. Here are a few examples:

- Context – It can be challenging to generate an accurate caption that conveys the intended message of an image. A single image can be perceived in different ways, leading to ambiguity in generating a descriptive caption. For example, a picture of a person riding a bicycle might be captioned differently depending on the specific details in the picture. The caption could vary from "a man riding a bicycle", "a woman riding a bicycle" or "men riding a vehicle" depending on the contextual information in the image.

Evaluation matrices	Results
BLEU-1	85.4
BLEU-2	67.6
BLEU-3	51.0
BLEU-4	38.8
METEOR	50.8

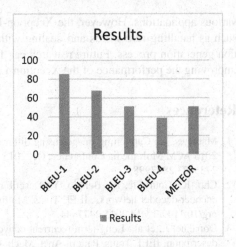

Fig. 3. Evaluation matrices on Flicker30k dataset using Xception-LSTM

- Ambiguity – Creating a precise and descriptive caption for an image can be a difficult task due to the ambiguity and subjectivity of visual content. Images can be interpreted in various ways, making it difficult to generate a single caption that accurately represents the content. Additionally, there may be more than one valid caption for a single image due to the different interpretations that people may have.
- Rare or Unseen Words – In some cases, image captioning models may generate captions that contain infrequent or unfamiliar words, making them difficult for people to comprehend. This issue can be particularly troublesome for individuals who do not have expertise in the language utilized in the caption.
- Data Bias – The process of training image captioning models involves using extensive datasets of image-caption pairs. However, these datasets may occasionally exhibit a bias towards particular types of images or captions. Consequently, the trained model may generate less accurate or less descriptive captions for certain types of images due to this bias.
- Evaluation – Assessing the quality of image captions lacks a single standard metric, and the suitability of various metrics varies based on the specific application. For instance, certain metrics may prioritize accuracy, whereas others may prioritize the diversity or originality of the generated captions.

6 Conclusion and Future Work

Compared to other models like VGG-LSTM and ResNet-LSTM, the Xception-LSTM model offers several advantages. For one, it boasts greater computational and memory efficiency, which makes it more suitable for training on larger datasets. Additionally, the LSTM-based language decoder employed by the Xception-LSTM model is capable of modeling long-term dependencies during the caption generation process. This is a crucial factor in generating coherent and semantically meaningful captions. Moreover, the Xception-LSTM model can be fine-tuned on other tasks such as visual question answering and image retrieval, which demonstrates its versatility and effectiveness in various applications. However, the Xception-LSTM model still faces some challenges such as handling rare words and dealing with the ambiguity and diversity in the caption generation process. Future research can focus on addressing these challenges and improving the performance of the Xception-LSTM model on image captioning.

References

1. Mathews, A.: Captioning images using different styles. In: MM 2015 – Proceedings of the 2015 ACM Multimedia Conference, Oct 2015, pp. 665–668 (2015). https://doi.org/10.1145/2733373.2807998
2. Cho, K., Courville, A., Bengio, Y.: Describing multimedia content using attention-based encoder-decoder networks. IEEE Trans. Multimedia **17**(11), 1875–1886 (2015). https://doi.org/10.1109/TMM.2015.2477044
3. Donahue, J., et al.: Long-term recurrent convolutional networks for visual recognition and description. IEEE Trans. Pattern Anal. Mach. Intell. **39**(4), 677–691 (2017). https://doi.org/10.1109/TPAMI.2016.2599174
4. Karpathy, A., Fei-Fei, L.: Deep visual-semantic alignments for generating image descriptions. IEEE Trans. Pattern Anal. Mach. Intell. **39**(4), 664–676 (2017). https://doi.org/10.1109/TPAMI.2016.2598339
5. Gao, L., Guo, Z., Zhang, H., Xu, X., Shen, H.T.: Video captioning with attention-based LSTM and semantic consistency. IEEE Trans. Multimedia **19**(9), 2045–2055 (2017). https://doi.org/10.1109/TMM.2017.2729019
6. Yang, Y., et al.: Video captioning by adversarial LSTM. IEEE Trans. Image Process. **27**(11), 5600–5611 (2018). https://doi.org/10.1109/TIP.2018.2855422
7. Wu, Q., Shen, C., Wang, P., Dick, A., van den Hengel, A.: Image captioning and visual question answering based on attributes and external knowledge. IEEE Trans. Pattern Anal. Mach. Intell. **40**(6), 1367–1381 (2018). https://doi.org/10.1109/TPAMI.2017.2708709
8. Lu, X., Wang, B., Zheng, X., Li, X.: Exploring models and data for remote sensing image caption generation. IEEE Trans. Geosci. Remote Sens. **56**(4), 2183–2195 (2018). https://doi.org/10.1109/TGRS.2017.2776321
9. Han, M., Chen, W., Moges, A.D.: Fast image captioning using LSTM. Clust. Comput. **22**, 6143–6155 (2019). https://doi.org/10.1007/s10586-018-1885-9
10. Xu, N., et al.: Multi-level policy and reward-based deep reinforcement learning framework for image captioning. IEEE Trans. Multimedia **22**(5), 1372–1383 (2020). https://doi.org/10.1109/TMM.2019.2941820
11. Yu, J., Li, J., Yu, Z., Huang, Q.: Multimodal transformer with multi-view visual representation for image captioning. IEEE Trans. Circuits Syst. Video Technol. **30**(12), 4467–4480 (2020). https://doi.org/10.1109/TCSVT.2019.2947482
12. Turkerud, I.R., Mengshoel, O.J.: Image captioning using deep learning: text augmentation by paraphrasing via back translation (2021). https://doi.org/10.1109/SSCI50451.2021.9659834
13. Chen, N., et al.: Distributed attention for grounded image captioning. In: MM 2021 – Proceedings of the 29th ACM International Conference on Multimedia, Oct 2021, pp. 1966–1975. https://doi.org/10.1145/3474085.3475354
14. Mahalakshmi, P., Fatima, N.S.: Summarization of text and image captioning in information retrieval using deep learning techniques. IEEE Access **10**, 18289–18297 (2022). https://doi.org/10.1109/ACCESS.2022.3150414
15. Ji, J., Ma, Y., Sun, X., Zhou, Y., Wu, Y., Ji, R.: Knowing what to learn: a metric-oriented focal mechanism for image captioning. IEEE Trans. Image Process. **31**, 4321–4335 (2022). https://doi.org/10.1109/tip.2022.3183434

16. Bae, J.W., Lee, S.H., Kim, W.Y., Seong, J.H., Seo, D.H.: Image captioning model using part-of-speech guidance module for description with diverse vocabulary. IEEE Access **10**, 45219–45229 (2022). https://doi.org/10.1109/ACCESS.2022.3169781
17. Ramos, R., Martins, B.: Using neural encoder-decoder models with continuous outputs for remote sensing image captioning. IEEE Access **10**, 24852–24863 (2022). https://doi.org/10.1109/ACCESS.2022.3151874
18. Vinyals, O., Toshev, A., Bengio, S., Erhan, D.: Show and tell: a neural image caption generator. In: Proceedings of the IEEE Conference on Computer Vision and Pattern Recognition (CVPR), pp. 3156–3164 (2015)
19. Karpathy, A., Fei-Fei, L.: Deep visual-semantic alignments for generating image descriptions. In: Proceedings of the IEEE Conference on Computer Vision and Pattern Recognition (CVPR), pp. 3128–3137 (2015)
20. Xu, K., et al.: Neural image caption generation with visual attention. In: Proceedings of the IEEE Conference on Computer Vision and Pattern Recognition (CVPR), pp. 3121–3129 (2015)
21. Anderson, P., et al.: Bottom-up and top-down attention for image captioning and visual question answering. In: Proceedings of the IEEE Conference on Computer Vision and Pattern Recognition (CVPR), pp. 6077–6086 (2018)
22. Li, X., Yin, X., Li, C., Hu, Z., Zhang, H., Sun, F.: VLP: vision-language pre-training for image captioning. In: Proceedings of the IEEE/CVF Conference on Computer Vision and Pattern Recognition (CVPR), pp. 10976–10985 (2020)
23. Cornia, M., Stefanini, M., Baraldi, L., Cucchiara, R.: Meshed-memory transformer for image captioning. In: Proceedings of the IEEE/CVF Conference on Computer Vision and Pattern Recognition (CVPR), pp. 1635–1644 (2020)
24. Lu, J., Batra, D., Parikh, D., Lee, S.: Unicoder-VL: a universal encoder for vision and language. In: Proceedings of the IEEE/CVF Conference on Computer Vision and Pattern Recognition (CVPR), pp. 2238–2247 (2021)
25. Huang, L., Li, Y., Shen, J., Wu, J.: Parallel decoding of hierarchical structure for image captioning. In: Proceedings of the IEEE/CVF Conference on Computer Vision and Pattern Recognition (CVPR), pp. 11168–11177 (2021)
26. Hodosh, M., Young, P., Hockenmaier, J.: Framing image description as a ranking task: data, models and evaluation metrics. J. Artif. Intell. Res. **47**, 853–899 (2013)

Performance Analysis of Different Classifiers Using HOG and LBP for Traffic Sign Detection

Abhisek Panigrahi[✉], Bibhudatta Nayak, Soumya Sahoo, OmPrakash Sahoo, and Santanu Kumar Prusty

Department of Computer Science and Engineering, C.V. Raman Global University, Bhubaneswar, India
abhisekp2001@gmail.com

Abstract. A critical requirement for a traffic assistance system that can detect and recognize traffic signs in challenging circumstances has arisen as a result of the increased use of electrically powered autonomous cars. In order to overcome this difficulty, we provide a real-time method that greatly improves the effectiveness of traffic sign identification. We examine two widely utilized modern object detectors, Faster RCNN and YOLO, for traffic signal detection. Our experimental study shows that the faster RCNN model and the YOLO model give better accuracy, but they cannot meet real-time performance. In this paper, we evaluate a method for detecting traffic signs that has the potential to greatly enhance traffic safety and avoid accidents, especially in the era of electrically powered automobiles. The suggested solution has undergone extensive testing on the GTSDB and is based on reliable and effective techniques for feature extraction, ROI detection, and subclass sign recognition. Our method represents a significant advancement in the creation of traffic assistance systems that can swiftly and reliably detect and recognize traffic slgns.

Keywords: Gabor filter · HOG transform · Linear binary pattern · SVM · Random Forest · KNN

1 Introduction

The control of traffic and promotion of road safety are both greatly aided by traffic signs. Accidents and traffic jams are still frequently brought on by people disobeying traffic laws, such as failing to follow traffic signs. A driver alert system with the ability to precisely detect and recognize traffic signs in real-time is necessary to address this problem.

In this study, we offer a unique method for detecting and recognizing traffic signs in real-time in a busy traffic environment. The proposed method involves converting images of the road scene to grayscale and filtering them with simplified Gabor wavelets (SGW) to enhance the edges of traffic signs for improved detection [1]. The Maximally Stable External Regions (MSER) approach is then used to extract the area of interest (ROI). We have explored the use of multiple machine learning algorithms with various

S. Kadry and R. Prasath (Eds.): MIKE 2023, LNAI 13924, pp. 34–44, 2023.
https://doi.org/10.1007/978-3-031-44084-7_4

feature extraction techniques to classify the detected ROI from the MSER. This was done to compare the performance of different approaches and determine which is best suited for our proposed system. A classifier is used to classify the superclass of traffic signs.

Extensive tests were carried out on actual traffic scenarios to assess the detection accuracy, recognition accuracy, and processing speed of our suggested method. The findings demonstrate that our approach performs competitively in terms of accuracy and processing speed, qualifying it as a viable option for real-time traffic sign detection and identification.

In addition, we have explored the use of multiple machine learning algorithms with various feature extraction techniques to classify the detected ROI from the MSER [2]. This was done to compare the performance of different approaches and determine which is best suited for our proposed system. The results of these tests are included in our findings, providing a comprehensive evaluation of the effectiveness of our proposed approach.

The ultimate objective of the proposed method is to create a driver warning system that, by giving timely and precise notifications to drivers when they approach traffic signs, will dramatically minimize traffic rule breaches. A system like this has the potential to significantly improve traffic flow, lessen accidents, and improve road safety, resulting in more effective and sustainable transportation systems.

Authors Contribution

1. We have recorded videos in busy local street to get some real life environment situation.
2. We have created a database of images from the videos for testing purpose.
3. We have compared different feature descriptor like HOG and LBP.

2 Literature Review

The process of recognising traffic signs typically consists of two stages: the initial stage is traffic sign detection, which looks at the position and size of the signs in the images of the traffic scene, and the second is traffic sign recognition, which is focused on classifying the signs into the correct category. Traffic sign detection often relies on the colour and form characteristics of the signs, while traffic sign identification frequently uses classifiers like convolutional neural networks (CNN) and SVM with discriminative features.

Red, blue, and white make up the limited colour palette used on traffic signs, which helps us recognise them from the surroundings [3]. The colour-based traffic sign detection approach is demonstrated to be the most direct and basic method in the current study. However, colour-based detection techniques frequently fall short of expectations owing to intense light, weak light, and other unfavourable weather circumstances.

Many researchers have recently used technologies in this field, as deep learning approaches have prominently displayed representation ability and produced remarkable performance in traffic sign identification. With a hinge-loss stochastic gradient descent approach and a convolution neural network based on deep learning, it was able to attain a high recognition rate. Offered a multi-column deep neural network operating on a graphics processing unit (GPU) for traffic categorization and achieved a recognition rate that was superior to that of a person. In order to categorise traffic signs, Qian employed CNN as the classifier and learned the discriminative feature of maximum pooling locations. He achieved a performance that was equivalent to that of the state-of-the-art technique [4]. Nevertheless, to satisfy the needs of real-time applications, it is required to further investigate the selection of discriminative features and explore the network topology in order to increase classification accuracy and processing speed.

For a long time, there were no tough public datasets in this domain, but that changed in 2011. Larsson and Felsberg, as well as Stallkamp et al., presented difficult datasets, containing annotations for traffic sign detection categorization. The German Traffic Sign Recognition Benchmark (GTSRB), the German Traffic Sign Detection Benchmark (GTSDB), and the Belgian Traffic Sign Classification (BTSC) were among the datasets used. More academics have been drawn to the GTSDB and, in particular, the GTRSB, and some of them have discovered novel techniques to verify utilizing this database.

3 Methodology

3.1 Dataset

The dataset is an important part of this strategy. A substantial amount of traffic sign data is required to train and validate a deep convolutional neural network traffic sign recognition model, which is currently included in well-known traffic sign databases. The GTSRB and GTSDB traffic sign datasets are used in this work. These two datasets contain a wide range of sophisticated traffic signs, including sign tilt, uneven illumination, traffic signs with distraction, occlusion, and comparable backdrop colours, as well as genuine scene maps (Fig. 1).

3.2 Proposed Method

Our proposed method has three steps: feature extraction, detection, and classification. First, we extract the road images from the streaming video and convert them to grayscale. Then comes the feature extraction phase. For edge detection and smoothing of the image, we use the simplified Gabor filter. By using different parameters, the image has been smoothed. The borders of the traffic signs were reinforced, and the regions inside the traffic signs were smoothed in the resulting Gabor feature map picture. Then comes the detection phase. In this step, we used the maximally stable extremal regions (MSERs) method to determine the regions of interest (ROI) and filtered out places where there was a minimal chance of traffic signs existing based on our stated filter rules.

The proposed regions were then categorised using a variety of different processor and classifier combinations to determine the super-class of traffic signs to which each area belonged (Fig. 2).

4 Feature Extraction

Feature extraction is the process of converting raw data into numerical features that may be processed while retaining the information in the original data set. It generates better results than merely utilising machine learning on raw data. It comprises features like

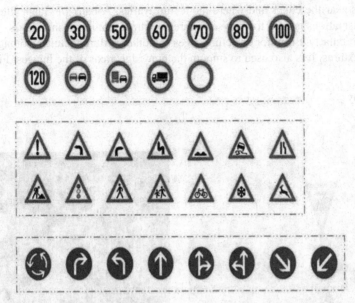

Fig. 1. Different Traffic Signs in Data-set

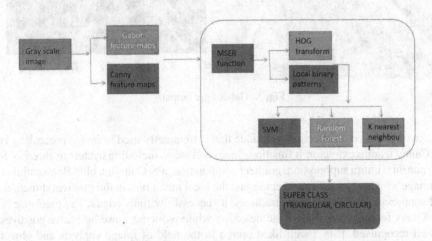

Fig. 2. Flowchart of the Methodology

corners, edges, ROI, and so on. The system's precise feature is to distinguish the traffic sign from the backdrop image by using image colour, shape, size, and orientation.

4.1 Gabor Filter

A Gabor filter is a linear filter that is used in image processing to identify edges, classify textures, and extract features. When a Gabor filter is applied to a picture, it produces the best results near the edges and where the texture varies. A bank of Gabor filters with varying orientations is used to analyse texture or extract features from images.

It is particularly useful for detecting edges and other texture features in images with complex textures. It is also used to smooth the non-edge areas of the images (Fig. 3).

4.2 Canny

Fig. 3. Gabor filter output

An effective edge detection technique that is frequently used in image processing is the Canny feature extractor. It functions in several steps, including hysteresis thresholding, non-maximum suppression, gradient computation, and Gaussian blur. By examining an image's intensity gradient, keeping just the local maximum in the gradient direction, and suppressing non-maximum gradients, it successfully finds edges. The precision of the Canny feature extractor's edge detection while reducing noise and false positives is well recognised. It is a well-liked option in the field of image analysis and object identification since it generates binary edge maps, which are commonly used in several computer vision applications (Table 1).

When compared to the Canny and traditional Gabor wavelet approaches, the SGW-based strategy produced the best computational results.

The Canny method produced a binary image, which resulted in a significant loss of image information and could not guarantee adequate detection and classification results, despite the SGW being slightly faster than the Canny.

Table 1. Comparison of the computational complexities of Canny, TWG and SWG algorithm

Algorithms	No. of Additions	No. of Multiplications
Canny	$40N^2$	$17N^2$
TGW	$48N^2\log2N^2 + 16N^2$	$32N^2\log2N^2 + 32N^2$
SGW	$18N^2$	$16N^2$

5 Traffic Sign Detection

Traffic signs are strictly classified based on shape and colour, so that they can be identified by these attributes by any AI. As a result, traffic sign recognition is dependent on its shape or shade.

In our suggested technique, we improved the shape information of traffic signs by using simpler Gabor filters to smooth and enhance the edges. Then the gabor maps have been processed by the MSER algorithm for finding the region of interest. Then the ROI's have been classified into different signs.

5.1 MSER Algorithm

MSER is a powerful tool for extracting regions of interest (ROI) from an image and has many practical applications in computer vision and image processing. Maximally Stable Extremal Regions (MSER) is an image segmentation algorithm that identifies and extracts regions of interest from an image. The algorithm works by identifying regions that are both stable and extremal with respect to a threshold parameter [5].

In this study, the MSER algorithm was employed to identify the regions of interest (ROI) in the image. By utilising a threshold value, specifically set at a delta of 18, in our implementation, the algorithm determined the stable and extremal regions based on their intensity values. A region in MSER is defined as a connected component of pixels with similar intensity values. Initially, the algorithm calculates the intensity threshold that produces the largest region and then progressively decreases the threshold until the region breaks into smaller pieces. The stable regions are the ones that persist across multiple thresholds, while the extremal regions are the ones that have the most distinct intensity values compared to their surroundings.

The chosen threshold value of 18 was determined empirically, based on the characteristics of the specific image dataset used in this context. It enabled the algorithm to capture the regions of interest with optimal performance. MSER demonstrates several advantages over other image segmentation methods, such as its robustness to noise and its ability to handle images with varying illumination and contrast. Its versatility has been leveraged in various applications, including object detection and text detection in images.

5.2 Image Pre-processing

In order to prepare the image data before feeding it through the different classifiers, we have implemented both HOG and LBP image per-processors. These techniques extract relevant features from the grayscale image data, which are then used as input for the subsequent classification steps. By using both HOG and LBP, we are able to capture complementary information about traffic signs, improving the accuracy of our system.

5.2.1 HOG

To determine which superclasses the suggested regions should be assigned to, we first extracted the Histogram of Oriented Gradient (HOG) features from the SGW feature map, which we dubbed SGW-HOG features [6]. We will be using the built-in function HOG from the Python library Skimage. By analysing the orientation and distribution of gradients, the HOG technique is used to extract features from the detected region of interest. The gradient strengths must be locally normalised to take into account variations in illumination and contrast, which necessitates combining the cells into bigger, spatially connected blocks. The road sign blob is separated into smaller sections termed 'cells' in order to discover the HOG characteristics. The cells were then gathered into blocks, and the resulting histograms were subjected to block normalisation to address the issue of illumination variance. It offers a condensed representation of the image that is reasonably resistant to changes in lighting, contrast, and image distortion. Better invariance to variations in illumination and shadowing is the effect of this normalisation. Values for parameters of HOG are orientations 9, pixels_per_cell (6, 6), and cells_per_block (6, 6).

skimage Compute the HOG by:

1. computing the gradient image in x and y
2. computing gradient histograms
3. Normalizing across blocks

5.2.2 LBP

Local Binary Patterns (LBP) is a widely used feature extraction technique that describes the local texture of an image by comparing the intensities of its pixels with their neighbors. In our proposed system, we use LBP as one of the image processors to extract features from the grayscale images of traffic signs [7]. Specifically, we have divided the image data into small parts, or regions, and for each region, we calculate a histogram of the LBP values of its pixels. These histograms are combined to generate a feature vector, which reflects the image's texture information. By using LBP as a feature extraction technique, we are able to capture the local texture and pattern information of the traffic signs, which can be useful for accurate classification. LBP is initially defined in a neighborhood of eight pixels, and a threshold is established using the centre pixel's grey value. All neighbors are assigned a value of 1 if their values are greater than or equal to those of the centre pixel; otherwise, they are set to 0. The values for parameters of LBP are radius = 1, n_points = 8 * radius, and method = 'uniform'.

6 Traffic Sign Classification

We have explored the use of multiple machine learning algorithms with various feature extraction techniques to classify the detected ROI from the MSER. Specifically, we have used three classifiers, namely Support Vector Machines (SVM), K-Nearest Neighbors

(KNN), and Random Forest, and compared their performance in terms of accuracy, precision, recall, and F1 score. These tests were conducted to determine the best classifier for our proposed system and to provide a more comprehensive evaluation of the effectiveness of our approach. A classifier is used to classify the superclass of traffic signs.

6.1 SVM

SVM is a strong and extensively used classification technique that has proven to be useful in a wide range of real-world applications. SVM has several advantages over other classification methods, including the ability to handle high-dimensional data, noise robustness, and effectiveness in dealing with non-linear data distributions. SVM has been utilised effectively in a range of applications, including image classification, due to its adaptability and dependability [8].

In our technique, we used a multi-class SVM classifier to divide data into two or more groups depending on input data attributes. Because the traffic signs had two super classes (triangular and circular), we trained a three-class SVM because there was a third super class known as the super class of negative data. This enabled us to efficiently categorise traffic signs, ensuring precise detection and identification.

SVM's capacity to handle high-dimensional data is one of its primary benefits. This was especially useful in our approach because the input photos comprised a significant number of elements that required to be analysed in order to identify and recognise traffic signs effectively. Furthermore, SVM is very resistant to noise and can deal successfully with non-linear data distributions, which improves its performance in real-world applications.

Overall, the use of SVM in our approach proved to be highly effective, achieving high levels of accuracy in both detection and recognition tasks. We believe that SVM will continue to play an important role in the development of transportation systems, providing a reliable and efficient method for traffic sign detection and recognition.

6.2 Random Forest

Random Forest is a well-known and mostly used classification and regression algorithm. It uses multiple decision trees and takes the majority vote for classification [9]. Random forest has the advantage of being able to accommodate missing values and a high number of features without overfitting. It is basic, adaptable, and capable of handling binary and categorical data. The random forest method is an ensemble learning strategy that combines several classifiers to improve the performance of a model.

6.3 KNN

A non-parametric supervised learning technique for classification and regression applications is the k-nearest neighbor (k-NN) classifier. It operates by categorising fresh observations in a feature space according to the majority class of their k-nearest neighbors. Euclidean distance, Manhattan distance, and Minkowski distance are a few of the often-used distance metrics for gauging how closely two observations match each other. Through cross-validation or other tuning techniques, the value of k, which stands for the number of nearest neighbors to take into account, is frequently calculated. The k-NN algorithm is easy to understand, can handle high-dimensional data, and is straight-forward. For large datasets, it can be computationally expensive and subject to the dimensional curse [10].

It is also known as a lazy learner algorithm since it keeps the training dataset instead of learning from it right away. When categorizing data, it instead utilizes the dataset to execute an action.

7   Experimental Results

Based on our proposed approach for real-time traffic sign detection and recognition using simplified Gabor wavelets (SGW) and Maximally Stable Extremal Regions (MSER), we conducted extensive tests to evaluate the performance of our system using various machine learning classifiers and feature extraction techniques.

We specifically evaluated the performance of our suggested method, which uses Histogram of Oriented Gradients (HOG) and Local Binary Patterns (LBP) as feature extraction techniques, against three classifiers: Support Vector Machines (SVM), K-Nearest Neighbors (KNN), and Random Forest (Table 2, Table 3).

Table 2. Comparision of SVM, Random Forest and KNN with LBP

LBP	SVM	Random Forest	KNN
Accuracy	0.802	0.841	0.826
Precision	0.986	0.845	0.868
Recall	0.829	0.992	0.935
F1 score	0.900	0.967	0.897
Specificity	0.389	0.684	0.461

Table 3. Comparison of SVM, Random Forest and KNN with HOG

HOG	SVM	Random Forest	KNN
Accuracy	0.961	0.944	0.954
Precision	0.991	0.974	0.972
Recall	0.965	0.961	0.975
F1 score	0.977	0.967	0.973
Specificity	0.952	0.863	0.870

The experimental results showed that HOG and LBP both yielded high accuracy, precision, recall, and F1 score for all three classifiers. However, we observed that HOG generally outperformed LBP in terms of accuracy and F1 score for all three classifiers.

SVM had the greatest accuracy and F1 score among the three classifiers for HOG. For LBP, on the other hand, the Random Forest classifier fared best, with the greatest accuracy and F1 score.

Overall, our experimental results demonstrate the effectiveness of our proposed approach for real-time traffic sign detection and recognition, with HOG as the preferred feature extraction technique and SVM as the preferred classifier for optimal performance (Fig. 4).

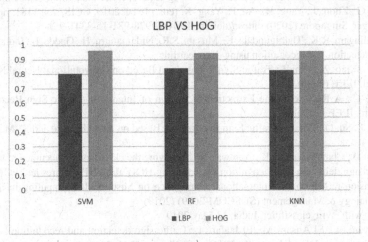

Fig. 4. LBP VS HOG

8 Conclusion

Our proposed approach using Gabor filter feature maps and SVM on the HOG feature proved to be a promising method for real-time traffic sign detection and recognition. The results showed high accuracy and an F1 score for circular and triangular shapes, which are common shapes for traffic signs. Additionally, we compared the performance of three classifiers (SVM, KNN, and Random Forest) with two feature extraction techniques (HOG and LBP) to further evaluate the effectiveness of our proposed approach.

For future improvement, we can explore the use of deep learning algorithms, such as Convolutional Neural Networks (CNNs), which have been shown to achieve state-of-the-art performance in image classification tasks. We can also investigate the use of more advanced feature extraction techniques, such as Scale-Invariant Feature Transform (SIFT) and Speeded Up Robust Features (SURF), to further enhance the detection and recognition of traffic signs.

Furthermore, we can extend the scope of our proposed approach to other road-related objects, such as pedestrian detection, lane detection, and vehicle detection, to develop a more comprehensive and advanced driver assistance system. Ultimately, the implementation of our proposed approach and future improvements can lead to the development of more effective and sustainable transportation systems by reducing traffic rule breaches, improving traffic flow, and enhancing road safety.

References

1. Umarov, M., Muradov, F., Azamov, T.: Traffic Sign Recognition Method Based on Simplified Gabor Wavelets and CNNs. IEEE (2021)
2. Yin, S., Xu, Y.: Fast traffic sign detection using color-specific quaternion Gabor filters. In: Zhai, G., Zhou, J., Yang, H., An, P., Yang, X. (eds.) IFTC 2019. CCIS, vol. 1181, pp. 3–12. Springer, Singapore (2020). https://doi.org/10.1007/978-981-15-3341-9_1
3. Megalingam, R.K., Thanigundala, K., Musani, S.R., Nidamanuru, H., Gadde, L.: Indian traffic sign detection and recognition using deep learning
4. He, Z., Xiao, Z., Yan, Z.: Traffic Sign Recognition Based on Convolutional Neural Network Model. IEEE (2020)
5. Deng, C.: A Review on the Extraction of Region of Interest in Traffic Sign Recognition System. IEEE (2020)
6. Tang, J., Su, Q., Lin, C.: Traffic Sign Recognition Based on HOG Feature and SVM. EITCE (2021)
7. Sonia, D., Chaurasiyab, R.K., Agrawal, S.: Improving the classification accuracy of accurate traffic sign detection and recognition system using HOG and LBP features and PCA-based dimension reduction. In: International Conference on Sustainable Computing in Science, Technology & Management (SUSCOM-2019) (2019)
8. HOG_with_svm_classififier, India, 13 May 2019
9. Ellahyani, A., El Ansari, M., El Jaafari, I.: Traffic sign detection and recognition based on random forests. https://doi.org/10.1016/j.asoc.Elsevier B.V (2016)
10. Narayana, M., Bhavani, N.P.G.: Detection of traffic signs under various conditions using Random Forest algorithm comparison with KNN and SVM. IEEE (2022)

Classification of the Class Imbalanced Data Using Mahalanobis Distance with Feature Filtering

S. Karthikeyan[1(✉)], T. Kathirvalavakumar[1], and Rajendra Prasath[2]

[1] Research Centre in Computer Science, V.H.N. Senthikumara Nadar College, Virudhunagar,
India
rgskarthi@gmail.com
[2] Department of CSE, Indian Institute of Information Technology, Sri City, Chittoor, India

Abstract. Class imbalance exists in many real-life applications. Classification of
an imbalanced dataset without pre-processing is more complex. Existing literature has numerous solutions to get rid of this problem either through data level
or algorithm level modification, but the imbalanced data classification still needs
significant improvements. In the proposed work, a new method has been formulated to reduce the misclassification in the classes. The feature filtering technique
is applied iteratively in an imbalanced dataset to use only significant features for
the classification. The chi-square statistics is used for the filtering process and
Mahalanobis distance (MD) is used for classification. The performance of the
MD is being compared with the state-of-the-art classifiers and observed that the
Mahalanobis distance performs better in 6 datasets and gives the same result in
3 datasets out of the 12 experimented datasets. The observations show that the
association between the features is a major deciding factor during the classification. If each feature is independent, then the classification region is separable even
though the dataset is imbalanced. The combination of the MD and the chi-square
feature filtering helps in achieving a better AUC score.

Keywords: Imbalanced Data · Classification · Feature Filtering · Mahalanobis
Distance

1 Introduction

Imbalanced data distribution between the classes exists in many fields. The importance
of solving this problem with respect to classification is still an open research problem.
From the literature, it is found that this problem can be solved either with data level or
algorithm level approach. The data re-sampling improves the performance of the model
but often leads to overfitting or under-fitting issues, and there are conflicting views and an
overall lack in research regarding the data level modification that needs to be addressed
properly [1], so instead of data re-sampling it has been decided to perform algorithm
level approach to solve the class skewed classification.

S. Kadry and R. Prasath (Eds.): MIKE 2023, LNAI 13924, pp. 45–53, 2023.
https://doi.org/10.1007/978-3-031-44084-7_5

MD [2] is a metric which takes into account the correlation between variables and provides a more accurate measure of the difference between two observations than the Euclidean distance. MD is better than the Euclidean Distance [3] since it considers the mean and covariance to calculate the distance of each pattern with the available patterns. Buda et al., [4] have used the MD in handling the extreme imbalanced data especially in anomaly detection. Abdi and Hashemi [5] have used the MD to oversample the minority class data; the generated synthetic samples have the same class average of the samples in the minority class that helps in reducing the sample overlapping in different class regions. Bennin et al., [6] have proposed the MAHAKIL algorithm that divides the minority class data into two parent groups and each new synthetic sample inherits the properties of the parents which creates diversity in the data distribution and avoids overfitting issues. Arun and Lakshmi [7] have shown that the initial discrimination measure for the minority class data is calculated using the MD and the minority class samples are separated into clusters based on MD. The parent samples for the genetic algorithm (GA) are chosen from different clusters to form the synthetic samples. Taguchi and Rajesh [8] have presented the Mahalanobis-Taguchi System (MTS) classifier, the MD is used to calculate the similarity of unknown samples from the known samples. Taguchi method is used for an optimized classification performance, but MTS lacks in finding a perfect threshold for classification. El-Banna [9] has proposed the Mahalanobis Genetic Algorithm (MGA) by combining MD and GA. Training data is converted into the balanced form by applying SMOTE on the minority class data and the MD is considered as a critical threshold value (MDcr) to separate the major and minor classes. During the classification, if the MD of the test pattern is greater than MDcr then it is considered as a majority class sample, otherwise it is considered as a minority class sample. El-Banna [10] has proposed the Modified Mahalanobis Taguchi System to enhance the performance of MTS by applying the ROC curve to determine the threshold that helps in separating the classes. Yao and Lin [11] have proposed an evolutionary MD oversampling which integrates the multi objective particle swarm optimization with Gustafson-Kessel algorithm to improve the learning ability of the classifier in a multi class environment. Siddappa and Kampalappa [12] have proposed the local mahalanobis distance learning method to improve the performance of the K-nearest neighbour algorithm for classification. Here the MD is used to extract the necessary information that helps in finding the separable region of the classes and avoids the risk of overfitting.

Feature filtering is based on the important relationship between the input features and the output. After removing the irrelevant features, better classification accuracy is obtained with minimal overfitting issues. Some popular methods used for feature filtering are chi-square, Pearson correlation coefficient, recursive feature elimination and analysis of variance tests [13].

In this research, a new approach is proposed to classify the imbalanced data by using MD after filtering insignificant features but without data re-sampling. It extracts the significant features based on the chi square test. It increases the performance of the classification and reduces the misclassification in both major and minor classes. The rest of the sections are organized as follows: Sect. 2 describes the proposed method, Sect. 3 presents the findings of the data analysis, and Sect. 4 concludes with the major findings.

2 Proposed Method

Significant feature selection, MD calculation, probability calculation for the calculated distance and data classification are involved in this method. The proposed method filters the insignificant features using the chi-square statistics and performs classification using the MD. Following the feature filtering, MD is calculated between the testing pattern and the major and minor class of the training data separately. With the obtained MD value, the probability scores are calculated for both major and minor classes and a class with the highest probability is considered as class label for the test pattern. The predicted labels are compared with the actual labels of the testing pattern to find the misclassification in both major and minor classes and AUC score.

3 Mahalanobis Distance

MD is a statistical distance measure that measures the distance between a test pattern and the mean of the training patterns by taking into account the correlation between the features. The MD is used to determine how a test pattern is similar or dissimilar with the training patterns. Initially covariance (C) for the training patterns is calculated. Mean of the training patterns is calculated (μ). MD is calculated using

$$D^2 = (x - \mu).C^{-1}.(x - \mu)^T$$

where x, μ, C^{-1} represents a test pattern, mean of the training patterns and the inverse of the covariance of the training patterns respectively. The steps in calculating the MD are as follows:

1. Generate mean pattern with the means of each feature of the training dataset
2. Calculate difference between test pattern and mean pattern
3. Find Inverse covariance for the training patterns
4. Transpose the values obtained in step-2
5. Multiply the results obtained in step 2, step 3 and step 4
6. Find Square root for the value obtained in step 5.

3.1 Feature Filtering

Feature filtering can significantly improve the performance of the classifier because of its significance. Scikit-learn [14] is a python based tool that assists in performing classification, regression and many other tasks. Significant features for our experiment are extracted using the "selectKBest" function available in scikit-learn. The score functions available in scikit-learn for the feature filtering are ANOVA test (f_classif), mutual information for a discrete target (mutual_info_classif), Pearson correlation coefficient (f_regression), mutual information for a continuous target (mutual_info_regression) and chi-square statistics (chi2). These score functions are applied on the selectKBest method to identify the function that performs better on the dataset. From the trials made, it is observed that the chi-square statistics score function performs better than others. So, the chi-square function is used for filtering the significant features. The steps in feature filtering are as follows:

1. Compute the chi-square statistics for each feature
2. Each feature is assigned with a score
3. Use selectKBest method to pick the required number of representative features based on the highest score obtained in step-2

3.2 Proposed Procedure

1. Perform feature filtering on the original data to select the k-significant features by selectKBest function with chi-square statistics.
2. After the feature filtering, divide the original dataset randomly into Training (70%) and the Testing (30%) datasets.
3. Divide the Training dataset into Major and Minor class data.
4. Find MD for the test data with Major and Minor training data separately.
5. If the covariance matrix has zeros in the diagonals during MD calculation, then the matrix is singular, and cannot compute its inverse. To overcome this issue, add a small amount of noise with the dataset.
6. The MD value of the Major and Minor classes are used for probability prediction.
7. Divide the MD value of the Major class by the sum of MD values of the Major and Minor classes. Obtained value is subtracted from 1 and the result is the score. Do the same division for the Minor class data.
8. Class of the highest probability score is the class of the test pattern.
9. Compare the predicted label with the actual label to evaluate the performance.
10. Do the steps 1 to 9 by varying k from 2 to the number of features in the dataset.

4 Results and Discussion

The proposed method is experimented with 12 imbalanced datasets with different Imbalance Ratio (IR) which are collected from Keel repository [15] and UCI repository [16]. The information about the datasets is given in Table 1. The original dataset is divided into training dataset (70%) and testing dataset (30%) and no data leakage is present in both training and testing datasets.

The classification accuracy is not a proper measure to evaluate the performance of the classifiers in an imbalanced dataset since the class with lesser number of instances is not reflected in accuracy score because of the large volume of instances in other classes [17]. Area under the curve (AUC) is a proper evaluation measure for the imbalanced data, since it allocates equal importance to both the classes. So, the AUC score is considered in the rest of the discussion.

The experiment starts without pre-processing and the results of the classification are observed and found that the classifiers strive to differentiate the classes when the features of different classes are highly correlated and that leads to unstable or inaccurate classification. Yeast-1–2-8-9_vs_7 is such a dataset with more correlation and is shown in Fig. 1. It gives challenges in determining the variables that are truly playing a role to obtain the outcome. When there is a lesser correlation between the features then the classification is done without difficulty even if the classes are imbalanced, which is the Iris0 dataset and is shown in Fig. 2.

Table 1. Dataset Information

Dataset	# Patterns	IR	Category
Abalone9-18	731	16.4	Real
Abalone19	4174	129.4	Real
Ecoli-0–1-3-7_vs_2-6	281	39.1	Real
Glass5	214	22.7	Real
Iris0	150	2	Real
Page-blocks-1-3_vs_4	472	15.8	Real/Integer
Shuttle-c2-vs-c4	129	20.5	Integer
Yeast-1–2-8-9_vs_7	947	30.5	Real
Yeast-2_vs_8	482	23.1	Real
Yeast5	1484	32.7	Real
Yeast6	1484	41.4	Real
Diabetes	390	18.18	Real/Integer

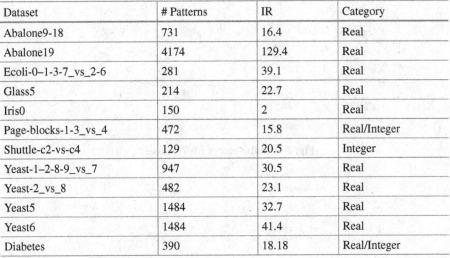

Fig. 1. Pairwise plot of Yeast-1–2-8-9_vs_7 Dataset

The dataset is also experimented with C4.5, K-Nearest Neighbour (KNN), Support Vector Machine (SVM) and Multi-Layer Perceptron (MLP) classifiers for comparison. The AUC score of each classifier is recorded for each filter size and the scores are plotted in Figs. 3 and 4. In each figure, x-axis and y-axis represents filter size and AUC score respectively.

Each classifier performs better at a certain filter size which is dependent on its characteristics. The best results of each classifier are tabulated to analyze the classifier that works better with the imbalanced data. The AUC scores of the original dataset

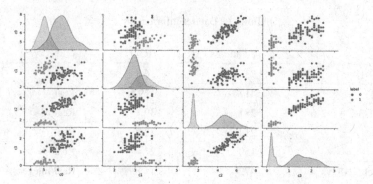

Fig. 2 Pairwise plot of Iris0 Dataset

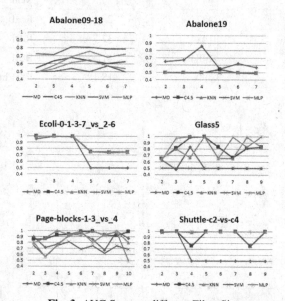

Fig. 3 AUC Score at different Filter Size

without feature filtering and the best AUC result obtained with feature filtering are shown in Table 2. In Table 2, –F along with the classifier name represents the results after applying the filters and –O represents the results without applying the filters. From Table 2, it is clear that the original data without pre-processing does not give better results in 10 out of 12 datasets. Table 3 shows the best AUC score obtained from the proposed method on different classifiers.

From Table 3, it is observed that the Mahalanobis distance with feature filtering works better than C4.5, KNN, SVM and MLP classifiers in 6 datasets and gives same result in 3 datasets and in other datasets the observed results are closer to the better results of other classifiers. The problem in calculating the inverse covariance arises in

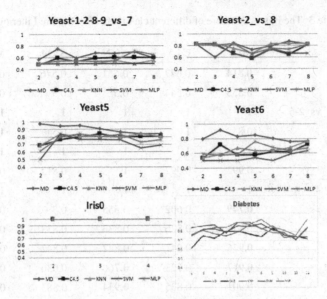

Fig. 4 AUC Score at different Filter Size

Table 2 AUC score of different classifiers with and without Feature Filtering

Dataset	MD-F	MD-O	C4.5-F	C4.5-O	KNN-F	KNN-O	SVM-F	SVM-O	MLP-F	MLP-O
Abalone9-18	**0.814**	0.79	**0.677**	0.629	**0.651**	0.538	**0.576**	0.5	**0.761**	0.723
Abalone19	**0.855**	0.569	**0.544**	0.493	**0.5**	0.499	**0.5**	**0.5**	**0.5**	**0.5**
Ecoli-0–1-3-7_vs_2-6	1	0.5	1	0.743	1	0.743	1	0.75	1	0.75
Glass5	**0.833**	0.5	1	0.833	1	0.833	**0.5**	**0.5**	1	1
Iris0	1	1	1	1	1	1	1	1	1	1
Page-blocks-1-3_vs_4	**0.992**	0.875	**0.992**	0.988	1	0.804	**0.812**	0.687	**0.988**	0.5
Shuttle-c2-vs-c4	1	0.5	1	1	1	1	1	1	1	1
Yeast-1–2-8-9_vs_7	**0.749**	0.592	**0.602**	0.598	**0.555**	0.553	**0.5**	**0.5**	**0.714**	0.648
Yeast-2_vs_8	**0.873**	0.822	**0.826**	0.815	**0.833**	0.829	**0.833**	0.663	**0.833**	0.818
Yeast5	**0.976**	0.838	**0.841**	0.805	**0.842**	0.804	**0.842**	0.688	**0.841**	0.762
Yeast6	**0.917**	0.766	**0.723**	0.723	**0.766**	0.766	**0.636**	0.636	**0.677**	0.672
Diabetes	**0.901**	0.737	**0.835**	0.717	**0.934**	0.934	**0.883**	0.833	**0.868**	0.84

the datasets, ecoli-0–1-3-7_vs_2-6, glass5, shuttle-c2-vs-c4, yeast-1–2-8-9_vs_7, yeast-2_vs_8, yeast5 and yeast6, and are resolved by adding a small amount of noise to the original dataset.

Table 3 The Best AUC score of different classifiers after Feature Filtering

Dataset	MD	C4.5	KNN	SVM	MLP
Abalone9-18	**0.814**	0.677	0.651	0.576	0.761
Abalone19	**0.855**	0.544	0.5	0.5	0.5
Ecoli-0–1-3-7_vs_2-6	1	1	1	1	1
Glass5	0.833	1	1	0.5	1
Iris0	1	1	1	1	1
Page-blocks-1-3_vs_4	0.992	0.992	1	0.812	0.988
Shuttle-c2-vs-c4	1	1	1	1	1
Yeast-1–2-8-9_vs_7	**0.749**	0.602	0.555	0.5	0.714
Yeast-2_vs_8	**0.873**	0.826	0.833	0.833	0.833
Yeast5	**0.976**	0.841	0.842	0.842	0.841
Yeast6	**0.917**	0.723	0.766	0.636	0.677
Diabetes	0.901	0.835	**0.934**	0.883	0.868

5 Conclusion

Chi-square statistics is used in feature filtering to accelerate the classification process. The key findings of the experiment are each classifier exhibits better results at different filter sizes; Strong association between the features makes the classification process more complex; and less association between the features and the wide dispersion of feature values help the classifier to extract the significant features. The combination of the MD and the chi-square feature filtering outperforms the selected state-of-the-art classification algorithms and provides a better classification to the imbalanced classification problems.

References

1. Fernández, A., García, S., Galar, M., Prati, R.C., Krawczyk, B., Herrera, F.: Learning from imbalanced data sets. Vol. 10. Springer, Cham (2018)
2. Mahalanobis, P.C.: On the generalized distances in statistics: Mahalanobis distance. Journal Soc. Bengal **26**, 541–588 (1936)
3. Ghorbani, H.: Mahalanobis distance and its application for detecting multivariate outliers, pp. 583–595. FactaUniversitatis, Series, Mathematics and Informatics (2019)
4. Buda, M., Maki, A., Mazurowski, M.A.: A systematic study of the class imbalance problem in convolutional neural networks. Neural Netw. **106**, 249–259 (2018)
5. Abdi, L., Hashemi, S.: To combat multi-class imbalanced problems by means of over-sampling techniques. IEEE trans. Knowl. Data Eng. **28**(1), 238–251 (2015)
6. Bennin, K., Keung, J., Phannachitta, P., Monden, A., Mensah, S.:Mahakil: Diversity based oversampling approach to alleviate the class imbalance issue in software defect prediction. IEEE Trans. Softw. Eng. **44**(6), 534–550 (2017)
7. Arun, C., Lakshmi, C.: Genetic algorithm-based oversampling approach to prune the class imbalance issue in software defect prediction. Soft Computing **26**(23), 12915–12931 (2022)

8. Taguchi, G., Rajesh, J.: New trends in multivariate diagnosis. Sankhyā: The Indian Journal of Statistics, Series B, pp. 233–248 (2000)
9. El-Banna, M.: A novel approach for classifying imbalance welding data: Mahalanobis genetic algorithm (MGA). The Int. J. Adv. Manuf. Technol. **77**, 407–425 (2015)
10. El-Banna, M.: Modified Mahalanobis Taguchi system for imbalance data classification. Comput. Intelli. Neurosci. 2017 (2017)
11. Yao, Leehter, Lin, Tung-Bin.: Evolutionary mahalanobis distance-based oversampling for multi-class imbalanced data classification. Sensors **21**(19), 6616 (2021)
12. Siddappa, N., Kampalappa, T.: Imbalance data classification using local mahalanobis distance learning based on nearest neighbour. SN Comput. Sci. **1**, 1–9 (2020)
13. Sarker, I.H.: Machine learning: Algorithms, real-world applications and research directions. SN Comput. Sci. **2**(3), 160 (2021)
14. Pedregosa, F., et al.: Scikit-learn: Machine learning in Python. J. Mach. Learn. Res. **12**, 2825–2830 (2011)
15. Derrac, J., Garcia, S., Sanchez, L., Herrera, F.: Keel data-mining software tool: Data set repository, integration of algorithms and experimental analysis framework. J. Mult. Valued Logic Soft Comput. 17 (2015)
16. Dua, D., Graff, C.: UCI machine learning repository [http://archive.ics.uci.edu/ml]. University of California, School of Information and Computer Science, Irvine, CA (2019)
17. Kulkarni, A., Chong, D., Batarseh, F.A.: Foundations of data imbalance and solutions for a data democracy. data democracy, pp. 83–106. Academic Press (2020)

Towards Data-Centric Approaches to Lung Cancer Classification

Mark Movh and Isah A. Lawal(✉)

Noroff University College, Kristiansand, Norway
`isah.lawal@noroff.no`

Abstract. There is an ever-growing need to review Artificial Intelligence and its corresponding implementation methodology in medical image analysis. The discussion of optimizing code versus improving data is of prime importance when maximizing model performance in medical image classification. Recently, a majority of studies have been model-centric. It is crucial to investigate data-centric methodologies and how medical image quality impacts a model's learning capabilities. This study opts toward data-related modifications for model improvement in lung cancer classification, acting as a proof of concept for developing data-centric AI. The proposed data-centric approach (DCA) modifies CT-scan images of the lung through 3 stages; image preprocessing, image segmentation, and feature extraction. The modified images were used to train a simple Convolutional Neural Network (CNN) for the classification task. We evaluate the performance of the proposed method using a publicly available real-world dataset of lung CT scans. Our method achieves a classification score (F1 score) of up to 0.889. This performance is superior to that reported using a model-centric approach on the same dataset, which conducted automatic hyperparameter optimization using the random search algorithm.

Keywords: Lung Cancer Classification · Data-Centric AI · Medical Image Analysis

1 Introduction

Lung cancer is the most prominent and deadly variant of cancer, indicating the need for accurate identification and diagnosis. Computed Tomography (CT) scans are a method for capturing images of the lungs, to identify clumps of abnormal cells. CT is considered the most common, due to its accuracy and benefit of perspective without having features overlap one another [3]. However, even with the technological development of these machines, human limitations are often the cause of inefficient information evaluation. Manual reading, understanding, and overall analysis of the scans could become unreliable without an experienced radiologist [4]. Therefore, to combat these issues, the process of automating detection and diagnosis in healthcare has been researched and developed through Machine Learning.

© The Author(s), under exclusive license to Springer Nature Switzerland AG 2023
S. Kadry and R. Prasath (Eds.): MIKE 2023, LNAI 13924, pp. 54–66, 2023.
https://doi.org/10.1007/978-3-031-44084-7_6

Machine Learning (ML) is an area within Artificial Intelligence that revolves around computers and how they can learn in order to process data, find more complex patterns, and present information collected from these patterns. The process mainly consists of training a model on real-world examples, such that it is capable of learning their distinctive characteristics. Three major ML types are supervised, semi-supervised, and unsupervised learning [9]. Each category has various problem-solving capabilities, however this study will primarily focus on supervised ML as it focuses on lung cancer classification. In supervised ML there are features (inputs) and labels (outputs); the goal of the algorithm is to learn how it can map these features to their respective labels [2]. Once an ML model has been created and trained, it can be further improved through specific optimization. However, these optimizations may come in many forms based on the two important components that build up the ML model; the data, which would be some type of input (images etc.), and the code, that is, the algorithm that undergoes the learning described above [6].

Recently, there has been a large discussion as to which part should be optimized for a more accurate model; the data, or the code. Andrew Ng [7] popularized this discussion through his presentation of model-centric and data-centric AI concepts, bringing to light an important consideration as ML applications continue growing. A model-centric approach refrains from adjusting the data while continuously optimizing a model's structure by tuning the model parameters and hyperparameters to maximize performance (see Fig. 1) [6]. A data-centric approach, on the other hand, is concerned with keeping the model structure fixed while improving the quality of data through pre-processing (see Fig. 2) [8].

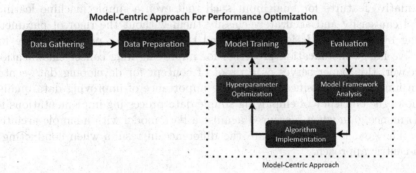

Fig. 1. Model-centric approach focused on model optimization. The dotted region encapsulates the overarching stages of this optimization for performance improvement.

The problem with the model-centric approach for medical image classification is that the model sometimes becomes too complex due to excessive parameter tuning to fit the raw medical images. Thus potentially leading to a model with poor classification accuracy that is undesirable for medical applications. Additionally, a lack of available data in large amounts and a lack of refined solutions

Data-Centric Approach For Performance Optimization

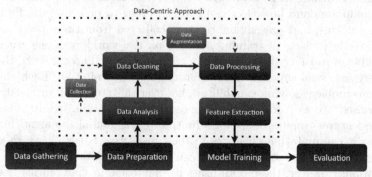

Fig. 2. Data-centric approach focused on data enhancement. The dotted region encapsulates the primary stages to achieve higher-quality data for performance improvement.

within the medical sector should pave the way for an increase in data-centric methodologies [6,12].

The main idea of this study is to improve the quality of medical images and data preparation stages through a data-centric approach, thus achieving a more accurate lung cancer classification model. We improve the performance of a simple CNN model implementation through modification of image pre-processing, image segmentation, and feature extraction stages. Firstly, we extract prominent information through data analysis. Secondly, we undergo pre-processing stages to transform the data into more suitable input. Finally, determining the most informative features for final input such that even a simple machine learning model can easily and accurately classify them without the rigor of parameter tuning by an expert. It is acknowledged that this paper presents simple but effective image modification processes for improving lung cancer classification. Moreover, this paper serves as a proof of concept for developing data-centric AI in lung cancer classification and the importance of improving data quality. It shows the efficiency of employing simple data processing implementations for performance gain when measured against a base model with a simple architecture. The goal is to demonstrate the difference in results when conducting a data-centric approach.

2   Related Work

There has been a growing interest in research on the concept of data-centric AI since the term became popularized. Still there exists a large knowledge gap within the field due to a lack of available studies, especially in the medical sector.

Research by Angela Zhang et al. [13] in 2022 expressed the importance and limitations of transforming model practices into data practices in healthcare. The paper outlines several concepts that could prove useful in the future, placing heavy emphasis on dataset creation and improvement, along with model optimizations. While they do praise data-centric AI, they rightly point out the limitations. Large, quality and standard datasets are hard to gain access to, much less create. In the medical sector, this may be especially difficult, as data is not so easily available, either due to the privacy of individuals or the differences in data structures within institutions.

Overall, there is a lack of available literature on data-centric approaches in medical image applications, mainly in lung cancer classification. Even so, despite the term being newly formed, the concept of enhancing data and images has been heavily researched. Chaturvedi et al. [3] presented the intricate nature of image modification and how it can improve the performance of models. This review of current research highlights a variety of available and refined techniques. Three primary stages are outlined for a data-centric approach in medical image applications: image preprocessing, image segmentation, and feature extraction.

Studies such as the one conducted by Vas and Dessai [10] in 2017 on small medical images, go into further detail about the effectiveness of these stages. Vas and Dessai first cropped the images to reduce unnecessary parts, then applied 3×3 median filters to remove impulse noise. Moreover, the images were segmented through morphological operations such that only the Region of Interest (ROI), the lungs in this case, are kept. For feature extraction, the Gray-Level Co-Occurrence Matrix (GLCM) was used, scanning for Haralick features. The artificial neural network was chosen for the classification algorithm for the classification task.

Vero and Srinivasan [11] more recently in 2020, also conducted such an approach by applying image pre-processing, segmentation, feature extraction, and furthermore feature selection. First, Histogram Equalization was used to make image intensity differences clearer, followed by an Adaptive Bilateral Filter to remove any noise present. For image segmentation, the nodules were segmented using the Artificial Bee Colony method. Then various techniques were tested to locate the nodules within an image. Through ROI feature extraction, the following features were considered: volumetric, texture, intensity, and geometric.

Similarly to the [11] method, our proposed method (DCA) employs basic but rigorous data-related stages allowing for a comprehensive comparison of techniques, and providing insight into the potential of data-centric approaches in lung cancer classification and other healthcare applications.

3   Methodology

3.1   Data Processing

For the lung cancer classification DCA modeling, we used the Lung Nodule Analysis (LUNA16) dataset described in Sect. 4.1. Figure 3 shows samples of some of the raw CT scan slices and the following discussion explains the data processing techniques that were used to improve the overall quality of the image and in preparing them as input to a simple convolutional neural network.

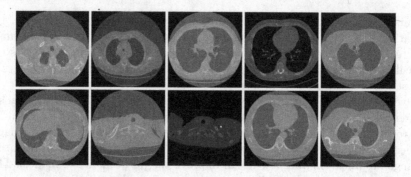

Fig. 3. 10 randomly selected samples from the LUNA16 dataset. Each sample is a single slice from a computer tomography scan.

Firstly, simple filtering techniques, such as the median filter, were used to reduce any potential noise in the slices. The prime reason for using median filters is that they are better at detail conservation, such as edges. The core implementation of this is that it selects the median value for a pixel, depending on its neighbors. The filter size determines the number of these neighbors, and a 5×5 filter was selected for our study. The transformation of one of the raw images can be seen in Fig. 4. For a larger testing environment, the Mean and Gaussian filters were later employed to measure the effectiveness of implementing other basic filters for performance changes.

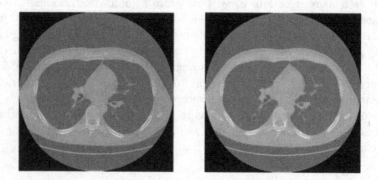

Fig. 4. Figure showing how the 5×5 median filter modifies the original raw image (left) causing smoothing and blur (right).

Secondly, we performed image segmentation. Regions of interest are first identified, then followed by morphological operations. Every slice in every CT scan was individually segmented, going through 8 steps. Firstly, the image is transformed into black and white, followed by the second step, border refinement. The third step is to label the different regions of the image and then, in step 4, remove those deemed irrelevant. This is done by only keeping the largest two areas. The next step conducts binary erosion to create a distinction between the lungs and blood vessels. Binary closing comes after, filling in black gaps in the regions. The 7th step ensures any black holes left over are filled, and finally in step 8, the binary mask created is superimposed on the original slice. The stages and their transformations are shown in Fig. 5.

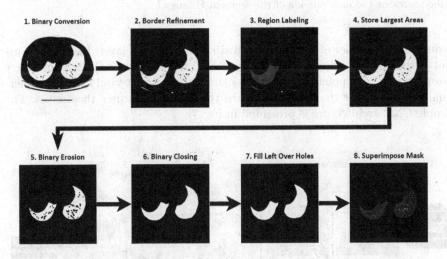

Fig. 5. Procedure of segmenting CT scan slices to prepare them for feature extraction.

Finally, the feature extraction stage. In the case of the dataset used, the images contained labeled nodule objects. The CT scans were stored as 3D arrays, with each one of the 3D arrays accompanied by real-world coordinates, allowing training and testing data generation. Thus in the feature extraction stage, the nodules within the 3D arrays would be extracted based on regions of interest. Meaning a nodule would be found given its respective coordinates, and then cubic voxels of size $36 \times 36 \times 36$ would be cut around these coordinates. These voxels would then have a corresponding label marked as 1 (nodule). Non-nodule voxels were generated by randomly picking coordinates and slicing the array at these coordinates. A sample of these is shown in Fig. 6. Not all of these cubes will be perfectly sliced. Nodules or randomly cut voxels could be taken on the border of the segmented lung, causing much of the black background to be cut along with it. While this randomness potentially results in some overlapping, it was considered reasonable to allow the model to generalize and avoid overfitting.

3.2 Classification Algorithm

The architecture of the CNN that was built for the proposed lung cancer classification is inspired by design in [1] except that for our DCA approach, relatively fewer convolutional layers are enough for distinguishing cancerous and non-cancerous lung scan images. We also modified the fully connected layer,

Fig. 6. 5 voxels cut from random segmented CT scan slices. The black values on the cube represent the area outside of the segmented lungs.

reducing the number of filters and adjusting the dropout layer. Lastly, the output activation was changed to sigmoid, and the loss was adjusted to binary cross-entropy. The point was to create a simple model that would rely on higher-quality data rather than trying to learn the complex features themselves. The simple CNN architecture is presented in Fig. 7.

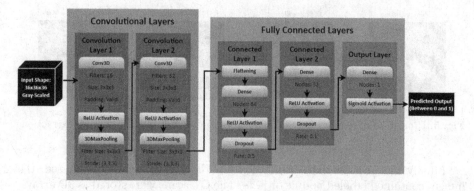

Fig. 7. Schematic of the CNN used for the classification of the processed CT scan images.

The CNN is created by putting the layers in sequential order. The left part of the diagram shows the convolutional layer, whereas the right shows the fully connected part of the CNN. The input layer accepts a 3D array as input of size $36 \times 36 \times 36$, noting that these are grayscaled. Following is a 3D convolution containing 16 filters of size $3 \times 3 \times 3$ with the ReLU activation function. The next layer contains the pooling layer, implemented through 3D max-pooling, with a kernel size of $3 \times 3 \times 3$, and the stride is the default. The second 3D convolution layer is similarly implemented, consisting of 32 $3 \times 3 \times 3$ filters and a ReLU activation function, followed by a 3D pooling layer with a $3 \times 3 \times 3$ kernel size. The results of this layer are then flattened, before going into the fully connected layer which is responsible for the classification of labels. The first Dense layer contains 64 nodes and a ReLU activation function. A dropout function is utilized, with a value of 0.5. The following Dense layer is very similar to the previous, except it contains 32 nodes and outputs into the Dropout layer set to 0.1. The final layer contains only one node and applies the sigmoid function for the binary problem. Only one output is given, this being a predicted output between 0 and 1. Closer to 0 means higher certainty that it is not a nodule, whereas values closer to 1 indicate higher certainty of a nodule.

4   Experiments and Discussion

4.1   Dataset

The LUng Nodule Analysis (LUNA16) dataset [5] was used to evaluate our proposed method. The open-access dataset contains CT scan images of patients' lungs, published by Zenodo as a challenge for lung cancer classification through ML algorithms. 486 data points are identified within these images and then extracted for model training and testing. The training and testing data is generated as discussed in the methodology. These 486 data points are divided approximately into an 80-20% split, having 388 data points in the training set and 98 data points in the testing set.

From every data point, 96 voxels (48 positively labeled for nodule, 48 negatively labeled for non-nodule) are extracted, resulting in 37248 training entries and 9408 testing entries. The model is trained through batches, going through all 388 data points and training the model on the 96 voxels present in each batch. Predictions are generated in batches, taking one data point at a time, and are prepared accordingly such that they are comparable to the testing set. We assessed the performance of the DCA approach on the testing set using the following metrics: accuracy, precision, sensitivity, specificity, and F1 score. The experiment is repeated 5 times to gain an average for each metric.

4.2   Experimental Setup

The CNN model presented in Sect. 3.2 was developed using Tensorflow in Python version 3.9.13. The training and testing of the model were done on a PC containing the following components: Ryzen 7 3700X (3.6 GHz) CPU, ASUS ROG Strix 2060 GPU, and 32 GB (3200 MHz) RAM. Several experiments were conducted however, the model's parameters remained fixed throughout these experiments. The model is trained, and its predictions are evaluated against the processed data discussed in Sect. 3.1. The model is compiled with binary cross-entropy for the loss function and the adam optimizer with a default learning rate (0.001).

4.3   Results and Discussion

Once the model had been built and compiled, the evaluations began using the test set. A total of 6 different evaluations were conducted, and their results were collected. The first evaluation involved using only the raw, unmodified CT scan data. The second used only median filters to transform the data. The third involved using only fully segmented CT scan images. The fourth, fifth, and sixth evaluations involved using the proposed data-centric approach (DCA) with Median, Mean, and Gaussian filters, respectively. The various evaluations were conducted to view how the lung cancer classification performance would be impacted depending on which data processing techniques were implemented.

Table 1 summarizes the result of the evaluations. As seen in the results, using only the unmodified data produces classification scores of at most 81% in almost all scoring metrics, with each additional data modification process improving the results. The proposed DCA with the Median filter achieves the highest classification score (F1 score) of 88.9%, confirming that modifying the data quality is a valid approach to improving a model's performance. Interestingly, using only correctly segmented CT scan images scored almost as high as the full data-centric approach on the test set, with sensitivity being higher at 92%. From this, it can be said that proper image segmentation can increase classification performance for nodules.

Table 1. Results of raw data, application of single processing methods, and three variants of the proposed data-centric approach, utilizing various filters. IS - Image Segmentation, DCA - Data-Centric Approach (Filter Applied).

	Raw Data	Filter Only	IS Only	DCA (Median)	DC (Mean)	DCA (Gaussian)
Accuracy	81.0%	84.0%	87.0%	88.8%	85.6%	85.6%
Precision	81.0%	85.4%	84.6%	89.4%	87.4%	85.6%
Sensitivity	80.0%	85.2%	92.0%	88.4%	87.4%	85.0%
Specificity	81.0%	83.0%	81.8%	89.4%	83.8%	87.0%
F1-Score	81.0%	84.0%	87.4%	88.9%	85.2%	86.2%

The DCA approach showed a decrease in sensitivity over applying image segmentation on the lungs. However, this decrease is resultant of the model having more stable and robust learning since the full DCA approach remained unbiased as sensitivity and specificity were similar. Additionally, with the DCA (median) approach, the results of all the scoring metrics are very close to each other, which means the model is more balanced, predicting true positives as well as true negatives equally. This is highly desirable for lung cancer classification using CT scan images as in healthcare, it is especially important to attain as high performance as possible since people's lives are considered.

Considering the results and advantages of the proposed data-centric approach, there was further investigation taken into the classification results of the best-performing approach, this being the DCA (Median) implementation. A confusion matrix was constructed, which is shown in Fig. 8. The confusion matrix of the model's predictions further confirms that the model is balanced in terms of predictions and is not biased towards one class. However, with a large number of false positives and false negatives, it would not be sufficient to employ this solution in real-life systems. Nevertheless, the improvement of quality in data allowed for an increase in performance across all measured metrics.

		Actual Values	
		Positive Nodule 1	Negative Non-Nodule 0
Predicted Values	Positive Nodule 1	88.4%	10.6%
	Negative Non-Nodule 0	11.6%	89.4%

Fig. 8. Confusion Matrix of the DCA (Median) implementation, showing the percentages of true positives (top left), false positives (top right), false negatives (bottom left), and true negatives (bottom right).

This study favored a data-centric approach in medical image analysis due to the general problem of lack of the availability of quality data. Contrarily, model-centric approaches were also tested to confirm the effectiveness of the proposed approach further. The raw data was kept fixed while the CNN model discussed in Sect. 3.2 was iteratively improved. Various testing was conducted, ranging from manual hyper-parameter optimization to automatic hyper-parameter tuning. From dozens of models generated, the best scoring (average across all metrics) was selected for comparison as shown in Fig. 9. For the model-centric implementation, the number of layers remains the same. However, the optimizer is changed to Ftrl (Follow the regularized leader) as it presented more stabilized results. Moreover, each layer was automatically hyper-tuned through the use of the random search algorithm. The first convolutional layer had a range of 16 to 64 filters, whereas the second convolutional layer had a range of 32 filters to 128. Each of these layers had a step of 8 in these ranges and tried filter sizes of 3 and 5. The first fully connected layer had a range of 128 to 256 nodes, and the second fully connected layer had a range of 64 to 128 nodes. Each fully connected layer had a step of 16 in these ranges.

As shown, the model-centric approach scored lower than the proposed data-centric approach, except in sensitivity, where it scored similarly. It should be noted that the model-centric results were obtained after an exhaustive hyperparameter search that was time-consuming and computationally complex. Whereas the data-centric approach was limited in its applied techniques, a variety of image modifications can still be applied for an improved classification rate in the scope of image enhancement applications in healthcare. This additional evaluation demonstrates the importance of improving data quality and its impact on performance that can still be built upon. Overall, the discussion shows that by minimizing the number of uninformative artifacts in an image, the model could focus better on identifying the prominent features present and improving the classification rate.

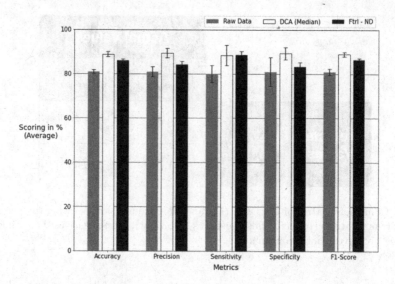

Fig. 9. Performance of model-centric approach (MCA) in black vs. proposed DCA in white. The gray bars are the results of the baseline model trained with the unmodified data only. Error bars measure standard deviation.

As a comparison of the proposed DCA to similar studies, Ge Zhang, Lin, and Wang in 2021 [14] conducted their 3D CNN implementation for lung nodule detection on CT scans in the LUNA16 dataset. Similar to our approach, the paper underwent data preprocessing where 3D nodule patches were segmented and extracted, with the additional stage of labeling the malignancy suspiciousness of said patch. Through the application of these preprocessing stages for data augmentation and a 3D CNN based on DenseNet architecture for classification, the implementation by Zhang, Lin, and Wang achieved an accuracy score of 92.4%, a sensitivity score of 87.0%, and a specificity score of 96.0%. Despite not having an identical experimental setup, our proposed data-centric approach scored similarly while having a much smaller and less complex CNN architecture. Complex architectures, such as the compared method, are often less favorable as they become increasingly tedious to optimize and train, making them less comprehensible to medical experts. In healthcare, it is important for an implementation to match or complement the opinion of a human expert. Our data modification methodology conducting simple changes allowed us to improve the quality of data, avoiding this reliance on complex models to learn difficult features and reduce interpretability. Our solution, using a simple 5-layered CNN trained on the preprocessed data, provided results that are balanced between the identification of nodules and non-nodules, having similar sensitivity and specificity scores.

Our results stand as a proof-of-concept of the potential of building data-centric AI. By exploring the area of data modification, data-centric approaches can be refined, and the quality of data can be improved until it reaches sufficient performance for real-life applications in healthcare.

5   Conclusions

This study demonstrated the benefits of modifying CT scan images to achieve higher-quality input for lung cancer classification. Due to the extensive nature of these data modification stages, the changes improved performance in all measured metrics. The study contributes to the concepts of data-centric AI by extensively reviewing available methodologies and presenting a simple, but effective data-centric approach for computational intelligence in healthcare. This approach acts as a proof-of-concept that employing simple implementations of image preprocessing, image segmentation, and feature extraction, allow for attaining the most important features of CT scans and inputting those into a simple CNN model to achieve good classification results. Exploring concepts of model hyperparameter optimization (i.e., model-centric AI) also indicated that achieving higher results for such a problem was more effective with our proposed data-centric solutions.

While this study promotes the concept of data-centric AI, and its importance in achieving high performance, we noted that our study was limited to the exploration of a single dataset, experimenting with few data processing techniques for lung cancer classification. Nevertheless, the results demonstrated that data-centric AI is worth investing time in to gain informative insights about medical images and how their quality will impact a model's learning capabilities. Future work would explore other image applications in healthcare, particularly where data is often limited.

References

1. ArnavJain: Candidate generation and luna16 preprocessing. Kaggle (2017). https://www.kaggle.com/code/arnavkj95/candidate-generation-and-luna16-preprocessing#Reading-a-CT-Scan. Accessed 20 Nov 2022
2. Brownlee, J.: Master Machine Learning Algorithms: Discover How They Work and Implement Them From Scratch, pp. 11–12. Machine Learning Mastery (2016)
3. Chaturvedi, P., Jhamb, A., Vanani, M., Nemade, V.: Prediction and classification of lung cancer using machine learning techniques. In: IOP Conference Series: Materials Science and Engineering, vol. 1099, no. 1, p. 012059 (2021). https://doi.org/10.1088/1757-899x/1099/1/012059
4. Gao, J., Jiang, Q., Zhou, B., Chen, D.: Convolutional neural networks for computer-aided detection or diagnosis in medical image analysis: an overview. Math. Biosci. Eng. **16**(6), 6536–6561 (2019). https://doi.org/10.3934/mbe.2019326
5. van Ginneken, B., Jacobs, C.: Luna16 part 1/2. Zenodo (2016). https://doi.org/10.5281/zenodo.3723295. Accessed 20 Nov 2022
6. Hamid, O.H.: From model-centric to data-centric AI: a paradigm shift or rather a complementary approach? In: 8th IEEE International Conference on Information Technology Trends (ITT), pp. 196–199 (2022). https://doi.org/10.1109/ITT56123.2022.9863935

7. Ng, A.: A chat with Andrew on MLOps: from model-centric to data-centric AI (2021). https://www.youtube.com/watch?v=06-AZXmwHjo&t. Accessed 20 Nov 2022

8. Polyzotis, N., Zaharia, M.: What can data-centric AI learn from data and ml engineering? In: 35th NeurIPS Conference on Neural Information Processing Systems, pp. 1–2 (2021). https://doi.org/10.48550/ARXIV.2112.06439

9. Rajkomar, A., Dean, J., Kohane, I.: Machine learning in medicine. N. Engl. J. Med. **380**(14), 1347–1358 (2019). https://doi.org/10.1056/NEJMra1814259

10. Vas, M., Dessai, A.: Lung cancer detection system using lung CT image processing. In: 2017 International Conference on Computing, Communication, Control and Automation (ICCUBEA), pp. 1–5 (2017). https://doi.org/10.1109/ICCUBEA.2017.8463851

11. Vero, A., Srinivasan, A.: Deep learning for lung cancer detection and classification. Multimed. Tools Appl. **79**, 7731–7762 (2020). https://doi.org/10.1007/s11042-019-08394-3

12. Whang, S.E., Roh, Y., Song, H., Lee, J.G.: Data collection and quality challenges in deep learning: a data-centric AI perspective. arXiv preprint arXiv:2112.06409 (2021). https://doi.org/10.48550/ARXIV.2112.06409

13. Zhang, A., Xing, L., Zou, J., Wu, J.C.: Shifting machine learning for healthcare from development to deployment and from models to data. Nat. Biomed. Eng. **149**, 104420 (2022). https://doi.org/10.1038/s41551-022-00898-y

14. Zhang, G., Lin, L., Wang, J.: Lung nodule classification in CT images using 3D densenet. J. Phys: Conf. Ser. **1827**(1), 012155 (2021). https://doi.org/10.1088/1742-6596/1827/1/012155

Comparative Analysis of Machine Learning Approaches for Classifying Erythemato-Squamous Skin Diseases

Bhavana Kaushik[1]([⊠]) [iD], Ankur Vijayvargiya[2], Jayant Uppal[3], and Ankit Gupta[4]

[1] School of Computer Science, University of Petroleum and Energy Studies, Dehradun, India
kau.bhavana@gmail.com
[2] Accenture, Noida, India
[3] People Strong, Noida, India
[4] Samsung R&D, Noida, India

Abstract. In the recent era, Machine Learning and Artificial Intelligence have come to a very great development point as we can use ML algorithms to predict the type of Erythemato-Squamous (Skin) diseases of the skin. In Dermatology, differential diagnosis of skin diseases is quite challenging in real life because most skin diseases share many histopathological features. And in this work, Psoriasis, Lichen Planus, Seborrheic Dermatitis, Chronic Dermatitis, Pityriasis Rosea, and Pityriasis Rubra Pilaris are among the skin illnesses for which eight different algorithm analytical comparison is done. Moreover, each classifier algorithm is discussed in detail with its pros and cons. The machine learning algorithms like Support Vector Machine, Decision tree, Random Forest, KNN, Naïve Bayes, Gradient Boosting, XGBoost, and Multilayer Perception have been proven to be successful in preserving state information through exact segmentation/classification. Random forest, Gradient Boosting, and XGBoost outperform all other methods and give an accuracy of 100% on the given ESD dataset. While Support Vector Machine gives the least accuracy of 72.97%. The paper also discusses the difficulties connected with skin disease segmentation or categorization. Furthermore, the study proposes future potential directions that include real-time analysis.

Keywords: Erythemato-Squamous Diseases · Machine Learning · Classification · Skin Diseases · Comparative Analysis

1 Introduction

In recent times we observed that many people are suffering from skin diseases or skin cancer which can be curable at an early stage but now they are not curable thus after watching the continues advancement and development in technology and especially in AI & ML we decided to combine them and take the help from various machine learning algorithms to predict the skin diseases at the earliest stage so that patient can be cured within time and it can also help all medical field and especially in dermatology to predict the diseases at the earliest stage. Many times doctor is not able to get the type of skin

© The Author(s), under exclusive license to Springer Nature Switzerland AG 2023
S. Kadry and R. Prasath (Eds.): MIKE 2023, LNAI 13924, pp. 67–77, 2023.
https://doi.org/10.1007/978-3-031-44084-7_7

disease in the earlier stage because at the beginning stage all types of skin diseases show the same symptoms and also share the same histopathological features so identifying in the earliest stage is also a very challenging task for the doctors thus this time is wasted and can cause the disease to grow very quickly and can cause to death and cancer. This wasted time is very crucial for the patient and in this time if our algorithm can make the accurate and right decision then we can save the life of the patient. One of the hardest challenges facing today's health care organisations (hospitals, medical facilities) is the provision of high-quality services at fair pricing. Providing patients with accurate diagnoses and efficient treatments are examples of quality care [1]. The majority of hospitals presently use a hospital information system to manage their patient or healthcare data. Typically, these systems output enormous amounts of data in the form of statistics, text, charts, and images [2]. Regrettably, these data are rarely used to guide clinical decisions. Utilising pertinent computer-based information and/or decision assistance technologies can yield the desired results. A critical question is raised in this context: "How can we transform data into useful knowledge that enables clinicians to make informed therapeutic decisions?" This is the main motivation behind studying. Erythemato-squamous diseases (ESDs) are very prevalent skin conditions. The six different varieties are psoriasis, seboreic dermatitis, lichen planus, pityriasis rosea, chronic dermatitis, and pityriasis rubra pilaris. By a little margin, all of them exhibit the medical symptoms of erythema and scaling [3]. **Psoriasis -** It is believed that psoriasis is an immune system issue. As skin cells accumulate to form scales, itchy, dry areas start to appear. Triggers include illnesses, stress, and the common cold. The most typical sign is a rash on the skin, but it can also affect the joints or nails. The treatment's goals are to get rid of scales and slow down skin cell growth. Medication, light treatment, and topical ointments can all help [4]. **Seborrheic dermatitis -** This skin ailment is characterised by flaking patches and red skin, mainly on the scalp. It can also appear on oily body parts like the face, upper chest, and back. Seborrheic dermatitis can create obstinate dandruff in addition to scaly patches and red skin. Self-care and medicinal shampoos, creams, and lotions are used in the treatment. Treatments may need to be repeated [5]. **Lichen planus -**An inflammatory skin and mucous membrane disease. When the immune system mistakenly targets skin or mucous membrane cells, lichen planus develops. Lichen planus shows on the skin as reddish, itchy pimples with a flat top. It creates lacy, white patches on mucous membranes, such as the mouth, and sometimes severe ulcers. Lichen planus is frequently self-resolving. Topical treatments and antihistamines may help if symptoms are bothersome [6]. **Pityriasis Rosea -** A hasty that twitches as a huge spot on the abdomen, chest, or back and then spreads out into a pattern of smaller lesions. The cause of pityriasis rosea is unknown, but it is thought to be caused by a viral infection. The illness creates a rash on the torso, upper legs, and upper arms that is mildly irritating. Pityriasis rosea is frequently self-resolving. Antihistamines, steroid cream, and, in rare situations, antiviral medicines can all assist [7]. **Chronic dermatitis -** Dermatitis is a word used to define a group of itchy, inflammatory skin disorders marked by epidermal abnormalities. Dermatitis affects one out of every five people at some time in their livesIt can have many different patterns and many different causes. Eczema and dermatitis are commonly used interchangeably. The term "eczematous dermatitis" is occasionally used. Acute, chronic, or a combination of the two types of dermatitis are possible. Dermatitis, another name for acute eczema, is a rapidly growing red rash that can blister and enlarge. Eczema (also known as dermatitis) is a chronic, uncomfortable skin condition. Typically, it is thicker, darker than the surrounding skin, and heavily scraped [8]. **Pityriasis Rubra Pilaris (PRP)** – It is a term used to describe a collection of rare skin illnesses characterized by scaling patches that are reddish orange in colour and have well-defined edges. They can cover the entire body or convinced parts like the prods and laps, palms, and soles. Islands of sparing are patches of uncomplicated skin, especially on the stem and limbs, which are frequently seen. The palms and soles are commonly affected, becoming swollen and yellowish in appearance (palmoplantar keratoderma). PRP is frequently misdiagnosed as psoriasis or another skin disorder [9].

Skin cells (squamous) deteriorate, and erythema (redness of the skin) is caused by infection with one of these skin illnesses. Dermatologists frequently examine patients both clinically and based on histological features [10]. Clinical examinations involve looking at the colour, presence of zits, their size, location, and other symptoms. For each person/patient, the aforementioned examinations result in 12 clinical and 22 histological variables. Investigating these variables may have ambiguous and illogical results since they may cross paths, particularly in the early stages of ESD. So, there is a need to identify the appropriate classification technique to solve this problem and give a better result/prediction. This paper reviews all conventional techniques present to perform the differential diagnosis of ESD and also to identify the challenges existing with approaches. The contributions of this study are as follows:

 (i) Analytical comparison of all relevant conventional machine learning techniques for differential diagnosis of ESD is done.
 (ii) Also, the challenges and issues associated with each technique is identified.
 (iii) The study discusses the future potential directions that include real-time analysis.
 (iv) Discuss the difficulties connected with skin disease segmentation or categorization.

The structure of this paper is as follows: The description of material and methods used in this comparative analysis is given in Sect. 2. Section 3 contains the result and discussions of the analysis. Prominent challenges and future scope are described in the Sect. 4 and lastly Sect. 5 concludes the study.

2 Material and Methods

2.1 Materials

We used a standardized dermatology data set from the "University of California, School of Information and Computer Science's machine learning repository, or UCI. It has 34 properties, 12 of which are clinical and 22 of which are histological". Age and family history are continuous characteristics in the data set, with values ranging from 0–1. Every additional clinical and histological feature was given a degree from 0 to 3, where 0 meant the feature wasn't present, 3 meant it was present to its fullest extent, and 1, 2 meant it was present to a relatively moderate level. Naive Bayes, Random Forest, Support Vector Machines, XGBoost, Multi-layered perceptron, K-nearest neighbors, Decision tree, Gradient boosting DT are among the ML Classification Algorithms investigated in this paper [11] Table 1 shows the six classes of ESD.

Table 1. Six classes of ESD

Keys	Values (Class Labels)
1	Psoriasis
2	Seborrheic Dermatitis
3	Lichen Planus
4	Pityriasis Rosea
5	Chronic Dermatitis
6	Pityriasis Rubra Pilaris

2.2 Methods

2.2.1 Support Vector Machine (SVM)

The objective of the SVM, where n is the number of variables, is to identify the hyper plane in n-dimensional space. The hyper-plane is selected so that there is as little space as feasible between the closest data points and support vectors of the two distinct modules. The hyper plane can be pictured as both a plane and a line in three dimensions. Hyper plane separates the data points of two different classes [12]. Numerous of the prevailing (non)convex soft-margin losses can be observed as one of the substitutes of the L0/1 soft-margin loss. SVM have gained huge consideration for the last two decades due to its wide-ranging usage, so many researchers have established optimization procedures to solve SVM with various soft-margin losses [13].

For the prediction corresponding to a new input can be obtained using Eq. 1:

$$f(x) = B0 + sum(ai * (x, xi)) \tag{1}$$

where $f(x)$ is used to calculate the inner dot product which is the sum of the multiplication of each pair of the input values i.e., x as the new input and xi as each of the support vectors present in the training set. $B0$ and ai are the coefficients evaluated from the training data [14]. Figure 2 shows the visuals of the Support Vector Machines (SVMs) Model with all necessary features like absolute hyperplane with maximum margin, hyperplane with positive and negative trends and the nearest data points as the support vectors.

2.2.2 Random Forest

It is, as the name implies, a group of various decision trees, each of which predicts some class, and the class having the most votes is accepted as the predicted class. These decision trees produce distinct outcomes relatively. This concept is highly effective in the reduction of prediction errors if predicted through a single decision tree. In this approach, an individual tree may be in the wrong direction, but the common direction could be in the right direction [15].

2.2.3 Naive Bayes

This classifier is based on the Bayes theorem and approaches the probabilistic strategy in classification through prediction. Equation 2 states the approach of Bayes theorem:

$$P(A|B) = P(B|A)P(A)/P(B) \qquad (2)$$

where we discover the chance of happening of A assuming that B had already occurred. In this concept A is considered as the hypothesis and B is considered as the evidence. This approach is best when the features are not affected by each other [16].

2.2.4 Decision Tree

As the name suggests, we can find an analogy between a tree and a result tree. A result tree is similar to a tree by having split conditions as a node, directing edges as branches, and decisions as leaves. The formation of a tree involves feature decision and branching conditions and holds the decision by preventing further branching. This approach follows the greedy concept by splitting the branches with lower prediction cost i.e., the class with the maximum data points should be classified initially at 0 level/root node [17].

2.2.5 K-nearest Neighbours

As the name suggests, this algorithm finds the separate clusters of data points present in proximity i.e., near to each other based on the distance between the two data points. In this classification approach, the K refers to the number of neighbors as the class labels and the mode of k labels is considered as the predicted outcome. The efficiency of this algorithm decreases with an increase in the number of predictor variables [18]. For a new input having real values, the distance is most likely to be measured through Euclidean distance given by Eq. 3:

$$EuclideanDistance(x, xi) = sqrt(sum((xj - xij)^2)) \qquad (3)$$

where x is the new input and xi is the existing point covering all the j input attributes [19].

2.2.6 Gradient Boosting DT

As the name suggests, in this approach small steps are initiated from a point in a direction by enhancing the weak learners to make them strong. It consists of a cost function, feeble leaner, and preservative sequential approach to improving the presentation of the predictive model [20]. This classifier algorithm is highly used to optimize the user-defined cost functions by using the gradients in the loss function to make them controlled and realistic [21].

2.2.7 Multi-Layered Perceptron

As the name suggests, it refers to the neural networks or system of input, output layers, and various hidden layers between them with multiple neurons connected. A perceptron is referred to as a neuron with a random activation function. This algorithm uses the technique of backpropagation, a repetitive approach of combining the weights and the inputs which are achieved through the threshold function to minimize the cost function [22].

2.2.8 XGBoost

This is known as an extreme gradient boosting algorithm. In this approach the framework of gradient boosting is conserved. It is a highly optimized algorithm in terms of software as well as hardware resources usage for supercilious prediction outcomes in a quick time with minimal computing cost. This approach involves Gradient descent methodology as gradient boosting for strengthening the weak learners like CARTs [23].

3   Results and Discussions

This section will analyse all the classifier accuracy and performance using the confusion matrix and will give the proper comprehension regarding the best classifier algorithm. To do this analysis few pre-processing steps is to be applied on used dataset. First the count of null values of attributes will be checked it is seen that there are 8 rows of age attribute with null values. Now, describe the dataset to check the composition of the dataset. After this, the descriptive statistic of the dataset is collected, which is shown in Table 2.

Table 2. Description of dataset without replacing null values.

Parameters	Values
count	358.000000
mean	36.296089
std	15.324557
min	0.000000
25%	25.000000
50%	35.000000
75%	49.750000
max	75.000000

Now, in lieu of the null values with the median value of the attribute and describe the dataset again. Table 3 describes the altered dataset, after replacing the null values present in the age attribute.

Now, it can be clearly seen that there is only a slight difference in the mean frequency of the dataset. Mean frequency difference percentage = 0.18%. Now, import all the 8 classifier algorithms i.e., Naive Bayes, Support Vector Machines, Random Forest, XGBoost, Multi-layered perceptron, K-nearest neighbors, Decision tree, Gradient

Table 3. Description of dataset after replacing null values.

Parameters	Values
count	366.000000
mean	36.363388
std	15.037366
min	7.000000
25%	25.000000
50%	35.000000
75%	48.000000
max	75.000000

boosting DT. By meeting the internal classes ration in both sets, divide the dataset into a train set and a test set with a test size of 0.1. Now, the next step is to check the distribution of classes in training and test set. Figure 1 shows that the distribution of both the training and test set are in proportion.

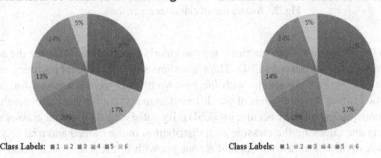

Fig. 1. Class distribution in training and test set

The internal ratio of classes is same in both the sets as shown in Fig. 10. After training these models with various classifier algorithms mentioned above and test for the accuracy scores. Figure 2 describes the accuracy scores of the classifier algorithms. It is calculated as the number of accurate forecasts produced divided by the total number of predictions, then multiplied by 100 [24]. Plot the confusion matrices for each of the classification algorithms. The confusion matrix serves as a performance metric for categorisation using machine learning. The output of the machine learning classification problem can be two or more classes, hence it is a performance evaluation for that problem [24]. The Figs. 3 shown below are the confusion matrices of the classifiers algorithms which compares the predicted class of ESD by classifiers algorithm to the actual class of ESD for a particular set of attribute values from test set.

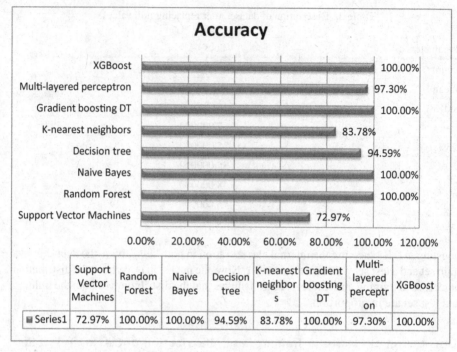

Fig. 2. Accuracies of classifier algorithms

From the above confusion matrices, we can clearly see predictions as per the accuracy scores of all the classes of ESD. The algorithms with 100 percent accuracy or the Ensemble of classifier algorithms with the best accuracies were employed for better classification about the prediction of the differential analysis of erythemato-squamous diseases constructed on their accuracies (ESD). By satisfying the internal classes ratio in both sets and checking the classification distribution in the trained and trial sets, the dataset was divided into a train set and a trial set with a test size of 0.1. After that, we can see that the internal class ratio is the same in both sets, and we also trained these models using the various classifier techniques discussed above, and we tested their accuracy: "Support Vector Machines (72.97%), Random Forest (100.0%), Naive Bayes (100.0%), Decision Tree (94.59%), K-nearest neighbors (83.78%), Gradient Boosting DT (100.0%), Multi-layered perceptron (97.3%), and XGBoost (100.0%) were the most popular".

4   Challenges and Future Scope

Challenges in the current study is as follows: (i) In the data set, there is 2.2 percent missing data for the age attribute. The mean frequency was used to replace missing data with true values. As a result, training M.L. model would have been more effective if we had used real data. Our dataset size is small so it can lead to many problems like overfitting, Measurement errors, Missing values, Sampling Bias, etc. Due to this model

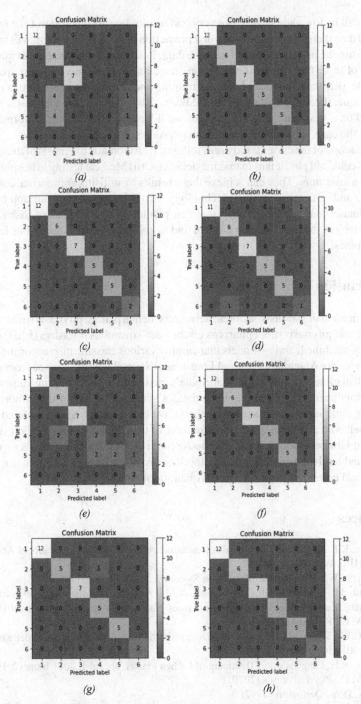

Fig. 3. (a): SVM Confusion matrix, Figure (b): Random Forest Confusion matrix, Figure (c): Naive Bayes Confusion matrix, Figure (d): Decision tree Confusion matrix, Figure (e): KNN Confusion matrix, Figure (f): Gradient boosting DT Confusion matrix, Figure (g): MLP Confusion matrix, Figure (h): XGBoost Confusion matrix

accuracy will be low and can produce very bad results also at sometimes. Like in if have the biased data then it can lead to the worst prediction [25]. (ii) The system will take time even if we use the best method with massive data. In some circumstances, this may result in the use of more CPU power. Furthermore, the data may take more storage space than is available. (iii) Vast quantity of data for training and testing is acquired. As a result of this technique, data inconsistencies may emerge. This is due to the fact that some data is updated on a frequent basis. As a result, we'll have to wait for more information. If this is not the case, the old and new data may produce contradictory results.

Future scope of the current study is as follows: (i) Automatic diagnosis of these illness groupings could aid physicians in making decisions. (ii) Medical testing in hospitals must be kept to a minimum. They can achieve these results by utilising appropriate computer-based info and/or decision-making technology. (iii) We can improve our app by using a larger dataset and creating an app that can predict a huge number of diseases. (iv) We can establish the link between clinical and histological features using our feedback methodologies.

5 Conclusion

The classification of psoriasis, seboreic dermatitis, lichen planus, pityriasis rosea, chronic dermatitis, and pityriasis rubra pilaris as erythemato-squamous disorders (ESD) is made possible by machine learning models that employ various classification algorithms, such as Support Vector Machines, Random Forest, and Naive Baye. We can see from the preceding calculations that utilizing the attribute's median to replace missing values results in a very small percentage change in the dataset's mean frequency. Furthermore, while comparing eight other classification methods, we can find that XGBoost fared exceptionally well with a 100% accuracy rate for three of them, namely Random Forest, Naive Bayes, and Gradient Boosting DT. XGBoost could be useful since it uses a mixture of software and hardware enhancement methods to provide supercilious prediction results with minimal computer assets in a small quantity of period.

References

1. Greiner, K.E.A.C.: Health Professions Education: A Bridge to Quality. National Academies Press (US), Washington (2003)
2. Brook, C.: What is a Health Information System? (2020)
3. Elsayad, M., Al-Dhaifallah, M., Nassef, A.M.: Analysis and diagnosis of erythemato-squamous diseases using CHAID decision trees. Yasmine Hammamet, Tunisia (2018)
4. Staff, M.C.: Psoriasis (2020)
5. Gary, C.W., Sara, P.M., Jaboori, K.A.: Diagnosis and treatment of seborrheic dermatitis, no. Feb 1, 2015 Issue (2015)
6. Usatine, R.P.: Diagnosis and Treatment of Lichen Planus. no. Jul 1, 2011 Issue (2011)
7. Staff, M.C.: Pityriasis rosea (2020)
8. Oakley, D.A.: Dermatitis (1997)
9. Vanessa Ngan, D.A.O.: Pityriasis rubra pilaris (2015)
10. Verma, S.S.P., Kumar, S.: Comparison of skin disease prediction by feature selection using ensemble data mining techniques. Inf. Med. Unlocked, 1–16 (2019)

11. Ilter, N., Guvenir, H.A.: Dermatology Data Set. (01 January 1998). [Online]. Available: https://archive.ics.uci.edu/ml/datasets/dermatology

12. Gandhi, R.: Support Vector Machine — Introduction to Machine Learning Algorithms. (7 June 2018). [Online]. Available: https://towardsdatascience.com/support-vector-machine-introduction-to-machine-learning-algorithms-934a444fca47

13. Wang, H., Shao, Y., Zhou, S., Zhang, C., Xiu, N.: Support Vector Machine Classifier via Soft-Margin Loss. IEEE Trans. Pattern Analy. Mach. Intell. 1–11 (2021)

14. Brownlee, J.: Support Vector Machines for Machine Learning. (20 April 2016). [Online]. Available: https://machinelearningmastery.com/support-vector-machines-for-machine-learning/

15. Yiu, T.: Understanding Random Forest. (12 June 2019). [Online]. Available: https://towardsdatascience.com/understanding-random-forest-58381e0602d2

16. Gandhi, R.: Naive Bayes Classifier. (5 May 2018). [Online]. Available: https://towardsdatascience.com/naive-bayes-classifier-81d512f50a7c

17. Gupta, P.: Decision trees in machine learning (18 May 2017). [Online]. Available: https://towardsdatascience.com/decision-trees-in-machine-learning-641b9c4e8052

18. Harrison, O.: Machine Learning Basics with the K-Nearest Neighbors Algorithm (11 September 2018). [Online]. Available: https://towardsdatascience.com/machine-learning-basics-with-the-k-nearest-neighbors-algorithm-6a6e71d01761

19. Brownlee, J.: K-Nearest Neighbors for Machine Learning (15 April 2016). [Online]. Available: https://machinelearningmastery.com/k-nearest-neighbors-for-machine-learning/

20. Brownlee, J.: A Gentle Introduction to the Gradient Boosting Algorithm for Machine Learning (9 September 2016). [Online]. Available: https://machinelearningmastery.com/gentle-introduction-gradient-boosting-algorithm-machine-learning/

21. Singh, H.: Understanding Gradient Boosting Machines (November 4 2018). [Online]. Available: https://towardsdatascience.com/understanding-gradient-boosting-machines-9be756fe76ab

22. Bento, C.: Multilayer Perceptron Explained with a Real-Life Example and Python Code: Sentiment Analysis (21 September 2021). [Online]. Available: https://towardsdatascience.com/multilayer-perceptron-explained-with-a-real-life-example-and-python-code-sentiment-analysis-cb408ee93141

23. Morde, V.: XGBoost Algorithm: Long May She Reign! (8 April 2019). [Online]. Available: https://towardsdatascience.com/https-medium-com-vishalmorde-xgboost-algorithm-long-she-may-rein-edd9f99be63d

24. Agrawal, R.: The 5 Classification Evaluation metrics every Data Scientist must know (17 September 2019). [Online]. Available: https://towardsdatascience.com/the-5-classification-evaluation-metrics-you-must-know-aa97784ff226. Accessed 9 Decemeber 2021

25. EduPristine: Problems of Small Data and How to Handle Them (2016)

Automatic Detection of Waterbodies from Satellite Images Using DeepLabV3+

Seifedine Kadry[1]([✉]) [iD], Mohammed Azmi Al-Betar[2], Sahar Yassine[1],
Ramya Mohan[3], Rama Arunmozhi[3], and Venkatesan Rajinikanth[3] [iD]

[1] Department of Applied Data Science, Noroff University College, 4612 Kristiansand, Norway
skadry@gmail.com, sahar.yassine@noroff.no
[2] Artificial Intelligence Research Center (AIRC), College of Engineering and Information Technology, Ajman University, Ajman, United Arab Emirates
[3] Department of Computer Science and Engineering, Division of Research and Innovation, Saveetha School of Engineering, SIMATS, Chennai 602105, India

Abstract. In recent times, it has become increasingly popular to examine a wide range of environmental and earth data using remote sensing schemes supported by satellite images (SI). Due to the complex nature of spatial, spectral, and temporal characteristics of SI, it is quite difficult for automatic analysis to be performed as it requires specially designed algorithms. As part of the research proposal, an enhanced deep-learning scheme will be implemented in order to extract the waterbodies from the chosen SIs. The phases involved in this scheme includes; (i) the collection and resizing of images, (ii) Shannon's Entropy preprocessing, (iii) DeepLabV3+ extraction of waterbodies, (iv) the comparison of extracted sections with ground truth (GT) and the calculation of performance metrics, and (v) validation of the effectiveness of the implemented scheme. In the proposed work, the proposed multi-thresholding approach is combined with DeepLabV3+ in order to get the waterbodies to be extracted from the chosen test images. DeepLabV3 is demonstrated to have excellent segmentation performance when it is compared to UNet and SegNet. The experimental results of this scheme indicate that DeepLabV3+ results in higher Jaccard value ($>89\%$), Dice value ($>93\%$), and segmentation accuracy value ($>97\%$) when compared to other pretrained segmentation schemes.

Keywords: Earth data · Satellite image · Waterbodies · DeepLabV3+ · Segmentation

1 Introduction

The rapid improvement in technology and the avalaibility of various computing facilities helped to implement the Artificial Intelligence (AI) schemes to analyse a variety of the database. Examination of the satellite data is one of the common procedure in geoscience studies and the outcome of this assesment helps to provide a variety of vital information regarding various key geoscience information for necessary monitoring and decision making tasks [1, 2].

Detection of waterbodies from the Satellite İmagery (SI) is one of the key reaseach area and this helps to achieve the following;

S. Kadry and R. Prasath (Eds.): MIKE 2023, LNAI 13924, pp. 78–86, 2023.
https://doi.org/10.1007/978-3-031-44084-7_8

- Water resource management: Accurate mapping of waterbodies is important for the management of water resources such as rivers, lakes, and reservoirs. It helps in monitoring changes in water levels, identifying areas of water scarcity or abundance, and planning for the allocation of water resources.
- Flood monitoring: Waterbody detection is crucial in monitoring floods, which can cause significant damage to property and human lives.
- Environmental studies: It can help in the identification of wetlands, which are critical habitats for wildlife and provide several ecosystem services such as water purification and carbon sequestration.
- Climate change studies: Changes in waterbodies, such as shrinking of glaciers, melting of sea ice, and changes in river flow patterns, can provide insights into the impacts of climate change. Detection of these changes can help in developing mitigation and adaptation strategies.
- Navigation and transportation: It helps in the planning of shipping routes, construction of ports, and monitoring of maritime activities.

In the literature, there exists a variety of procedures to detect the waterbodies from SIs, such as image segmentation, thresholding, and machine learning algorithms. Compared to the traditional image processing schemes, the AI supported methods are widely adopted in the literature for automatic examination and assesment of the SI.

Convolutional neural networks (CNNs) are a type of deep learning algorithm that has shown great promise in waterbodies analysis from satellite imagery. CNNs are well-suited for image analysis tasks as they can automatically learn and extract relevant features from images without the need for manual feature engineering. CNNs can be trained using labeled satellite images to detect and classify waterbodies at different scales. The labeled images are typically manually annotated by experts to indicate the location of waterbodies. The CNN then learns to recognize patterns in the labeled images and applies them to new images to detect waterbodies [3, 4].

The CNNs are a powerful tool for waterbodies analysis from satellite imagery. They can be used for detection and classification of coastal waterbodies, mapping of inland waterbodies, flood monitoring, and detection of water quality parameters. The information obtained from CNNs can be used for various applications such as water resource management, environmental studies, and disaster response planning. In the proposed research, DeepLabV3+ is considered to extract and evaluate the waterbodies from the choen SI. The various stages of this scheme involves in; image collection and resizing, preprocessing with Shannon's Entropy (SE), DeepLabV3+ supported segmentation and computing the necessary performance measures by comparing the mined region with ground truth (GT). The experimental outome of this study confirms that the proposed scheme is efficient in achiving a better segmentation outcome on the chosen image database.

This work includes of five sections; Sects. 2 and 3 demonstrates the literature review and methodology, Sects. 4 and 5 present experimental outcomeand conclusion of this study.

2   Literature Review

Recently, deep-learning based examination of SIs are widely discussed by the researchers. The work of Zhang et al. (2021) proposes a method for waterbodies segmentation using deep convolutional neural networks (DCNNs). The proposed method achieved high accuracy in waterbodies segmentation in high-resolution satellite images [5]. The study of Gao et al. (2022) proposes an unsupervised method for waterbodies segmentation using generative adversarial networks (GANs) and region merging. The proposed method achieved high accuracy in waterbodies segmentation in Landsat 8 satellite images [6]. The research by Yuan et al. (2020) proposes a deep learning-based method for waterbodies segmentation using data augmentation techniques such as rotation, flipping, and scaling. The proposed method achieved high accuracy in waterbodies segmentation in multispectral satellite images [7].

The work of Rana and Babu (2022) compares the performance of different waterbodies segmentation techniques, including thresholding, segmentation using morphological operations, and deep learning-based methods [8]. The review work by Babu and Rajam (2020) provides an overview of different machine learning techniques for waterbodies segmentation from satellite images, including thresholding, clustering, and deep learning-based methods [9]. The paper also discusses the challenges and future directions in waterbodies segmentation. Based on this motivation, proposed research also implements automatic evaluation of waterbodies from the chosen satellite images.

To achieve a better segmentation accuracy, this research considers the preprocessed SIs and implements the DeepLabV3+. The experimental results of this work confirms that the proposed scheme works well on the chosen images.

3   Methodology

This section of the research presents the methodology proposed in this research to examine the SIs. Figure 1 depicts the various stages existing in the waterbodies detection scheme with DeepLabV3+.

The necessary SI database is collected from [10] and then every image and the GT are resized to $512 \times 512 \times 3$ pixels. These images are then enhanced using the Shannon's entropy supported tri-level thresholding along with the Mayfly-Algorithm (SE + MA) [10]. The waterbodies region from the enhanced image is then mined using the DeepLabV3+ and the extracted section is compared with the Ground-Truth (GT) and the necessary performance metrics are computed. Based on the mean value of these measures, the merit of the proposed technique is verified.

3.1   Database

The necessary test images for this study is obtained from [10]. The considered image and the ground-truth (GT) is resized to 512×512 pixels and in this work a total of 1000 images are considered for the examination. Among the considered images, 80% of the database is chosen for training the pretrained segmentation system and 20% is considered for testing the performance. The sample test image and its related GT is depicted in Fig. 2.

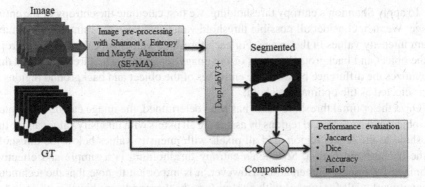

Fig. 1. Proposed satellite image evaluation approach

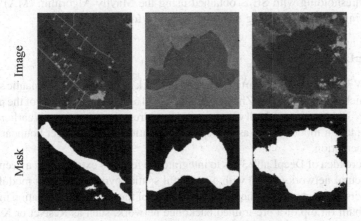

Fig. 2. Benchmark satellite image database

3.2 Thresholding

The chosen RGB scaled test images are preprocessed with a tri-level thresholding to enhance the visibility of the waterbody section for better segmentation. The related thresholding considered in this work can be found in [11–13].

Shannon's Entropy (SE) thresholding is a technique used in image processing to automatically determine a threshold value for image segmentation. This technique uses SE formula to calculate the optimal threshold value for a given image. The basic idea behind SE thresholding is that the threshold value should be chosen in such a way that it maximizes the difference in entropy between the object and background regions of the image. The entropy of an image can be calculated using the formula shown in Eq. (1):

$$H = -\sum (p\_i * \log 2(p\_i)) \tag{1}$$

where p_i is the probability of the intensity level i occurring in the image. The sum is taken over all possible intensity levels in the image.

To apply Shannon's entropy thresholding, we first calculate the entropy of the entire image. We then consider all possible threshold values between the minimum and maximum intensity values in the image. For each threshold value, we calculate the entropy of the object and background regions of the image separately. The threshold value that maximizes the difference between the entropies of the object and background regions is then selected as the optimal threshold.

Once the optimal threshold value has been determined, the image can be segmented into object and background regions by assigning all pixels with intensity values above the threshold to the object region, and all pixels with intensity values below the threshold to the background region. Shannon's entropy thresholding is a simple and effective technique for image segmentation. However, it is important to note that the technique may not work well for images with non-uniform backgrounds or objects with complex shapes. In such cases, more sophisticated thresholding techniques may be required. The optimal thresholding with SE is obtained using the Mayfly Algorithm (MA) and the complete information regarding this scheme can be found in [14, 15].

3.3 DeepLabV3+

DeepLabV3+ is a state-of-the-art deep neural network architecture for semantic segmentation, which was proposed by Zhang et al. in 2020 [16]. It is an extension of the previous DeepLabV3 architecture, which was designed to address the limitations of earlier semantic segmentation models, such as low spatial resolution, coarse object boundaries, and over-segmentation.

The core idea of DeepLabV3+ is to integrate the features extracted by a deep convolutional neural network (CNN) with a powerful spatial pyramid pooling module and a decoder network that restores the spatial resolution of the output segmentation map. The model is built on top of a pre-trained backbone network, such as ResNet or Xception, which is used to extract high-level features from the input image.

The spatial pyramid pooling module in DeepLabV3+ is designed to capture features at multiple scales and at different levels of granularity. This module takes the output feature maps from the backbone network and applies pooling operations at different scales, with each scale capturing features at a different level of detail. This allows the model to capture both global and local context information and better distinguish between objects that are close together.

The decoder network in DeepLabV3+ is used to restore the spatial resolution of the output segmentation map, which is reduced by the pooling operations in the spatial pyramid module. The decoder network upsamples the feature maps to the original input resolution and fuses them with the corresponding feature maps from the backbone network, using a skip-connection mechanism similar to that used in U-Net [17, 18].

The final output of DeepLabV3+ is a pixel-wise segmentation map, where each pixel is assigned a label indicating the object or background class. The model is trained using a cross-entropy loss function, which measures the discrepancy between the predicted and ground-truth segmentation maps.

Overall, DeepLabV3+ has demonstrated state-of-the-art performance on several benchmark datasets for semantic segmentation. Its combination of deep CNNs, spatial pyramid pooling, and decoder networks allows it to achieve highly accurate segmentation results, even in complex scenes with multiple objects and cluttered backgrounds.

3.4 Performance Evaluation

There are several commonly used measures for evaluating the performance of image segmentation algorithms. Some of the most widely used measures are considered in this research work and the necessary expressions are shown in Eqns. (2) to (5) [19, 20];:

$$Jaccard = TP/(TP + FP + FN) \tag{2}$$

$$Dice = (2 * TP)/(2 * TP + FP + FN) \tag{3}$$

$$Accuracy = (TP + TN)/(TP + TN + FP + FN) \tag{4}$$

$$Precision = TP/(TP + FP) \tag{5}$$

where TP is the number of true positives, TN is the number of true negatives, FP is the number of false positives, and FN is the number of false negatives.

4 Result and Discussions

This section of the research presents the experimental outcome achieved using the implemented pretrained schemes. The proposed scheme is trained using 800 images along with its GT and the merit of this scheme is verified using 200 testing images.

The experimental outcome of DeepLabV3+ is presented in Fig. 3. Figure 3(a) presents the outcome achieved during algorithm training and Figs. 3(b) and (c) shows the training and validation accuracy and loss for a chosen epochs of 120. This confirms that the proposed approach is efficient in acieving a better accuracy. After the training, the proposed scheme provides a binary segmentation result and it is depicted in Fig. 3(d). This picture also confirms that the segmented outcome of this scheme is better and it similatr to the GT.

The performance measures from these images needs to be computed by executing a pixel wise comparison among the segmented section and the GT and the mean value of these measures are then considered for the validation. The sample results and the related GT obtained during the study is presented in Fig. 4.

The result of DeepLabV3+ is compared and confirmed against other segmentation procedures, like UNet, SegNet, VGG-UNet and VGG-SegNet available in the literature and the mean ± standard deviation value is then considered for the assesment as shown in Table 1. This table confirms that, the proposed approach helps in getting a better result on the chosen database compared to other existing approaches.

The limitation of this research is, it implements a thresholding combined with the segmentation process. In the future, the SE can be replaced with other thresholding schemes existing in the literature. Further, the computational complexity of the proposed scheme can be reduce by considering the light-weight deep learning based segmentation achemes.

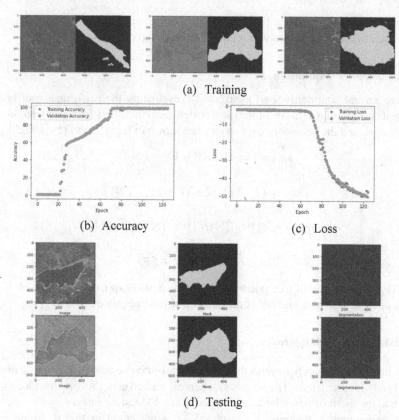

(a) Training

(b) Accuracy   (c) Loss

(d) Testing

Fig. 3. Experimental outcome for DeepLabV3+

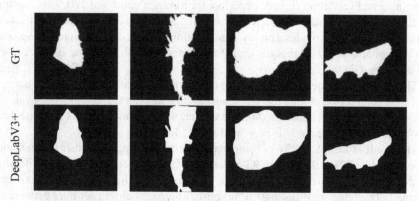

Fig. 4. Sample GT section and its related segmented waterbody region

Table 1. Computed performance values of the test images

Scheme	Jaccard (%)	Dice (%)	Accuracy (%)	Precision (%)
DeepLabV3+	89.28 ± 1.05	93.63 ± 0.85	97.38 ± 0.31	95.17 ± 0.22
UNet	87.17 ± 1.17	91.13 ± 0.82	96.06 ± 0.33	93.37 ± 0.41
SegNet	86.81 ± 1.08	90.28 ± 0.75	96.22 ± 0.28	93.09 ± 0.38
VGG-UNet	88.83 ± 1.15	92.53 ± 0.38	97.14 ± 0.27	94.85 ± 0.16
VGG-SegNet	88.39 ± 1.13	92.17 ± 0.52	96.74 ± 0.19	94.29 ± 0.23

5 Conclusion

This work implements a methodology by integrating the thresholding and dee-segmentation procedure to detect waterbodies in SI. The experimehtal outcome confirms that the DeepLabV3+ has demonstrated its ability to accurately detect waterbodies, even in complex scenes where the waterbodies are partially or completely occluded. By leveraging the powerful features of DeepLabV3+ and fine-tuning it on waterbody detection datasets, the model can accurately detect waterbodies at different scales, shapes, and orientations. Additionally, DeepLabV3+ can effectively distinguish waterbodies from other similar-looking features such as shadows, rocks, and vegetation, making it a robust tool for waterbodies detection. Overall, DeepLabV3+ based waterbodies detection has the potential to provide valuable insights into water resources management, environmental monitoring, and disaster response planning. However, like any other computer vision model, it has its limitations and requires careful evaluation and validation before deployment in real-world applications.

References

1. Kadhim, I.J., Premaratne, P.: A novel deep learning framework for water body segmentation from satellite images. Arabian J. Sci. Eng. 1–12 (2023)
2. Rambhad, A., Singh, D.P., Choudhary, J.: Detection of flood events from satellite images using deep learning. In: Intelligent Data Engineering and Analytics: Proceedings of the 10th International Conference on Frontiers in Intelligent Computing: Theory and Applications (FICTA 2022), pp. 259–268. Singapore: Springer Nature Singapore (2023)
3. Daniel, J., Rose, J.T., Vinnarasi, F., Rajinikanth, V.: VGG-UNet/VGG-SegNet supported automatic segmentation of endoplasmic reticulum network in fluorescence microscopy images. Scanning 2022 (2022)
4. Rajinikanth, V., Kadry, S., Damaševičius, R., Sankaran, D., Mohammed, M.A., Chander, S.: Skin melanoma segmentation using VGG-UNet with Adam/SGD optimizer: a study. In: 2022 Third International Conference on Intelligent Computing Instrumentation and Control Technologies (ICICICT), pp. 982–986. IEEE (2022)
5. Zhang, Z., Meng, L., Ji, S., Huafen, Y., Nie, C.: Rich CNN features for water-body segmentation from very high resolution aerial and satellite imagery. Remote Sensing 13(10), 1912 (2021)
6. Guo, Z., Lin, W., Huang, Y., Guo, Z., Zhao, J., Li, N.: Water-Body Segmentation for SAR Images: Past, Current, and Future. Remote Sensing 14(7), 1752 (2022)

7. Yuan, K., Zhuang, X., Schaefer, G., Feng, J., Guan, L., Fang, H.: Deep-learning-based multispectral satellite image segmentation for water body detection. IEEE Journal of Selected Topics in Applied Earth Observations and Remote Sensing **14**, 7422–7434 (2021)
8. Rana, H., Sivakumar Babu, G.L.: Object-oriented approach for landslide mapping using wavelet transform coupled with machine learning: a case study of Western Ghats, India. Indian Geotechnical Journal **52**(3), 691–706 (2022)
9. Aalan Babu, A., Mary Anita Rajam, V.: Water-body segmentation from satellite images using Kapur's entropy-based thresholding method. Computational Intelligence **36**(3), 1242–1260 (2020)
10. https://www.kaggle.com/datasets/franciscoescobar/satellite-images-of-water-bodies?select=Water+Bodies+Dataset
11. Kadry, S., Rajinikanth, V., Koo, J., Kang, B.-G.: Image multi-level-thresholding with Mayfly optimization. Int. J. Elect. Comp. Eng. **11**(6), 2088–8708 (2021)
12. Rajinikanth, V., Palani Thanaraj, K., Satapathy, S.C., Fernandes, S.L., Dey, N.: Shannon's entropy and watershed algorithm based technique to inspect ischemic stroke wound. In: Smart Intelligent Computing and Applications: Proceedings of the Second International Conference on SCI 2018, Vol. 2, pp. 23–31. Springer Singapore (2019)
13. Rajinikanth, V., Satapathy, S.C., Dey, N., Vijayarajan, R.: DWT-PCA image fusion technique to improve segmentation accuracy in brain tumor analysis. In: Microelectronics, Electromagnetics and Telecommunications: Proceedings of ICMEET 2017, pp. 453–462. Springer Singapore (2018)
14. Zervoudakis, K., Tsafarakis, S.: A mayfly optimization algorithm. Comput. Ind. Eng. **145**, 106559 (2020)
15. Bhattacharyya, T., et al.: Mayfly in harmony: a new hybrid meta-heuristic feature selection algorithm. IEEE Access **8**, 195929–195945 (2020)
16. Zhang, D., Ding, Y., Chen, P., Zhang, X., Pan, Z., Liang, D.: Automatic extraction of wheat lodging area based on transfer learning method and deeplabv3+ network. Comput. Electron. Agric. **179**, 105845 (2020)
17. Chen, L.-C., Zhu, Y., Papandreou, G., Schroff, F., Adam, H.: Encoder-decoder with atrous separable convolution for semantic image segmentation. In: Proceedings of the European conference on computer vision (ECCV), pp. 801–818 (2018)
18. Peng, H., Zhong, J., Liu, H., Li, J., Yao, M., Zhang, X.: ResDense-focal-DeepLabV3+ enabled litchi branch semantic segmentation for robotic harvesting. Comput. Electron. Agric. **206**, 107691 (2023)
19. Rajinikanth, V., Durai Raj Vincent, P.M., Srinivasan, K., Ananth Prabhu, G., Chang, C.-Y.: A framework to distinguish healthy/cancer renal CT images using the fused deep features. Frontiers in Public Health 11 (2023)
20. Manic, K.S., Rajinikanth, V., Al-Bimani, A.S., Taniar, D., Kadry, S.: Framework to Detect Schizophrenia in Brain MRI Slices with Mayfly Algorithm-Selected Deep and Handcrafted Features. Sensors **23**(1), 280 (2022)

Topic Classification of Text-Based Lesson Questions in Turkish with BERTurk

Ayşegül Albayrak Doğan[1]([✉]) [iD], Ahmet Sayar[1] [iD], and İlker Çetiner[2] [iD]

[1] Kocaeli University, 41380 Izmit, Kocaeli, Turkey
aysegulalbayrak1@gmail.com, ahmet.sayar@kocaeli.edu.tr
[2] ArgeLabs Information Technologies, 41275 Yeniköy, Başiskele, Kocaeli, Turkey
ilker.cetiner@argelabs.com.tr

Abstract. With the corona virus pandemic, social contact has been kept to a minimum, and education in schools can be carried out remotely. As a result of this, the concept of distance education has gained importance. In this study, natural language processing (NLP) and its effects on distance education are discussed, and by using NLP, a topic classification system is proposed. Classification is applied to text-based lesson questions in Turkish language. In this way, the questions asked by the students can be quickly directed to the teachers in the relevant specialty through a system to be designed, and the processes can be accelerated. In the data preparation phase, the real-world lesson questions were collected and converted from image to text using the EasyOCR library, and topic classification was performed on the data set using the Berturk model. Since the image-to-text method was used in the data set preparation phase, we encountered some noise in the data. To clean the data, different data preprocessing and cleaning techniques are applied. Finally, the training has been performed, and accuracy rates are presented.

Keywords: Natural Language Processing · Intention Classification · Data Preprocessing · Topic Classification · Bert Model · NLP

1 Introduction

In changing and developing world conditions; especially due to effects of the Covid-19 pandemic and the effects of the earthquake disaster centered in Kahramanmaraş, which occurred in Turkey on February 6, 2023, a decision was made to provide distance education until the major impact of the earthquake was overcome in order to continue educational activities. It has been experienced that the effects of distance education, which gained importance with the pandemic conditions and the effects of natural disasters experienced afterwards, continue. In such situations of necessity, it is inevitable to ensure sustainability and adapt quickly to changing conditions [3]. As a result of the aforementioned conditions, the world of distance education, which is defined as time and space independent learning, continues to occupy the agendas of teachers, students and families [12].

© The Author(s), under exclusive license to Springer Nature Switzerland AG 2023
S. Kadry and R. Prasath (Eds.): MIKE 2023, LNAI 13924, pp. 87–94, 2023.
https://doi.org/10.1007/978-3-031-44084-7_9

NLP, which is becoming more and more important in the age of the digitalizing world and has an impact in almost every field, can be used in many areas such as social media, banking transactions, appointment systems, chat applications, question and answer systems. In the world of education, it is widely integrated in many areas such as research, science, linguistics, e-learning, assessment systems and contributes to achieving positive results in other educational environments such as schools, higher education system and universities [1]. NLP develops methods that produce solutions by understanding human needs in different fields. One of these methods, BERT that was developed by Google in November 2018 to better understand human language, has remained popular in the field of natural language processing in recent years.

In this study, the factors observed in the process of automatically labeling which course class the questions asked in dialog systems, which will provide fast interaction in the field of education, are evaluated. The accuracy rates of the classification process using the BERTurk language model were observed on 4027 data belonging to History, Turkish, Philosophy, Geography, Biology, Mathematics, Physics, Chemistry, History, Turkish, Philosophy, Geography, Biology, Mathematics, Physics, Chemistry courses among the high school courses in Turkish language and in the National Education curriculum of the Republic of Turkey [15]. The data used in the training consists of a pool of data converted from image to text via EasyOCR library and course labeled during the conversion. It was observed some noisy data such as underscores (_) and hyphens (-) occurred in the image-to-text data depending on the quality. In this context, in order to examine the effect of the detected noisy data and the preprocessing work to be done on the questions on the classification of the data, different combinations were made and the training were repeated with the same hyper-parameters and the accuracy rates were examined. The same number and the same training set were used in each training. In the training, 20% test data was used. After the training, it was observed that the performance rates did not differ significantly even if cleaning operations were performed. Despite noisy data, successful classification was achieved at a rate of 0.97.

Related studies are presented in the second part to the general one in our study. The work done in the third section is summarized. Tests and evaluation studies were carried out in the fourth section. The fifth section contains the results and comments.

2   Relatedwork

Classification techniques have been used in many different scientific and application domains. Those classification studies can be grouped into three categories. The first group is applied to image and video data; the second group is a hybrid classification applied to text and image data together; and the third group's classification techniques are applied merely to text-based data.

The samples of image and video data classification are given as follows: [7] study the performance of the vehicle classification algorithms by using deep learning algorithms on video streams. Topçu et al. [17] have applied the classification to remote satellite image data by using capsule networks. Some of them are based on image data, and others are based on text data. Şentaş et al. [13] studied the performance of Support Vector Machine (SVM) and Convolutional Neural Network Algorithms (CNN) for real-time vehicle type classification. Tasiev et al. [16] presented a real-time vehicle type classification using CNN. The hybrid classifications are applied to text and image data together. Omurca et al. [11] proposed a document image classification system by fusing deep and machine learning models. Sevim et al. [14] studied multi-class document image classification by using deep visual and textual features. Yurtsever et al. [18] worked on a search technique enabling figure search by text in large scale digital document collections.

In this study, we focus on the third group of classification techniques, which are applied merely to text-based data. The topic classification problem is a Natural Language Processing (NLP) problem that is often studied in the literature. BERT, who has created a solution to this problem, has given strong results in most of the studies conducted in the field of subject classification in many languages. In this section, studies that use BERT modeling in the field of subject classification and evaluate the effect of preprocessing stages on performance rates are examined. The sources examined in the literature are usually in foreign languages and the number of studies on Turkish is in the minority. In our study, the studies that can guide the effect of the data consisting of a set of Turkish questions on the performance rate of the data preprocessing stage when using the BERT model are shared below in chronological order.

Hazrati et al. [5], in their study to detect irony in texts published on social media, stated that non-standard expressions are generally included in the texts used by social media users. The BERT model was trained before and after these detected expressions were cleaned and the success rates were shared. They stated that data cleaning has a great impact on the success rate in sentiment analysis. After the data cleaning process we performed on the Turkish dataset, it was observed that there were not very big differences in the success rates. In this context, it is seen that the data set in different fields has an effect on the success rate on BERT.

Kurniasih and Manik [8] investigated the effects of training with and without data preprocessing on deep learning. In this study, the effect of the processes such as correction of abbreviated words, removal of repeated syllables, removal of hashtags on the accuracy performance rate of the Bert model was examined.

Another study examining the effect of systematic practical perturbations on the performance of deep learning-based text classification models such as CNN, LSTM and BERT was conducted by Miyajiwala et al. [10]. In the study where punctuation marks, stop words, ineffective words, and unwanted words are expressed as perturbations, it is stated that BERT is a more sensitive model than other models when perturbations are added or removed.

Bayrak and Issifu [4] worked on two main tasks in their study. The first one is dialect recognition and the second one is sentiment analysis with BERT model on tweet data and they shared their classification results. In the data preprocessing stages, they applied cleaning steps such as Html tags, URLs, leaving spaces after Arabic numbers, removing Arabic-specific accent marks. They showed that preprocessing has a positive effect on Dialect Recognition unlike Sentiment Analysis.

In their study, Zhu et al. [19] showed that for text classification tasks with modern NLP models such as BERT, methods applied over various types of noise do not always improve performance and may even degrade it. For different types of noise in the dataset, they show that BERT is robust to injected noise, but not necessarily under weak supervision noise. In our study, it was observed that the removal of ineffective words relatively decreased the performance rate.

In this study for author profile detection, Alzahrani and Jololian [2] discussed the effects of data preprocessing techniques before Bert model training. In the preprocessing stages, training was carried out after the removal of stop words, reteweet tags, hashtags, mentions, and urls, and the success rates obtained were shared. It was observed that the highest performance rate was obtained on data trained without any data preprocessing.

In this study by Jiang et al. [6], a pre-trained and fine-tuned BERT model was used to classify OCR translated texts into book excerpts according to subject areas. As a result of the classification, the effect of OCR noise on the performance rate was analyzed.

Maharani [9] stated that data quality greatly affects the classification performance in classification using the Bert model. In this study, as a result of the study conducted to classify tweets related to emergencies, it was emphasized that data quality should be improved in order to avoid misclassification and it was stated that this issue will be studied in future studies.

3 Architecture

The general flow we followed during our study to conduct the training and evaluate the results is shown in Fig. 1. During the creation of the labeled data, the question set was converted from image to text using EasyOcr and a total of 4027 data were obtained. The dataset was split into training and test data by 20% before training. The same number of questions was used in all experiments. In the data preprocessing stages, the preprocessing method was presented with different combinations depending on the data set content and training was performed. The training was named according to the order in which the training was performed. For the training, library management was provided by installing Anaconda on a computer with GeForce RTX 3080 Nvidia graphics card. Hyperparameters used for each training are available in Table 1. Training was performed on a preprocessed dataset and took 4 min on average. Six different training trials were conducted, are listed below. The data sample used in the trainings can be seen in Table 2.

In the first training, "A), B), C), D), E), A) Yalnız/Yalnız, B) Yalnız/Yalnız, C) Yalnız/Yalnız, D) Yalnız/Yalnız, E) Yalnız/Yalnız" answer choices were removed from the sentence and the training was realized.

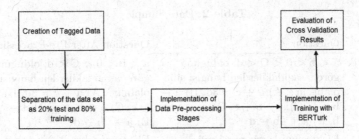

Fig. 1. Training Model Flowchart

Table 1. Model Parameters

Parameter	Value
Hugging Face Model	dbmdz/bert-base-turkish-uncased
Use Early Stopping	True
Early Stopping Delta	0.01
Early Stopping Metric	mcc
Early Stopping Metric Minimize	False
Early Stopping Patience	5
Evaluate During Training Steps	6000
fp16	False

In the second training, the Natural Language Toolkit (NLTK) library was used to remove stopwords from the dataset and training was performed.

In the third training, the training was performed by removing only the options starting with "Yalnız/Yalnız" from the sentence in the data set.

In the fourth training, the Stopwords belonging to the NLTK library on the dataset and the choices starting with "Yalnız/ Yalnız" were removed and the training was performed.

In the fifth training, no cleaning was performed on the data set.

The sixth training, words were merged by removing hyphens (-) and under-scores (_) from the data set. The method followed during word merging is as follows:

The hyphen represents subtraction in mathematical formulas, so there are rules to be considered during word merging. For example, since the hyphen in the formula "$x = a - b$" should not be removed and proceeded as an expression in the form of "ab", the merging method here is based on the rule of Turkish grammar that the word at the end of the line is divided while the word at the end of the line is divided, and no single letter is left at the end of the line and at the beginning of the line. In addition, it was checked that the word following the line does not correspond to a variable in the formula in Mathematics. For formula variables, a fixed list was prepared according to the content of the data set: ab, ba, abc, acb, bac, bca, cab, cba, a, b, c. If the word checked before and

Table 2. Data Sample

Lesson	Question	Question After Pre-Processing
Matematik	a < b < 0 < C < d, olduguna göre, asagidakilerden hangisi sifir olabi- lir?, A) a + c = 3d, B) a − b = c, C) a − b + c, D) d − c = b, E) a + 2b − d	a < b < 0 < C < d, olduguna göre, asagidakilerden hangisi sifir olabilir?, A) a + c = 3d, B) a − b = c, C) a − b + c, D) d − c = b, E) a + 2b − d
Fizik	Elektronun karsit parçacigi pozitron için; I Kütlesi elektronunki ile aynidir. II. Yük miktari elektronunki ile aynidir. III. Yükünün isareti elektroninki ile aynidir. yargilarindan hangileri doğrudur? 1, A) Yalniz I, B) Yalniz II D) ve III E) II ve III, C) I ve II	Elektronun karsit parçacigi pozitron için; I Kütlesi elektronunki ile aynidir. II. Yük miktari elektronunki ile aynidir. III. Yükünün isareti elektroninki ile aynidir. yargilarindan hangileri doğrudur?

after the hyphen in the word to be merged was present in the above list, the word was not merged.

4   Tests and Evaluation

The success rates of the experiments conducted within the scope of the study are available in Table 3. According to the data preprocessing stages performed on the data, the steps are indicated in the columns according to the trainings. The number of test data used for success rate calculation and the number of incorrectly predicted questions are shared. In the validation process performed on the data model obtained after the training, the course distributions of the incorrectly predicted questions are in Table 4.

Table 3. Training Cross Validation Results

Training	Accuracy	Test Question Count	False Prediction Count	Remove Stop-words	Remove "-, _"	Remove Start With "Yalnız"	Remove All Answers Choices
First	0.985472	826	12				✓
Second	0.984261	826	13	✓			
Third	0.979418	826	17			✓	
Fourth	0.978208	826	18	✓		✓	
Fifth	0.978208	826	18				
Sixth	0.970944	826	24		✓		

Table 4. Number of False Predicted Questions by Lessons

Training	Biyoloji	Coğrafya	Felsefe	Fizik	Kimya	Matematik	Tarih	Türkçe	TOTAL
First	1	3	2	3	0	0	1	2	12
Second	3	3	1	3	0	0	1	2	13
Third	3	2	1	5	1	0	1	4	17
Fourth	3	3	3	3	2	0	1	3	18
Fifth	7	2	0	6	2	0	1	0	18
Sixth	7	4	2	5	0	0	1	5	24

5   Conclusions

We investigated the performance rates of the pre-trained and fine-tuned BERTurk model on data converted from image to text using EasyOCR, and the effect of data preprocessing on classification by course categories on a dataset of high school questions. Our analysis shows that the pre-trained and fine-tuned BERTurk language model achieves good performance rates despite the OCR noise. The most successful result was obtained in the first training as 0.98. When the results are analyzed, it can be seen from Table 4 that the highest number of incorrect predictions was in the Physics course and there was no misclassification in the Mathematics course.

In the scope of our study the training accuracy performance rates did not show large differences in the data preprocessing steps. In particular, it was observed that the incorrect predictions were obtained based on similar question contents. Therefore, the number of incorrect predictions can be improved in future studies by providing diversity over the training data set. Different combinations can be added to the training steps by diversifying the data preprocessing stages. For example, although the word "ka- lem" was corrected as "kalem", no processing was performed on the word "ka - lem". Therefore, these words were included separately in the training. The training can be performed again by performing the cleaning process for these words.

References

1. Alhawiti, K.M.: Natural language processing and its use in education. Int. J. Adv. Comput. Sci. Appl. **5**(12) (2014)
2. Alzahrani, E., Jololian, L.: How different text-preprocessing techniques using the BERT model affect the gender profiling of authors. arXiv preprint arXiv:2109.13890 (2021)
3. Aras, K.S., Kocasaraç, H.: Eğitimin dijital boyutunda öğrenme-öğretme araçları. Uluslararası Karamanoğlu Mehmetbey Eğitim Araştırmaları Dergisi **4**(2), 117–134 (2022)
4. Bayrak, G., Issifu, A.M.: Domain-adapted BERT-based models for nuanced Arabic dialect identification and tweet sentiment analysis. In: Proceedings of the the Seventh Arabic Natural Language Processing Workshop (WANLP), pp. 425–430 (2022)

5. Hazrati, L., Sokhandan, A., Farzinvash, L.: Profiling irony speech spreaders on social networks using deep cleaning and BERT. In: CLEF, pp. 1613–0073 (2022)
6. Jiang, M., Hu, Y., Worthey, G., Dubnicek, R.C., Underwood, T., Downie, J.S.: Impact of OCR quality on BERT embeddings in the domain classification of book excerpts. In: CHR, pp. 266–279 (2021)
7. Kul, S., Eken, S., Sayar, A.: Trafik gözetim videolarında araç sınıflandırma algoritmalarının etkinliğinin Ölçülmesi (2016)
8. Kurniasih, A., Manik, L.P.: On the role of text preprocessing in BERT embedding-based DNNs for classifying informal texts. Neuron **1024**(512), 256 (2022)
9. Maharani, W.: Sentiment analysis during Jakarta flood for emergency responses and situational awareness in disaster management using BERT. In: 2020 8th International Conference on Information and Communication Technology (ICoICT), pp. 1–5. IEEE (2020)
10. Miyajiwala, A., Ladkat, A., Jagadale, S., Joshi, R.: On sensitivity of deep learning based text classification algorithms to practical input perturbations. In: Arai, K. (ed.) SAI 2022. LNNS, vol. 507, pp. 613–626. Springer, Cham (2022). https://doi.org/10.1007/978-3-031-10464-0_42
11. Omurca, S.İ, Ekinci, E., Sevim, S., Edinç, E.B., Eken, S., Sayar, A.: A document image classification system fusing deep and machine learning models. Appl. Intell. **53**, 1–16 (2022)
12. Sayılır, K., Sarı, Y.P., Pepele, H.R., Yetkin, S.G.: Uzaktan eğitim faaliyetlerine ilişkin ortaokul öğrencilerinin görüşlerinin değerlendirilmesi. Ulusal Eğitim Dergisi **3**(2), 417–435 (2023)
13. Şentaş, A., et al.: Performance evaluation of support vector machine and convolutional neural network algorithms in real-time vehicle type and color classification. Evol. Intell. **13**, 83–91 (2020)
14. Sevim, S., et al.: Multi-class document image classification using deep visual and textual features. Int. J. Comput. Intell. Appl. **21**(02), 2250013 (2022)
15. Sun, C., Qiu, X., Xu, Y., Huang, X.: How to fine-tune BERT for text classification? In: Sun, M., Huang, X., Ji, H., Liu, Z., Liu, Y. (eds.) CCL 2019. LNCS (LNAI), vol. 11856, pp. 194–206. Springer, Cham (2019). https://doi.org/10.1007/978-3-030-32381-3_16
16. Tashiev, İ., et al.: Konvolüsyonel sinir ağı kullanarak gerçek zamanlı araç tipi sınıflandırması real-time vehicle type classification using convolutional neural network
17. Topçu, M., Dede, A., Eken, S., Sayar, A.: Multilabel remote sensing image classification with capsule networks. In: 2020 International Congress on Human-Computer Interaction, Optimization and Robotic Applications (HORA), pp. 1–3. IEEE (2020)
18. Yurtsever, M.M.E., Özcan, M., Taruz, Z., Eken, S., Sayar, A.: Figure search by text in large scale digital document collections. Concurr. Comput.: Pract. Exp. **34**(1), e6529 (2022)
19. Zhu, D., Hedderich, M.A., Zhai, F., Adelani, D.I., Klakow, D.: Is BERT robust to label noise? A study on learning with noisy labels in text classification. arXiv preprint arXiv:2204.09371 (2022)

Highest Accuracy Based Automated Depression Prediction Using Natural Language Processing

S. V. Tharun[1], G. Saranya[1], T. Tamilvizhi[2], and R. Surendran[3](✉)

[1] Department of Computer Science and Engineering, Amrita School of Computing, Amrita Vishwa Vidyapeetham, Chennai, India
g_saranya@ch.amrita.edu
[2] Department of Computer Engineering, Panimalar Engineering College, Chennai, India
[3] Department of Computer Science and Engineering, Saveetha School of Engineering, Saveetha Institute of Medical and Technical Sciences, Chennai, India
surendran.phd.it@gmail.com

Abstract. In today's fast-paced society, psychological health issues such as anxiety, depression, and stress have become prevalent among the general population. Researchers have explored the use of machine learning algorithms to predict the likelihood of depression in individuals. As datasets related to depression become more abundant and machine learning technology advances, there is an opportunity to develop intelligent systems capable of identifying symptoms of depression in written material. By applying natural language processing and machine learning algorithms to analyze written text, such as social media posts, emails, and chat messages, researchers can potentially identify patterns and linguistic cues associated with depression. These patterns may include changes in word usage, tone, and sentiment. The dataset consists of text-based questions on this information channel. At present, machine learning techniques are highly effective for analyzing data and identifying problems. Researchers have conducted comparisons of the accuracy achieved by different machine learning algorithms using the complete set of attributes as well as a subset of selected attributes. In summary, while the potential for AI to aid in mental health diagnosis and treatment is exciting, it's important to proceed with care and consideration for the complexities of the field and the needs of patients.

Keywords: Machine Learning · Natural Language processing · Data Visualization · Artificial Intelligence · Decision Tree Making · Logistic Regression

1 Introduction

Depression is a significant contributor to global disability, affecting millions of people worldwide. While older adults may have a lower incidence of depression, late-life depression (LLD), also known as geriatric depression, still poses a significant public health concern due to its association with increased morbidity, suicide risk, cognitive and physical impairments, and self-neglect. As the global population continues to age,

S. Kadry and R. Prasath (Eds.): MIKE 2023, LNAI 13924, pp. 95–104, 2023.
https://doi.org/10.1007/978-3-031-44084-7_10

it becomes increasingly important to identify and treat LLD. Although the definition of Late Life Depression (LLD) can vary, it typically pertains to depression experienced by individuals who are aged 60 or older, and can encompass both early and late onset episodes. Like depression in younger individuals, LLD can manifest in various ways, with symptoms ranging from subtle changes in mood to major depression as defined in the Diagnostic and Statistical Manual of Mental Disorders (DSM-5). It is important to approach preliminary results with caution and acknowledge that further research is necessary to better integrate artificial intelligence (AI) in mental health research with clinical practice.

Depression is a prevalent mental health condition that can affect anyone. It is characterized by a persistent feeling of sadness, hopelessness, and low energy, often leading to a loss of interest in activities, difficulty sleeping, and changes in appetite. The causes of depression can be complex, with contributing factors ranging from genetics to environment and psychology. While depression can be a debilitating condition, it is treatable with the help of mental health professionals, and treatment options can include medication, psychotherapy, or a combination of both. Seeking help early is essential in preventing the condition from becoming more severe, as untreated depression can lead to serious consequences, including suicide. Depression can have significant impacts on daily functioning, including work, school, and relationships. Some individuals may experience only one episode of depression, while others may have recurrent episodes throughout their lives. It is important to remember that depression is a legitimate medical condition that should be taken seriously, and with proper treatment and support, many individuals with depression can achieve a better quality of life.

NLP can be used to analyze various aspects of text, such as sentiment, tone, and word choice, to identify patterns that may indicate the presence of depression. The project aims to develop a model that can accurately predict depression in individuals using their written text, such as social media posts or chat messages. The machine learning algorithms to predict the stress level during this quarantine with the following Methodology such as Acquisition of data, Pre-processing the data, Splitting the Training and Testing Data, Classification of data and Analysis on the performance of the model.

2 Related Works

Meera Sharma, Sonok Mahapatra, and Adeethyia Shankar employed a Classification and Regression Tree (CART) model, a type of decision tree frequently utilized in machine learning for information mapping. Although CART models have various applications, they used it for binary classification of their data. One issue highlighted in their study is that CART models may not be readily trainable in other languages due to insufficient datasets in those languages. To address the limitations in the dataset, future studies could consider utilizing more current and extensive samples from various industries [1]. Usman Ahmed et al., developed a method based on the standard PHQ-9 questionnaire to extract nine distinct behavior types that align with the Diagnostic and Statistical Manual of Mental Disorders 5 (DSM-V) for measuring depression symptoms from authored patient text. Their method effectively handles uncertainty and interprets the embedding working. One limitation of their approach is that constructing rules requires prior knowledge and expertise in the system.

In the future, research could focus on improving the accuracy of detecting triggering rules and identifying key words to help psychiatrists make accurate notes and diagnoses [2]. Konda Vaishnavi et al., compared the accuracy of five machine learning techniques, including Logistic Regression, Classifier, Random Forest, Decision Tree Classifier, K-NN and Stacking, in identifying mental health issues based on various criteria. Their results indicated that these classifiers were more accurate than other classifiers, with ROC areas between 0.8 and 0.9. However, a limitation of their study was the use of a minimal dataset. Future research could address this limitation by utilizing a larger dataset to improve accuracy [3]. Theodoros Iliou ét al., assessed the effectiveness of seven classification algorithms, which comprised the Nearest-neighbor classifier, J48, Random Forest, Multilayer Perceptron, Support Vector Machine, JRIP, and FURIA, on both the raw and preprocessed datasets. One limitation of these data preprocessing methods and classification algorithms is that they may face difficulties in the training phase if the dataset contains a high amount of irrelevant or redundant information, as well as noisy or unreliable data [4]. The proposed approach involves utilizing the strengths of a commonly-used classifier called "XGBoost" to achieve precise classification of data into four categories of mental disorders: Schizophrenia, Autism, OCD, and PTSD. The results of the experiments demonstrate that this approach is highly effective in accurately identifying various types of mental illness [5].

The approach consists of five modules, namely data acquisition, data pre-processing, training and testing data splitting, data classification, and analysis of model performance [6]. The primary benefit of this approach is its ability to precisely identify stress levels by utilizing the dataset obtained during the quarantine period. However, the limitations and issues of this method may become apparent when dealing with raw data and additional test cases. Further research may yield better results by exploring raw data and including more test cases. In the existing work, brain signals are used to find whether the person is in depression or not but it is a complex process since it uses wave signals. It is a physical device with complex mechanism and so, machine learning is not implemented. It is difficult to use in Telemedicine.

3   Proposed Work

The proposed model aims to develop a machine learning model with natural language processing capabilities that can accurately classify whether an individual is suffering from depression or not. Depression is a significant and prevalent issue in society, and managing it is challenging. Consequently, individuals are at a higher risk of experiencing mental distress. The machine learning approach is well-suited to address such complex tasks, as manual analysis of such data can be time-consuming [7]. One potential application of the model is to aid mental health professionals in identifying individuals who may be susceptible to depression and offering them suitable care and assistance. To achieve this goal, the project will entail collecting a dataset of textual samples from both depressed and non-depressed individuals, as well as creating and validating NLP algorithms for analyzing the text [8]. Machine learning can be employed to classify depression by leveraging past data to identify patterns, and enhance the model's accuracy by adjusting its parameters.

The model has potential applications in aiding mental health professionals to identify people who are at risk of developing depression and to provide them with suitable care and support. To accomplish this, the project will necessitate obtaining a collection of textual samples from both depressed and non-depressed individuals, as well as creating and validating NLP algorithms to analyze the text. Various algorithms can be tested, and the most effective model, with the highest accuracy, can be employed for classification purposes. Figure 1 depicts the entity relationship diagram. Figure 2 depicts the operations performed by the system and the user.

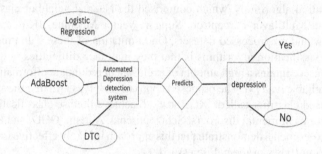

Fig. 1. ER diagram

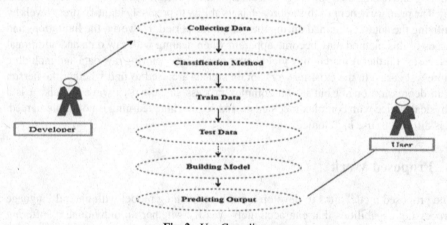

Fig. 2. Use Case diagram

Figure 3 depicts the sequence of activities happening during the prediction of depression when the user clicks the predict option in the website. Figure 4 depicts the Activity during the prediction of depression.

Figure 5 depicts the relationship between the users, interface and the model developed using the processes included as attributes. Both the training and testing data is used to develop the model.

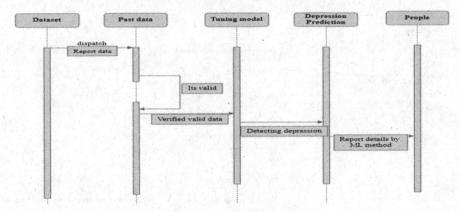

Fig. 3. Sequence diagram

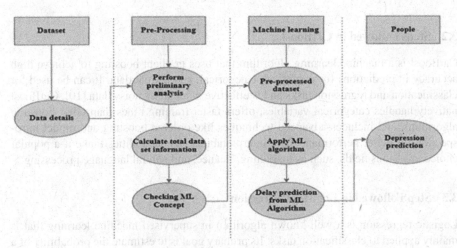

Fig. 4. Activity diagram

3.1 Steps Followed by Decision Tree Classifier

Decision Trees are a form of Supervised Machine Learning that involve splitting data based on certain parameters. This technique involves decision nodes and leaves as the two main entities that comprise a tree. Decision Trees have several real-life analogies and have significantly influenced machine learning, covering both classification and regression [9]. The decision tree classifier works by iteratively dividing the data into smaller subsets based on the most relevant features, until a specified stopping condition is met.

The algorithm selects the best feature to split the data at each node by maximizing the difference in impurity between the parent node and the child nodes. It continues until the leaf nodes are pure or the maximum depth of the tree is reached. The resulting tree can then be used to predict the target variable for new data by following the decision path from the root to a leaf node.

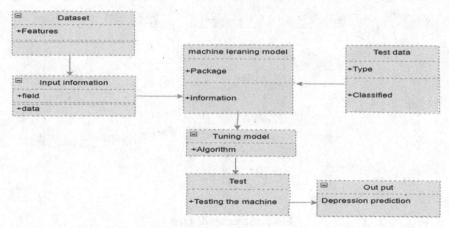

Fig. 5. Class diagram

3.2 Steps Followed in CatBoost

CatBoost is a machine learning algorithm that uses gradient boosting to achieve high accuracy in predictions for numerical, categorical, and textual data. It can be used for classification and regression tasks and is effective in handling noisy data [10]. CatBoost natively handles categorical variables, offers faster training times than other boosting algorithms, and includes advanced techniques like ordered boosting and model introspection for better performance and interpretability. Its capabilities make it a popular choice in various fields, such as marketing, finance, and natural language processing.

3.3 Step Followed in Logistic Regression

Logistic regression is a well-known algorithm in supervised machine learning that is mainly applied to classification tasks. Its primary goal is to estimate the probability of a binary target variable, where the target variable represents either a success/yes (denoted by 1) or a failure/no (denoted by 0). Logistic regression models are created to predict the likelihood of $P(Y = 1)$ based on the independent variable, X. This algorithm is effective and straightforward, and is frequently utilized in tasks such as spam detection, cancer detection, and diabetes prediction [11].

Data Preprocessing
Numpy and Pandas are commonly used Python libraries for data preparation and preprocessing. Numpy is used for mathematical operations and handling multi-dimensional arrays, while Pandas is used for data manipulation and analysis. Together, they provide a powerful toolkit for cleaning, transforming, and restructuring data before it is used for machine learning or other data analysis tasks [12]. To ensure that data can be efficiently processed by the algorithm, it is necessary to carry out data preprocessing, which involves transforming data into a format that is easy to comprehend and manipulate. In Python, the following import statements are commonly used for pre-processing data:

    import pandas as pd

import numpy as np

Data Visualization
Data visualization is the presentation of data in a pictorial or graphical format. It helps to identify trends, and relationships in the data. There are numerous libraries available in Python that enable the creation of visualizations, including but not limited to Matplotlib, Seaborn, Plotly, and Bokeh [13]. There are various libraries available that offer an extensive collection of functions and tools to create diverse types of graphs, charts, and plots such as line plots, scatter plots, bar graphs, histograms, and heat maps.These visualizations can be customized with different colors, labels, titles, and annotations to make them more informative and appealing to the audience in Fig. 6 and Fig. 7.

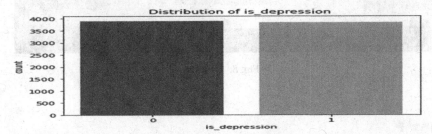

Fig. 6. Distribution Graph

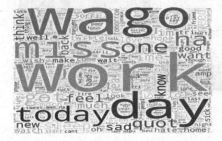

Fig. 7. Word Cloud Image

Algorithm Implementation
In order to effectively compare and evaluate the performance of various ML algorithms, it is essential to establish a standardized test harness. By developing such a test harness in Python, one can gain a template for comparing multiple machine learning models, and can add additional algorithms as needed [14, 15]. Since different machine learning models will exhibit unique performance characteristics, a test harness can be an effective tool for identifying the most suitable algorithm for a given problem [16, 17].

Deployment
In this module, the machine learning model that has been trained is typically saved in a

pickle file format (.pkl file). This file can then be used for deployment in order to create a more user-friendly interface and provide accurate predictions for Depression Prediction with help of Website in Fig. 8. In Fig. 9. Input as Text for the prediction of depression using NLP.

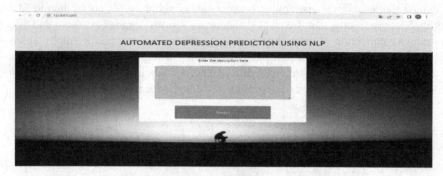

Fig. 8. Website

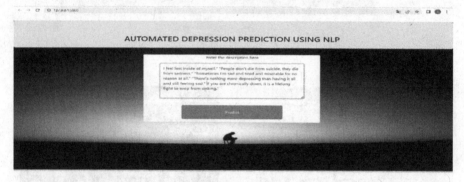

Fig. 9. Input as Text

The result of the proposed work is displayed in Fig. 10 as Predicted output (for depressed person) and Fig. 11. as Predicted output (for not depressed person).

In the proposed system, machine learning will be implemented for more accurate results. It utilized NLP based depression classification was predicted. This system can be implemented in telemedicine, Social Media Platforms to helps in analysing people's mental health. It is deployed in a web browser, so that retains the privacy of the patients. Brain signals are not required in this proposed system. It is a time efficient process.

Fig. 10. Predicted output (for depressed person)

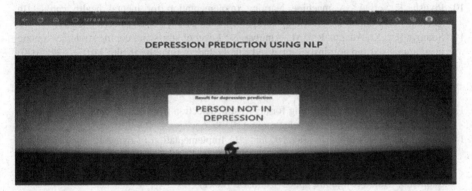

Fig. 11. Predicted output (for not depressed person)

4 Conclusions

The highest accuracy was achieved on a public test set using a high-accuracy scoring algorithm, which was then used in the application to detect depression in individuals. To enhance this system, future work could involve implementing it in multiple programming languages, deploying it on a cloud platform for better accessibility, incorporating more data to improve accuracy, and adding voice input functionality. These enhancements can provide an even more robust and effective tool for detecting depression in individuals.

References

1. Meera, S., Sonok, M., Adeethyia, S.: Predicting the utilization of mental health treatment with various machine learning algorithms. WSEAS Trans on Computers **19** (2019)
2. Usman, A., Jerry, C.L., Gautam, S.: Fuzzy explainable attention-based deep active learning on mental-health data. IEEE International Conference, pp. 6654–4407 (2019)
3. Konda, V., Nikhitha, K., Ashwath, R., Subba, R.N.V.: Predicting mental health illness using machine learning algorithm. Journal of Physics: Conference Series (2021)
4. Theodoros, I., Georgia, K., Mandani, N., Christina, L.: ILIOU Machine Learning Preprocessing Method for Depression Type Prediction, 257–263 (2019)

5. Kamal, M., et al.: Predicting Mental Illness using Social Media Posts and Comments **11** (2021). https://doi.org/10.14569/IJACSA.2020.0111271
6. Thanarajan, T., Alotaibi, Y., Rajendran, S., Nagappan, K.: Improved wolf swarm optimization with deep-learning-based movement analysis and self-regulated human activity recognition. AIMS Mathematics **8**, 12520–12539 (2023)
7. Anishfathim, B., Sreenithi, B., Trisha, S., Swathi, J., Sindhu, P.M.: The Impact of Mental Health due to Covid 19 – A Mental Health Detector Using Machine Learning. Second International Conference on Artificial Intelligence and Smart Energy (ICAIS), **147** (2022)
8. Cho, H.K.: Twitter Depression Data Set Tweets Scraped from Twitter, Depressed and Non-Depressed (2021). Available online: https://www.kaggle.com/hyunkic/twitter-depression-dat aset, accessed on 15 January 2022
9. Reya, P.R., Suchitra, S., Gopal, K.S.: The BMI and mental illness nexus: a machine learning approach. International Conference on Smart Technologies in Computing, Electrical and Electronics (ICSTCEE) (2020)
10. Piyush, K., et al.: A machine learning implementation for mental health care. 11th International Conference on Cloud Computing, Data Science & Engineering (2021)
11. Soumya Raj, K., Anagha Raj, M., Amulya, N.: Level of stress and coping strategies among institutionalised and non-institutionalised elderly. Indian J. Public Health **11**(03), 637 (2020)
12. Vaibhav, J., Dhruv, C., Piyush, G., Dinesh, K.V.: Depression and impaired mental health analysis from social media platforms using predictive modelling techniques. Fourth International Conference on I-SMAC (2020)
13. Amanat, A., et al.: Deep learning for depression detection from textual data. Electronics **11** (2022). https://doi.org/10.3390/electronics11050676
14. Surendran, R., Karthika, R., Jayalakshmi, B.: Implementation of dynamic scanner to protect the documents from ransomware using machine learning algorithms. In: 2021 International Conference on Computing, Electronics & Communications Engineering (iCCECE), Southend, United Kingdom, pp. 65–70. IEEE (2021)
15. Raymond, C., Gregorius, S.B., Sandeep, D., Fabian, C.: A textual-based featuring approach for depression detection using machine learning classifiers and social media texts. Computers in Biology and Medicine **135** (2021). https://doi.org/10.1016/j.compbiomed.2021.104499
16. Firoz, N., Beresteneva, O.G., Vladimirovich, A.S., Tahsin, M.S., Tafannum, F.: Automated Text-based Depression Detection using Hybrid ConvLSTM and Bi-LSTM Model. In: 2023 Third International Conference on Artificial Intelligence and Smart Energy (ICAIS), pp. 734–740. IEEE (2023)
17. Ramya, G.R., Bagavathi Sivakumar, P.: An incremental learning temporal influence model for identifying topical influencers on Twitter dataset. Soc. Netw. Anal. Min. **11**, 27 (2021). https://doi.org/10.1007/s13278-021-00732-4

Crime Prediction Using Modified Capsule Network with CrissCross Optimization on the Sentiment Analysis for Cyber-Security

Prem Kumar Singuluri[1], B. Jayamma[2(✉)], B. Divya Bharathi[1,2(✉)], T. Bhavya Sree[1,2(✉)], and B. Lakshmi Santhoshi[1,2]

[1] Department of CSE, Dean of Innovations, G. Pullaiah College of Engineering and Technology, Kurnool, India
dean@gpcet.ac.in
[2] G. Pullaiah College of Engineering and Technology, Kurnool, India
jayabana6@gmail.com

Abstract. Data mining from textual sources is known as social media text analytics. Social media, like any other text-based dataset, is amenable to text analysis. Today's society has a huge challenge in the form of crime, with the routes of which having shifted almost entirely to social media. Because of crime, both the standard of living and economic development have suffered. By sifting through the available data, we may identify trends in criminal behaviour and foresee potential incidents in the future. Unfortunately, not all crimes are recorded or solved due of a lack of evidence. Hence, tracking these offenders remains challenging. We can keep an eye on illicit activities on social media. For the simple reason that social media users sometimes make observations about their immediate environment. In this study, we propose a model for crime categorization on a Twitter dataset that combines the recently proposed Capsule Network with the tried-and-true Multi-layer Perceptron. The output of both networks is combined in a manner that makes the most of each network's strengths using a rule-based technique. Moreover, the Criss-Cross Optimization (CCO) method is used to fine-tune the capsule network's hyper-parameters. Crime statistics from reliable sources are used to make accurate comparisons between regions. We also evaluate recent changes in crime rates in both Jammu, India, and Ghaziabad, Uttar Pradesh. The most recent crime patterns during the last week have been documented (23, January 2019 to 30, January 2019). The studies show that the generated findings are consistent with the actual crime rate information. We anticipate that research of this kind will aid in gauging the current crime rate in various areas and in identifying criminal activity.

Keywords: Criss-Cross Optimization · Social-media · Capsule Network · Multilayer Perceptron · Criminal activity and in identifying criminal activity

1 Introduction

Criminal activity has far-reaching, deleterious effects on many elements of our society and culture. Recognizing high-crime areas and pinpointing the most recent incidents in a given area is a major problem for both local authorities and private citizens [1].

S. Kadry and R. Prasath (Eds.): MIKE 2023, LNAI 13924, pp. 105–116, 2023.
https://doi.org/10.1007/978-3-031-44084-7_11

On the other hand, city dwellers are always working to make their communities safer and friendlier places to live. The prevalence of crime is one of the most pressing issues confronting modern societies [2]. There have been much less research using social media to examine crimes and criminal behaviour than there have been on social crimes. Our primary data comes from social media. The real-time social networking platform Twitter has exploded in popularity throughout the globe. More than 300 million people throughout the globe use Twitter, and more than 500 million tweets are sent per day, according to data [3].

Using tweets for personal, professional, and business purposes is a terrific way to stay in touch with loved ones and co-workers [4]. Cyberstalking, cyberbullying, and other forms of cyber harassment are only some of the difficult problems that come up when people use Twitter. Data is collected from a social media platform by the use of a small number of sensitive terms [5, 6] associated with the crime, Assault. Among a significant volume of tweets, 3801 are selected to serve as datasets for the research. During the course of four years, a great deal of tweets are filtered out using keywords. The non-alphanumeric characters have been cleared out of the data [7]. Recognizable crime rates have also been acquired from security agencies, and a major component of our research involves comparing the datasheet built from Twitter data with the actual crime data conventional from security units [8, 9].

For seven predetermined locations in India (Ghaziabad, Chennai, Bangaluru, Chandigarh, Jammu, Gujarat, and Hyderabad), Twitter data represents the number of Tweets recorded using various keywords such as "Murder "Rape," and "Fight" between January 2014 and November 2018 [10]. A total of 3801 Tweets were gathered for this study and stored in a database. The correctness of the current study may be verified by comparing the predicted crime rates with actual crime rates obtained from security authorities (such as the National Crime Records Bureau, http://ncrb.gov.in/). There are a lot of people trying to figure out how to forecast a user's rank based on their social media account, and they're all working in various areas of social networks [12]. Measures of centrality and the structure of networks have been studied. Algorithms based on random walks and diffusion have been shown as viable methods for rating social media nodes [13, 14].

2 Related Works

Sivanantham et al. [16] Sentiment analysis' primary function is to examine customer feedback and assign ratings accordingly. The evaluations are generally unstructured, which means they need to be categorised or clustered in order to give useful data for the future. In order to improve the categorization accuracy in kid YouTube data emotional analysis, this study gives an overview of numerous machine learning algorithms. The sentiment analysis of children's videos on YouTube is improved by using a hybrid Support vector machine model trained with the ant colony optimisation approach. The paper examined Naive Bayes, SVM, and Adaboosting + SVM classification methods, and the findings were used to develop the suggested hybrid classifier. Each classifier's prediction for a set of test words submitted by children from YouTube is collected, and the classifier with the least detrimental and most secure prediction is proclaimed the winner. The suggested hybrid strategy outperforms both stand-alone machine learning algorithms and previously proposed hybrid approaches in terms of classification accuracy.

Using a hybrid method, Gautam and Bansal [17] established a system for real-time, automated cyberstalking detection on Twitter. First, lexicon-based, machine learning, and a hybrid approach were tested using recent, unlabeled tweets gathered through the Twitter API. All of the approaches that were utilised to extract the feature vectors from the tweets relied on the TF-IDF feature extraction technique. The best results from the lexicon-based procedure reached 91.1% accuracy, whereas the best results from the machine learning approach hit 92.4%. In contrast, when identifying unlabeled tweets retrieved through the Twitter API, the hybrid method showed the greatest accuracy at 95.8 percent. The machine learning method outperformed the lexicon-based method, and the suggested hybrid method performed very well. The real-time tweets acquired by Twitter Streaming were again classified and labelled using the hybrid technique with a new approach. The hybrid method again delivered the highest results as predicted, with an AUC of 98%, a precision of 94.6%, a recall of 94.1%, and an F-score of 94.1%. Machine learning classifiers' efficacy was evaluated across three distinct labelling strategies on each dataset. According to the study's experimental findings, the suggested hybrid technique outperformed previously applied methods in classifying both historical and real-time tweets. SVM outperformed competing machine learning techniques across the board.

2.1 Data Collection

2.1.1 Twitter Data Collection

From January 2014 to November 2018, information was culled from Twitter profiles by manually browsing through search results for profiles based in seven cities throughout India (Ghaziabad, Chennai, Bangaluru, Gujarat, and Hyderabad). Table 1 provides the latitude and longitude of potential study sites (extracted from Google Map Online).

Table 1. Seven different crime locations in India.

Location	Latitude and Longitude
Ghaziabad, India	28.6692°N, 77.4538°E
Chennai, India	13.0827°N, 80.2707°E
Bangaluru, India	12.9716°N, 77.5946°E
Chandigarh, India	30.7333°N, 76.7794°E
Jammu, India	32.7266°N, 74.8570°E
Gujarat, India	22.2587°N, 71.1924°E
Hyderabad, India	17.3850°N, 78.4867°E

The database now has 3801 tweets. We track down patterns of use on social networking sites that correspond to certain search terms. We are doing a process of labelling words. The ability to recognise essential phrases is crucial to the success of our investigation. We thus identify the key terms that pertain specifically to the criminal act. The terms "murder," "crime," "encounter," "hit and run," "rape," and "fight" are chosen for the social media crime rate study. For the first spot, 999 tweets are eliminated. On Twitter, the term "battle" is among the most frequently used keywords. Users often search for "crime" in the second location (Table 2).

Table 2. Statistics of tweets collected for the seven corruption metropolises

Location	Murder	Crime	Hit and run	Rape	Encouter	Fight
Jammu	41	43	1	44	78	99
Gujarat	62	50	8	91	37	101
Ghaziabad	117	228	43	222	121	268
Chennai	137	207	10	115	58	190
Bangaluru	170	112	32	111	43	158
Chandigarh	90	52	3	99	27	87
Hyderabad	52	98	12	80	48	156

As the safety of both domestic and foreign visitors is a top priority for the Indian government, these seven areas were chosen for the research. The National Crime Records Bureau found that five of these seven cities.

The most recent crime trends in India's two highest and lowest (Jammu) crime cities are measured using Twitter API as part of the validation of the suggested technique. Throughout the last week (January 23rd–January 30th, 2019), TAGS v6.1 has been used to keep track of crime statistics.

Table 3. Real crime data composed for seven crime cities

Location	Year 2014	Year 2015	Year 2016
Ghaziabad	240475	241920	282171
Jammu	23848	23583	24501
Gujarat	131385	126935	147122
Chennai	193200	187558	179896
Bangaluru	137338	138847	148402
Chandigarh	37162	37983	40007
Hyderabad	106830	106282	108991

2.1.2 Real Crime Data Collection

Each nation's ability to identify and deter crime depends on how well its criminal records are managed. The Indian police force is always trying new things in an effort to make its crime reporting system more effective. After forming a new task force in 1985, the Indian government established the National Crime Records Bureau to centralise and standardise crime statistics throughout the country (NCRB).

The NCRB database is a valuable resource for gathering accurate crime statistics in the targeted areas. We can verify the reliability of this study by comparing it to actual crime statistics that have been obtained by security authorities for the targeted areas. Table 3 displays actual crime statistics for the years 2014, 2015, and 2016 in seven different cities (Table 4).

Table 4. Joint dataset of tweets composed and NCRB crimes

Location	Total crime according to NCRB	Total tweets collected
Jammu	71932	306
Gujarat	405442	349
Bangaluru	424587	626
Chandigarh	115152	358
Ghaziabad	764566	999
Chennai	560654	717
Hyderabad	322103	446

2.2 Proposed Crime Detection Model

In the first stage, information is gathered from a number of users' Tweets using keyword searches. As this procedure generates a flood of tweets, a human filtering procedure is required to go through them and find the gems of information. Around four thousand tweets are collected and screened in this section. The second phase included gathering three years' worth of actual crime data from security authorities in the areas of interest. The final stage involves comparing the reported crimes with the actual ones. A comprehensive description of the model is provided below.

2.2.1 Data Pre-processing

Data pre-processing refers to the steps used to prepare the raw data for input into a neural network. Better results may be achieved if the data is properly structured. The dataset underwent the following preparation procedures:

- Tokenization was performed on all sentences in the dataset. The goal of tokenization is to separate phrases into their component parts, often words.
- A vocabulary is built out of the tokens, and it contains every single word in the dataset.
- A string of words is the input. A tokenizer is developed to transform the string of characters into a numerical series. As a result, each word has its own corresponding numeric value.

2.3 Classification Using Deep Learning

In this part, we provide the proposed model for the crime categorization issue based on the available research. Both the Capsule Neural Network and the Multilayer Perceptron Neural Network play important roles in the model. Whereas the Multilayer Perceptron uses the characteristics retrieved from the text sentences as its input, the Capsule Network uses the transformed sentences as input vectors of a defined length. The final sentence classification is based on the combined output of the two neural networks.

2.3.1 Embedding Layer

This is the model's first layer. This layer's job is to take integer-encoded information and transform it into vectors of a consistent length. Each word in the integer encoded data is represented by a distinct number. The layer is fed random weights and taught to embed the language as a whole. The layer produces dense vector embeddings of a constant size for each word.

2.3.2 Convolutional Layer

This layer is in charge of extracting characteristics from the input phrase at various locations within the text. The same may be said for convolution filters.

2.3.3 Primary Capsule Layer

The purpose of this initial capsule layer in the design is to convert the final result of the convolution layer into a capsule vector representation. This retains the semantic meaning of the words in the statement. This layer has 32-channel capsules that span 8 dimensions. A method known as routing by agreement determines which network layers to ascend to next. When numerous capsules in the current layer cast a vote for a certain capsule in the layer above, the aforementioned process activates that capsule.

2.3.4 Class Capsule Layer

The principal capsule layer's output serves as the layer's intake. Each capsule in the aforementioned layer communicates with the layer underneath it in a specific geographic area. Dynamic routing is the method that bridges this layer with the one below it. At this layer, the routings in the iterative dynamic routing technique are 3, and the capsules have a size of 16. A normalised text feature vector of size equal to the number of features is the network's input. Next, we'll break out how the network's training makes use of characteristics extracted from the data's context.

Sentence Length: This measures the total amount of words in the rumour post, omitting spaces, punctuation, and Hyperlinks.

Some individuals use all capital letters to convey strong emotions; this feature counts the number of words with just capital letters.

This is the number of times negation appears in the criminal report. A set of terms that may be used to convey denial or to symbolise a contradiction are compiled. Words like "never," "isn't," "barely," "no," and "shouldn't" are part of the lexicon.

Most of the time, individuals resort to foul language to express their emotions. Such terms like "moron," "fuck," "stupid," "bastard," and "bitch" are compiled into a set. The slang forms of these expressions, such as "stfu" and "wtf," are also included in this compilation. A peculiarity of the rumour post is the quantity of these terms. More often than asking a question or offering assistance, these words are used in response.

2.3.5 The Capsule Network Model

Although several deep neural network-based models have been presented as potential solutions to the issue of crime categorization, there is still room for expansion. In order to improve the precision of the criminal categorization system, this study explores the usage of a Capsule Network. Capsule Network's first publication demonstrated the model's superior performance over Convolutional Neural Networks when used to image classification on the MNSIT dataset. By picking up on certain visual cues, a Convolutional Neural Network can reliably identify things. Layers farther into the architecture are responsible for discovering more complex characteristics, such as the eyes or nose (in the case of face identification), whereas the first layers are responsible for recognising basic features like edges. As a result, the best forecast may be made by considering all of these factors together. We may deduce that CNN does not make use of spatial information and that the pooling function is used to link the layers. The fact that the pooling function utilised in Convolution Neural Networks is so effective is a tragedy, according to Geoffrey E. Hinton.

Inverse visuals are one of the goals of a Capsule Neural Network. Because of this, the approach makes an effort to discover the process that produced the desired picture and then replicate it. One of the most important aspects of the Capsule Network is its equivariance. The primary goal is to keep all data about an object's position and orientation in the network. If the target is rotated slightly, for instance, the activation vectors will shift somewhat as well.

2.3.6 Proposed Crisscross Optimization Algorithm (CCA)

Each iteration of the CCA, a population-based stochastic search method, consists of a horizontal crossover, followed by a vertical crossover. After each round of CCA's horizontal and vertical crossover and subsequent reselection by the competitive operation, the population is updated twice. Only the most effective solutions will be allowed to continue operating in a competitive environment. There are three reasons why it deserves to be considered excellent.

If you have an issue that spans more than one dimension, you may solve it by splitting the population into hyper cubes and searching the edges of each cube with a lower probability. This improves its capacity to search the whole world.Premature convergence may be prevented with the use of vertical crossover.The competitive process aids in leading to the ideal position and also speeds up the convergence rate.

A. Horizontal Crossover

It's the mathematical equivalent of two people swapping places throughout every dimension. Supposing the offspring of the pth parent is Y(p) and those of the qth parent are Y(q), we predict a horizontal crossover in the Th m dimensions. It is possible to mathematically describe the process of procreation.

$$MS_{hc}(p, m) = k_1.Y(p, m) + (1 - k_1).Y(q, m) + l_1.(Y(p, m) - Y(q, m)) \quad (1)$$

$$MS_{hc}(q, m) = k_2.Y(q, m) + (1 - k_2).Y(p, m) + l_2.(Y(q, m) - Y(p, m)) \quad (2)$$

where k_1, k_2 and l_1, l_2 are the uniformly distributed values between 0 to 1 and −1 to 1 respectively.

B. **Vertical Crossover**

It's also a crossing in all the dimensions of mathematics between two people. We will suppose that the vertical crossover occurs in the m 1th and m 2nd dimensions for each unique p. It is possible to mathematically describe the process of procreation.

$$[(MS)]\_vc(p, m\_1) = k\_1.Y(p, m\_1) + (1 - k).Y(q, m\_2)p \in N(1, M)$$

C. **Competitive Operator**

It's useful for stimulating rivalry between the next generation and its parents. To illustrate, in the case of a horizontal crossing, only the child that performs better than the parent will be evaluated for survival and further consideration.

D. **Crisscross Optimization Algorithm**

1. Initialize the size of the populations (N) and dimensions(M). After that randomly generate population vector 'P_{V_i}' and evaluate its function vector 'F_{V_i}'.
2. Execution of horizontal crossover with its competitive operator.
2.1 $B = permute(N)$ (For rearranging dimensions of N) for $p = 1$ to $N/2$.
2.2 Generate random numbers, $P \in (0, 1)$ and $P_1 = 1$.
    If $P = P_1$ then
2.3 $no1 = B(2p - 1)$ and $no2 = B(2p)$.
    for $j = 1$ to M.
2.4 Generate random numbers k, $k_2 \in (0, 1)$ and.
    $l_1, l_2 \in (-1, 1)$.
2.5 Calculate:

$$MS_{hc}(no1, j) = k_1.Y(no1, j) + (1 - k_1).Y(no2, j) + l_1.(Y(no1, j) - Y(no2, j)) \quad (4)$$

$$MS_{hc}(no2, j) = k_2.Y(no2, j) + (1 - k_2).Y(no1, j) + l_2.(Y(no1, j) - Y(no2, j)) \quad (5)$$

    End for
    End if
    End for
2.6 New '$P V_2$' will be obtained and calculate 'FV_2'
    If $FV_2 > FV_1$ then will obtain best FV and 'PV'
3. Perform vertical crossover with the competitive operator
3.1 Normalize
3.2 $B = permute(M)$
    for $p = 1$ to $M/2$.

3.3 Generate random numbers, $P \in (0, 1)$ and $P_2 = 1$

  If $P < P_2$ then

3.4 $no1 = B(2p - 1)$ and $no2 = B(2p)$

  for $j = 1$ to M

3.5 Generate random numbers $k \in (0, 1)$

3.6 Calculate:

$$MS_{vc}(j, no1) = k.Y(j, no1) + (1 - k).Y(j, no2) \tag{6}$$

  End for
  End if
  End for

3.7 Reverse normalizes MS_{vc} and updates PV with the new solution.

3 Results and Discussion

3.1 Measuring the Performance of Model

Recent tweets (obtained through the Twitter API) and real-time tweets (obtained via Twitter Streaming) were classified and labelled independently to evaluate the performance of classifiers with each applicable technique (lexicon-based, machine learning, and hybrid approach). Throughout both the training and testing phases of a model's development, its performance is evaluated using a variety of indicators. The confusion matrix is often used to establish the performance criteria.

A 2×2 truth table matrix, the confusion matrix in this research sums the values of True Pos, True Neg, False Neg, and False. Success is shown by True Pos (True Positive), which indicates the total number of cyberstalking tweets that were successfully identified, while failure is described by True Neg (True Negative). On the other hand, False Pos (False Positive) is a miss-hit that represents the total number of falsely detected cyberstalking tweets, and False Neg (False Negative) is the failure count that illustrates the total number of falsely identified non-cyberstalking tweets. The effectiveness of the cyberstalking detection system was evaluated using standard metrics such as accuracy, precision, f-score, and recall. During the real-time automated identification of tweets, the Area Under the Curve (AUC) was also determined.

3.1.1 Accuracy

Accuracy addresses the complete number of rights predictions anticipated by the classifier. Accuracy can be calculated using Eq. (7).

$$Accuracy = \frac{True\ Pos + True\ Neg}{True\ Pos + False\ Pos + False\ Neg + True\ Neg} \tag{7}$$

3.1.2 Precision

Precision shows the proportion between the true positives and the wide range of various positives. Precision can be calculated using Eq. (8).

$$Precision = \frac{True\ Pos}{True\ Pos + False\ Pos} \tag{8}$$

3.1.3 Recall

Recall describes the sensitivity and measures the proportion of true positive prediction to total positive. Recall can be determined using Eq. (9).

$$Recall = \frac{True\ Pos}{True\ Pos + False\ Neg} \tag{9}$$

3.1.4 F-Score

F-Score measures test accuracy and describe the harmonic average between precision and recall. F-score can be determined using the Eq. (10) (Table 5).

$$FScore = \frac{2Precision \times Recall}{Precision + Recall} \tag{10}$$

Table 5. Analysis of Various Pre-trained CNN Models

Algorithm	Accuracy	Precision	Recall	F-Score
DenseNet	0.918004	0.813725	0.447761	0.379863
ResNet	0.941113	0.426230	0.488060	0.360250
VGGNet	0.957229	0.953846	0.585075	0.310000
AlexNet	0.956454	0.965517	0.567164	0.384987
CapsNet	0.956454	0.636364	0.608955	0.414607
CapsNet C Co	0.96609	0.982456	0.67164	0.585714

All pre-trained models achieved better performance for predicting the crime rate, however the proposed model achieved high performance. The reason is that the hyper-parameters of CapsNet is optimized by CCO model and reaches the better accuracy. For instance, the proposed model achieved 96.60% of accuracy, 98% of precision, 67% of recall and 58% of F-score. But the existing pre-trained models such as DenseNet, ResNet, VGGNet. AlexNet and CapsNet achieved nearly 91% to 95% of accuracy, 63% to 95% of precision, 44% to 60% of recall and 30% to 41% of F-score analysis. Figures 1 and 2 presents the graphical representation of proposed model with existing pre-trained models.

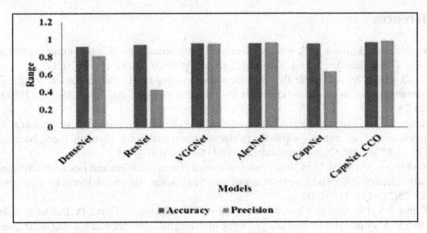

Fig. 1. Analysis of Various pre-trained models

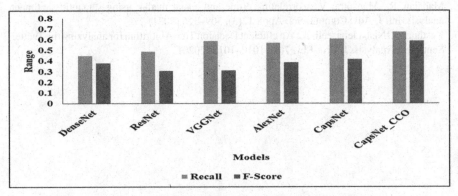

Fig. 2. Comparative analysis of Proposed model with existing techniques

4 Conclusion

The purpose of this research is to analyse crime statistics in real time using information from surveillance systems and public social media platforms. The case study chooses seven key cities from which to carry out the required research. It's common knowledge that many individuals share their thoughts and emotions with their followers all across the globe using the social networking service Twitter. Because of this, we choose to employ Twitter as a social media tool for our study's data collection on criminal activity. Since tweets represent the intents of users and may be detailed and in any format, the process of collecting data via Twitter is a job with a massive processing burden. The first step in our process is to gather the Tweets, then we filter them and lastly we compare them with actual crime statistics. We found that using social media to compare crime rates in different cities was a reliable method. In order to automatically identify criminality on live tweets on Twitter in real-time, a CapsNet model with CCO employing multiple ways on distinct segments was suggested in this article.

References

1. Sandagiri, C., Kumara, B.T., Kuhaneswaran, B.: Deep neural network-based crime prediction using Twitter data. Int. J. Syst. Serv.-Oriented Eng. (IJSSOE) **11**(1), 15–30 (2021)
2. Monika, Bhat, A.: Automatic Twitter crimeprediction using hybrid wavelet convolutional neural network with world cup optimization. Int. J. Pattern Recogn. Artif. Intell. **36**(05), 2259005 (2022)
3. Kasthuri, S., Nisha Jebaseeli, A.: An artificial bee colony and pigeon inspired optimization hybrid feature selection algorithm for twitter sentiment analysis. J. Comput. Theor. Nanosci. **17**(12), 5378–5385 (2020). https://doi.org/10.1166/jctn.2020.9431
4. Boukabous, M., Azizi, M.: Crime prediction using a hybrid sentiment analysis approach based on the bidirectional encoder representations from transformers. Indones. J. Electr. Eng. Comput. Sci. **25**(2), 1131–1139 (2022)
5. Kumar, V.S., Pareek, P.K., Costa de Albuquerque, V.H., Khanna, A., Gupta, D., Deepak Renuka Devi, S.: Multimodal sentiment analysis using speech signals with machine learning techniques. In: 2022 IEEE 2nd Mysore Sub Section International Conference (MysuruCon), pp. 1–8 (2022). https://doi.org/10.1109/MysuruCon55714.2022.9972662
6. Mahajan, R., Mansotra, V.: Correlating crime and social media: using semantic sentiment analysis. Int. J. Adv. Comput. Sci. Appl. **12**(3), 309–316 (2021)
7. Kasthuri, S., Nisha Jebaseeli, A.: An efficient Decision Tree Algorithm for analyzing the Twitter Sentiment Analysis. J. Crit. Rev. **7**(4), 1010–1018 (2020)

Sentiment Analysis Using Lexical Approach and Fuzzy Logic

Renjith V. Ravi[1] ⓘ, S. B. Goyal[2(✉)] ⓘ, Xiao ShiXiao[2,3],
Mustafa Muwafak Alobaedy[2], and Vladimir Kustov[4] ⓘ

[1] Department of Electronics and Communication Engineering, M.E.A Engineering College,
Malappuram, Kerala, India
renjithravi@meaec.edu.in
[2] City University, Petaling Jaya, Malaysia
drsbgoyal@gmail.com, mustafa.theab@city.edu.my
[3] Chengyi College Jimei University, Xiamen, China
[4] Saint Petersburg Railway Transport University of Emperor Alexander I, Saint Petersburg,
Russia

Abstract. Understanding the feelings expressed in a statement is the goal of
sentiment analysis. Depending on the thoughts expressed, a statement may be
good, neutral, or negative. Positive, neutral, or negative feelings might be present
in a statement. The reality is that the feeling expressed in each sentence whether
it be positive, negative, or neutral is not always obvious. The examination of
comparison, negation, intense, and sarcastic phrases, as well as the difficult work
of dealing with grammatical errors, provide enormous hurdles to this process. The
SA system is built in this study using a hybrid technique that combines dictionary-
based approaches with fuzzy logic to address these issues. The suggested system's
outputs may be classified as either high positive, positive, neutral, negative, or
high negative emotions. The analysis of customer satisfaction tweets about cloud
services from Google, Amazon, and Microsoft revealed that the suggested method
performs significantly, with results of 85.6% precision, 88.4% recall, and an 83.6%
F-score.

Keywords: Sentimental Analysis · Fuzzy Logic · SentiWordNet · SentiStrength

1 Introduction

Social media has transformed into another contact route between customers and organisations. Users have a platform to post and express their feelings, opinions, and preferences
regarding numerous subjects, people, goods, and services on social media platforms such
as Facebook, Twitter, and Tumblr [1]. Normally, questions created by the researcher are
used to gather text and reviews. This technique of data collection required a lot of time
and was difficult to control [2]. With the development of the web and technology, people
now use social media to post unstructured text reviews and comments. Views published
on social media may be categorised to establish the orientation of submitted text (negative, positive, neutral). To establish the user's opinion and emotion on a certain item or

S. Kadry and R. Prasath (Eds.): MIKE 2023, LNAI 13924, pp. 117–127, 2023.
https://doi.org/10.1007/978-3-031-44084-7_12

service, the sentiment intensity and strength of the post are assessed. Sentiment analysis is utilised to examine the content of both the text and the reviews [3].

Social networking sites must retain users to beat the competition [4]. Presenting and displaying items according to a client's interests, alerting them to future sales, and suggesting new products that are similar to user preferences discovered via social networking sites are all ways to increase customer happiness [5]. The content of the social web is dynamic and constantly evolving to reflect the changing social and emotional states of its users.

Sentiment analysis is a cutting-edge computer technique to enhance decision-making [6, 7]. Views are always conveyed in free text as evaluations, judgements, or remarks. It bases its summary on quotes from the reviews that refer to pertinent items and their features or qualities. To determine the attitude of a person who wrote about a product or event, sentiment analysis can be utilised to mine these online remarks [8].

A method for handling imperfect and diverse information is fuzzy logic [9]. It is a kind of multi-valued logic that focuses on reasoning that is approximate rather than precise and fixed. Traditional logic often has values between true and false; however, fuzzy logic variables might have a truth value that is between 0 and 1. By creating membership functions, fuzzy logic is used to categorise the emotions.

2 Related Works

Sentimental Analysis research is the subject of several new studies that are released annually. Regarding the creation, adaptation, or use of new sentimental lexicons [10–14], the use of traditional machine learning algorithms like nave Bayes [15, 16], regression techniques [17, 18], decision trees [17, 18], clustering [18, 19], ensemble classifiers and genetic algorithms [19, 20], or transfer learning [21], various new contributions can be found.

This study aims to organise the current understanding of the many uses of fuzzy logic in Sentimental Analysis. This study mainly focuses on the many activities that fall within the purview of sentiment analysis and other functions that rely on sentiment assessments to function. In each of these instances, fuzzy logic must act as the main approach used to carry out a crucial Sentimental Analysis-related activity, phase, or application in order to satisfy the criteria of this review; however, other techniques may also be used as part of the whole procedure or application.

3 Materials and Methods

Using a fuzzy logic framework, the suggested method divides tweets into five groups by pulling out and combining different kinds of information [22]. "High negative," "negative," "neutral," "positive," and "high positive" are the four categories. The suggested technique is composed of four processing steps, as shown in Fig. 1: data gathering, pre-processing, extraction of features, and model creation.

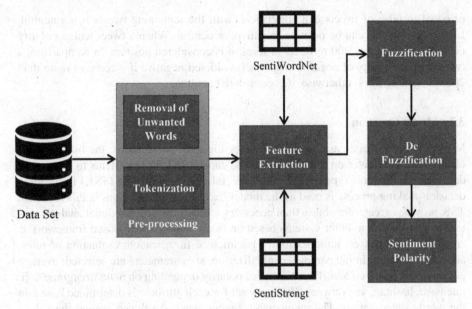

Fig. 1. Proposed Approach for Sentimental Analysis

3.1 Features of Dataset

The official Twitter accounts of Google, Microsoft, and Amazon provided the dataset for the suggested research project [22, 23]. Users who are pleased with the web services that these organisations provide have posted 9421 tweets about them. Punctuation, stop words, URLs, digits, foreign words, and Twitter keywords are all eliminated during the pre-processing phase. Then, the content is tokenized. Then there are words that make no sense together, hashtags, emojis, negative words, adverbs, and intensive words.

In feature extraction, the *SentiWordNet* dictionary polarity of words, the *SentiStrength* dictionary polarity of words, the polarity of emojis, adverb, intensive, negation, and hashtag, exclamation, and retweets are all retrieved from the input text. The *SentiWordNet* phrase "polarity" is symbolised by a number in the range [−1–1]. It has a linear impact on the output outcomes. Because of this, an output with a polarity of -1 is more likely to fall under the category of "very negative." The polarity corresponds to a result in the "high positive" class with a value of 1, and so on.

The word "polarity" from *SentiStrength* also refers to a value between −5 and 5. Moreover, it has a linear influence on the outcomes. The six groups of emoji are delight, grin, savouring, smirking, confuse, and wink. Each group has an impact on the output that may be favourable, negative, or neutral, as will be made clear throughout the model's design. The intense words, such as intensive, adverb, exclamation, and negation, each have an impact on the output results that may be either good, negative, or neutral. The effect of a hashtag may be positive, negative, or neutral, while the influence of a retweet might be positive, negative, or neutral. The percentage of uplifting tweets to total tweet volume is used to calculate the effect of a hashtag. More than 60% of the tweets using the positive hashtag are positive. In contrast, the negative hashtag contains more than or equal to 60% of tweets that are critical, with the remaining tweets being neutral. Likes and retweets can be positive, negative, or neutral. When a tweet with a polarity of positive words has 200 retweets or likes, it is considered positive. In comparison, a tweet with a polarity of negative words is considered negative if it receives more than 200 retweets or likes; otherwise, it is considered neutral.

3.2 Model Creation

Model creation is done as a method of fuzzy logic for two reasons: the fuzzy impact of the input parameter on the output and the capacity of fuzzy systems to accommodate inputs of diverse types. As a result, the Takagi-Sugeno-Kang (TSK) fuzzy logic decision-making process is used as the model for classifying emotions in this research. TSK provides greater flexibility than necessary in the proposed emotional analytic system in comparison to other systems based on fuzzy logic. The proposed framework is created in four phases: initialization of parameters, fuzzification, evaluation of rules, and consolidation. In the parameter initialization, six parameters are selected: average polarity depending on *SentiWordNet*, word polarity depending on *SentiStrength*, emoji, intensive, hashtag, and retweet. The fuzzy set for each attribute is determined based on the words stated earlier. The membership functions are developed around the values given earlier and employ a trial-and-error technique in which the final sets and processes are finished as the experiments are done. The fuzzification method is accomplished based on the TSK system by turning the crisp value for inputs having crisp values, such as polarity and hashtags, into fuzzy words. The criteria used in the rule assessment stage are produced via a trial-and-error method. A set of basic rules is constructed, as presented in Table 1. Nevertheless, the set gets refined as the trials are done. The aggregation procedure is based on using the rule's output with the greatest confidence.

Table 1. Rules for Fuzzy Controller

P\N	Negative	High Negative	Positive	High Positive
Negative	HN	HN	NN	NN
High Negative	HN	PP	HN	NN
Positive	NN	NN	HP	PP
High Positive	NN	NN	PP	HP

3.3 SentiWordNet

SentiWordNet [24] is a lexical resource for opinion mining and sentiment analysis. It is an extension of WordNet, a large lexical database of English language words that are grouped into sets of synonyms called synsets.

In *SentiWordNet*, each *synset* is assigned a score for three sentiment categories: positivity, negativity, and objectivity. These scores indicate the degree to which a word is associated with positive or negative sentiment, or if it is neutral and objective. The scores range from 0 to 1, with higher values indicating stronger sentiment.

SentiWordNet is often used in natural language processing (NLP) applications such as sentiment analysis, opinion mining, and text classification. It provides a valuable resource for analyzing text data and extracting sentiment information that can be used to make more informed decisions in various fields, including marketing, politics, and customer service. The algorithm of *SentiWordNet* is shown in Algorithm 1.

```
1. Function Label Propagation ()
2. {
3. Seed SentiWordNet
4. If sentiment present word;
5.       Label 'Opin'
6. Else
7.       Label 'Non-Opin'
8. End If
9. }
```

Algorithm 1: Algorithm for `SentiWordNet` [9]

3.4 SentiStrength

SentiStrength [25–27] is a sentiment analysis and natural language processing program that detects the sentiment of a given text on a scale of -5 to $+5$. Mike Thelwall, Kevan Buckley, and Georgios Paltoglou created it at the University of Wolverhampton in the United Kingdom.

SentiStrength uses a mix of linguistic rules and machine learning to figure out the tone of a piece of text. It can look at both short messages like tweets and long ones like news articles or book reviews. The program is very good at recognizing sarcasm and other types of subtle emotions.

SentiStrength employs the binary method of sentiment analysis, analysing the existence and intensity of positive and negative sentiment separately. A statement containing both positive and negative emotion, for example, will earn two different ratings, one for positive sentiment and one for negative sentiment.

SentiStrength is frequently utilized in academic and commercial research, notably in marketing, customer service, and social media analysis. It's accessible as both a free web tool and a software package for local installation.

```
1. Function Label Propagation ()
2. {
3. Seed SentiStrength
4. If sentiment present emojis, adverb, inten-
sive, negation, and hashtag, exclamation, or
retweets;
5.       Label 'Opin'
6. Else
7.       Label 'Non-Opin'
8. End If
9.}
```

Algorithm 2: Algorithm of SentiStrength [9]

3.5 Fuzzification

The first intitial phase in a fuzzy logic system is to find out what the inputs and outputs of the variables are. After the input variables and membership function are determined, the rule-base model (or matrix analysis of the fuzzy base of knowledge) must be comprised of expert IF-THEN rules [2, 9]. These regulations convert the variables used as inputs into an output. This will show if the likelihood of operational issues is normally low, typical, or high. Figure 2 depicts the fuzzification technique.

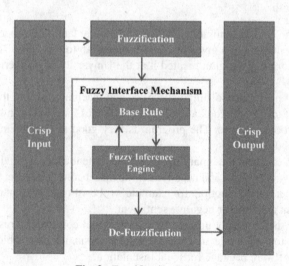

Fig. 2. Fuzzification Process

3.6 Membership Function Design

The Membership Function (MF) converts the input values of the linguistic variables into the supplied fuzzy sets [28–30]. Appropriate membership functions translate the inputs into the degree of membership for the partitions of linguistic variables. The many membership roles are varied. The right membership functions must be used for fuzzy sets in order for a fuzzy system to accurately reflect fuzzy modelling. In the proposed study, the categorization labels for the degree of semantic orientation and the sentence's positive or negative polarity are:

- High Negative (HN)
- Negative (NN)
- Positive (PP)
- High Positive (SP)

The sentence could be classified as objective or subjective. So, let's say it's seen as subjective. If so, it may be either negative or positive, or it might be either very positive, highly positive, or highly negative [1]. By carefully examining the lexical scores offered in *SentiWordNet*, trapezoidal and triangular membership functions are applied for the relevant fuzzy words in this study.

The MF can be defined using Eqs. (1) and (2) when the MF are trapezoids (in this case, 'Strong' or 'Weak').

$$\mu_{weak} (\text{trapezoid})(x; q_0, q_1, q_2, q_3) = max\left(min\left(\frac{x - q_0}{q_1 - q_0}, 1, \frac{q_3 - x}{q_3 - q_2} \right), 0 \right) \quad (1)$$

$$\mu_{strong} (\text{trapezoid})(x; q_3, q_4, q_5, q_6) = max\left(min\left(\frac{x - q_3}{q_4 - q_3}, 1, \frac{q_6 - x}{q_6 - q_5} \right), 0 \right) \quad (2)$$

The MF may be specified in the equation - if the MF are triangles (in this instance, "Positive" or "Negative") (3).

$$\mu_{Positive/Negative}(\text{triangle})(x, q_2, q_3, q_4) = max\left(min\left(\frac{x - q_2}{q_3 - q_2}, \frac{q_4 - x}{q_4 - q_3} \right), 0 \right) \quad (3)$$

3.7 Defuzzification

A fuzzy quantity is defuzzified, or changed into a known amount. The area of the generated figures on every MF is determined in three phases [2, 9]. Given that these areas are frequently triangles and trapezoids. If the membership degree isn't one, the area won't be in the shape of a triangle. Instead, it will be in the shape of a trapezium.

The Algorithm 3 shows the rules for defuzzification:

```
1. if (z ≤ 0.3) then
2. y='High Negative'
3. if (z ≥ 0.3 and z≤0.6) then
4. y=' Negative'
5. if (z ≥ 0.6 and z≤0.8) then
6. y='Positive'
7. if (z ≥ 0.8 and z≤1) then
8. y ='High Positive'
```

Algorithm 3: Rules for defuzzification [9]

4 Results and Discussion

As previously indicated, the information was gathered from the Twitter accounts of three different firms, and each message was individually annotated. In the studies, a total of 500 tweets, 100 in each category, were used. The outcomes of the suggested method are assessed using F-Score, recall, and precision. Table 2 contains the suggested approach's confusion matrix, whereas Table 3 contains the F-Score, recall, and precision [28–30].

Table 2. Confusion Matrix of the Proposed Approach

	High Positive	Positive	Neutral	Negative	High Negative
Neutral	6	4	95	24	25
Negative	0	0	2	69	0
High Negative	0	0	0	0	65
Positive	6	96	3	7	10
High Positive	88	0	0	0	0
Total	100	100	100	100	100

Table 3. Values of F-Score, recall, and precision from the Proposed Approach

Categories	Precision	Recall	F-score
High Positive	88%	100%	93%
Positive	96%	88%	91%
Neutral	95%	60%	75%
Negative	78%	96%	80%
High Negative	71%	98%	79%
Average	**85.6%**	**88.4%**	**83.6%**

The proposed approach has given an average precision value of 85.6, Recall of 88.4% and an F-Score of 83.6%. There was significantly reduced recall for the "positive" group (88%), and somewhat less recall for the "negative" category (96%). Also, the proposed approach had given a high recall value of 100% and 98% percentage in High Positive and High Negative category. There was somewhat less Recall for the "negative" group (96%), and slightly less for the "positive" category (88%). With just 55% of recall the "neutral" category, the suggested technique is seen as resilient since it was necessary to find both positive and negative comments when studying sentiment in this area in order to enhance the services. As a result, having poor "neutral" recall has no effect on how usable the suggested strategy is for SA of cloud service providers since it belongs to a less advantageous category. For all of the categories, the accuracy was good, except for "very negative," where some of the "neutral" categories were mistakenly labelled as negative by the proposed method. This is seen as a problem with the technique that was suggested, and it will be taken into account in future research.

5 Conclusion

In this essay, a novel approach to studying human emotions is proposed. It is based on a fuzzy logic system with many extracted characteristics. The suggested method, which can be categorized as a mix of machine learning and dictionary-based methods, combines the polarities of the emoji, hashtag, and intense words in fuzzy logic after extracting the polarities of the words in the input text using two dictionaries. As a result, the implementation of a fuzzy system involves three steps: fuzzification, rule evaluation, and aggregation. For each input, the system may produce one of the following five outputs: high positive, positive, neutral, negative, or high negative. Three well-known cloud service providers, Amazon, Google, and Microsoft, were included in a dataset measuring consumers' satisfaction with providers. The suggested approach's performance on the gathered data revealed that it had an F-score of 83.6%, an precision of 85.6%, and a recall of 88.4%. The outcomes also showed that 100% of the time, the suggested method correctly remembered both "high positive" and "high negative" categories. The recall for the "negative" group was somewhat lower (96%), while the recall for the "positive" category was slightly lower (88%). Despite the poor recall of the "neutral" category (60%), the suggested technique is regarded as robust since it was necessary to discover negative and sometimes positive feedback when studying the mood of such a domain in order to enhance their services. As the less advantageous category in the suggested method, having a poor "neutral" recall has no effect on the usefulness of the proposed technique for SA of cloud service providers. All categories, with the exception of "strong negative," had excellent results in terms of accuracy. Several of the "neutral" classes were given a negative label by the suggested technique, which is seen as a drawback and will be taken into account in further work.

References

1. Vashishtha, S.: Design And Development of Fuzzy Logic Based Sentiment Analysis System for Online Reviews & Social Media Posts (2022)
2. Mary, A., Jothi, J., Arockiam, L.: A framework for aspect based sentiment analysis using fuzzy logic. ICTACT J. Soft Comput. **8**, 1611–1615 (2018)
3. Verma, B., Thakur, R.S.: Sentiment analysis using lexicon and machine learning-based approaches: a survey. In: Tiwari, B., Tiwari, V., Chandra Das, K., Kumar Mishra, D., Bansal, J.C. (eds.) Proceedings of International Conference on Recent Advancement on Computer and Communication, pp. 441–447. Springer Singapore, Singapore (2018). https://doi.org/10. 1007/978-981-10-8198-9_46
4. Rodríguez-Penagos, C., Grivolla, J., Codina-Filba, J.: A hybrid framework for scalable opinion mining in social media: detecting polarities and attitude targets. In: Proceedings of the Workshop on Semantic Analysis in Social Media (2012)
5. Appel, O., Chiclana, F., Carter, J., Fujita, H.: A hybrid approach to sentiment analysis. In: 2016 IEEE Congress on Evolutionary Computation (CEC) (2016)

6. Wankhade, M., Rao, A.C.S., Kulkarni, C.: A survey on sentiment analysis methods, applications, and challenges. Artif. Intell. Rev. **55**, 5731–5780 (2022)
7. Ligthart, A., Catal, C., Tekinerdogan, B.: Systematic reviews in sentiment analysis: a tertiary study. Artif. Intell. Rev. **54**, 4997–5053 (2021)
8. Cui, J., Wang, Z., Ho, S.-B., Cambria, E.: Survey on sentiment analysis: evolution of research methods and topics. Artif. Intell. Rev. **56**, 8469–8510 (2023)
9. Mary, A.J.J., Arockiam, L.: ASFuL: aspect based sentiment summarization using fuzzy logic. In: 2017 International Conference on Algorithms, Methodology, Models and Applications in Emerging Technologies (ICAMMAET) (2017)
10. Xing, F.Z., Pallucchini, F., Cambria, E.: Cognitive-inspired domain adaptation of sentiment lexicons. Inform. Process: Manag. **56**, 554–564 (2019)
11. Bernabé-Moreno, J., Tejeda-Lorente, A., Herce-Zelaya, J., Porcel, C., Herrera-Viedma, E.: A context-aware embeddings supported method to extract a fuzzy sentiment polarity dictionary. Knowl.-Based Syst. **190**, 105236 (2020)
12. Wang, Y., Yin, F., Liu, J., Tosato, M.: Automatic construction of domain sentiment lexicon for semantic disambiguation. Multimed. Tools Appl. **79**, 22355–22373 (2020)
13. Ahmed, M., Chen, Q., Li, Z.: Constructing domain-dependent sentiment dictionary for sentiment analysis. Neural Comput. Appl. **32**, 14719–14732 (2020)
14. Mowlaei, M.E., Abadeh, M.S., Keshavarz, H.: Aspect-based sentiment analysis using adaptive aspect-based lexicons. Expert Syst. Appl. **148**, 113234 (2020)
15. Bi, J.-W., Liu, Y., Fan, Z.-P.: Representing sentiment analysis results of online reviews using interval type-2 fuzzy numbers and its application to product ranking. Inf. Sci. **504**, 293–307 (2019)
16. Li, Z., Li, R., Jin, G.: Sentiment analysis of Danmaku videos based on naïve bayes and sentiment dictionary. IEEE Access **8**, 75073–75084 (2020)
17. Bawa, V.S., Kumar, V.: Emotional sentiment analysis for a group of people based on transfer learning with a multi-modal system. Neural Comput. Appl. **31**, 9061–9072 (2018)
18. Al-Smadi, M., Al-Ayyoub, M., Jararweh, Y., Qawasmeh, O.: Enhancing Aspect-Based Sentiment Analysis of Arabic Hotels' reviews using morphological, syntactic and semantic features. Inform. Process. Manag. **56**(2), 308–319 (2019). https://doi.org/10.1016/j.ipm.2018.01.006
19. Riaz, S., Fatima, M., Kamran, M., Nisar, M.W.: Opinion mining on large scale data using sentiment analysis and k-means clustering. Clust. Comput. **22**, 7149–7164 (2017)
20. López, M., Valdivia, A., Martínez-Cámara, E., Victoria Luzón, M., Herrera, F.: E2SAM: evolutionary ensemble of sentiment analysis methods for domain adaptation. Inform. Sci. **480**, 273–286 (2019)
21. Saad, S.E., Yang, J.: Twitter sentiment analysis based on ordinal regression. IEEE Access **7**, 163677–163685 (2019)
22. Alharbi, J.R., Alhalabi, W.S.: Hybrid approach for sentiment analysis of twitter posts using a dictionary-based approach and fuzzy logic methods. Int. J. Semant. Web Inf. Syst. **16**, 116–145 (2020)
23. Alharbi, J.R., Alhalabi, W.S.: Sentimental analysis using fuzzy logic for cloud service feedback evaluation. Int. J. Inform. Comput. Technol. **8**, 1–10 (2018)
24. Husnain, M., Missen, M.M.S., Akhtar, N., Coustaty, M., Mumtaz, S., Prasath, V.S.: A systematic study on the role of SentiWordNet in opinion mining. Front. Comp. Sci. **15**, 154614 (2021)
25. Gouthami, S., Hegde, N.P.: Automatic sentiment analysis scalability prediction for information extraction using sentistrength algorithm. In: Brahmananda Reddy, A., Nagini, S., Balas, V.E., Srujan Raju, K. (eds.) Proceedings of Third International Conference on Advances in Computer Engineering and Communication Systems: ICACECS 2022, pp. 21–30. Springer Nature Singapore, Singapore (2023). https://doi.org/10.1007/978-981-19-9228-5_3

26. Khaira, U., Johanda, R., Utomo, P.E.P., Suratno, T.: Sentiment analysis of cyberbullying on twitter using SentiStrength. Indonesian J. Artif. Intell. Data Mining **3**, 21–27 (2020)
27. Sari, S., Khaira, U., Pradita, P.E.P.U., Tri, T.S.: Analisis sentimen terhadap komentar beauty shaming di media sosial twitter menggunakan algoritma sentistrength: sentiment analysis against beauty shaming comments on twitter social media using sentistrength algorithm. Indonesian J. Inform. Res. Softw. Eng. **1**, 71–78 (2021)
28. Jefferson, C., Liu, H., Cocea, M.: Fuzzy approach for sentiment analysis. In: 2017 IEEE international conference on fuzzy systems (FUZZ-IEEE) (2017)
29. Wang, Y., Subhan, F., Shamshirband, S., Asghar, M.Z., Ullah, I., Habib, A.: Fuzzy-based sentiment analysis system for analyzing student feedback and satisfaction. Comput. Mater. Continua **62**, 631–655 (2020)
30. Haque, M., et al.: Sentiment analysis by using fuzzy logic. arXiv preprint arXiv:1403.3185 (2014)

HTTP, WebSocket, and SignalR: A Comparison of Real-Time Online Communication Protocols

Nishant Sharma and Rekha Agarwal[✉]

Amity University, Noida, Uttar Pradesh, India
`ragarwal@amity.edu`

Abstract. Real-time web applications are widely implemented and used past few years. Different web applications like sports, taxi booking, stocks, tracking system needs to refresh itself without any human request. Different business and organization demand automate refresh of web pages to give users a better experience. Hyper Text Transfer Protocol (HTTP) is a basic and slow technology for information exchange between client and server. It is not appropriate for real time changes and updates. WebSocket is communication protocol with persistent connection. It has event driven architecture and duplex communication that makes it suitable for real time web. Further, SignalR is a library with automatic reconnection feature and fallback transports features. This research paper discusses HTTP, WebSocket and SignalR in terms of their data transfer, connection, performance, error handling, and reliability. This study gives an idea of three protocols performance and benefits in different use cases.

Keywords: Real time web applications · WebSocket · SignalR · Protocol

1 Introduction

A real-time web application enables user interaction in real-time such that any event and consecutive changes made by a user gets immediately effective in the system for access by others. Different real time applications like chat applications, collaborative document editing, real-time analytics, online gaming etc are widely used to support different tasks of users. When compared to conventional web applications, which often require users to refresh the page to see new content, real time wed applications offer a more interactive and engaging user experience. Real-time web applications like slack, Google Docs, Trello, and Twitch [1] etc are widely used by companies to customize intra-employee communication, document availability and collaborative updating, track tasks etc. In current era online applications, need constant communication between the server and client, the conventional HTTP request/response approach has shown to be inefficient. HTTP performs slow especially for applications demanding large data transfers. HTTP is a request-response protocol wherein each request and response must be communicated in its entirety before the next request can be generated. With this, the overall performance of an application may suffer due to delays caused by latency generated due to distance, network congestion or bandwidth limitations [2].

© The Author(s), under exclusive license to Springer Nature Switzerland AG 2023
S. Kadry and R. Prasath (Eds.): MIKE 2023, LNAI 13924, pp. 128–135, 2023.
https://doi.org/10.1007/978-3-031-44084-7_13

New technologies have consequently arisen with the goal to offer a more effective and trustworthy method of real-time communication. Using a single TCP(Transmission Control Protocol) connection, *WebSocket* is a technology that permits two-way communication among client and server. The ability of *WebSocket* to decrease network latency and enhance application speed is one of its key advantages to increase its adaptability by developers. This is because WebSocket uses initial handshake (for authorization and authentication), a WebSocket connection solely uses HTTP; thereafter, it uses a TCP connection to transfer data using its own protocol. More over, the WebSocket offers a faster method of data transport between client and server because it employs a binary data format that is more compact as compared to HTTP's text-based format, which is more compact. The amount of data that needs to be transmitted over the network is decreased which in result support performance and use less bandwidth [3].

Real-time communication in web applications is made simpler using the library SignalR, which is built on top of WebSocket. SignalR is the free and open-source that enables server to send asynchronous messages to client-side web application. SignalR shields from low-level facts so that it appears as though the client and server are connected permanently and persistently. SignalR offers an API for building server-to-client and client-to server remote procedure calls (RPCs) that may be used to provide data to clients and receive data from them. SignalR offers advantages like reduced latency, scalability, automatic reconnection etc. Web applications that require real-time updates, such as chat rooms, online games, real-time dashboards, and other applications, frequently employ SignalR [5].

The study in this paper presents an in-depth assessment between HTTP, WebSocket and SignalR based web systems in terms of functionality, scalability, usability, and security. This study presents pros and cons of each of three and discusses use cases in various situations. The research will assist architects and developers to choose the best approach for real time web application.

2 Literature Study

According to Alex Diaconu [11] limitation of HTTP since its inception HTTP has gone a long way. The issues using HTTP as the underlying transport protocol have forced WebSocket to evolve to achieve higher performance. HTTP was developed to provide hypermedia resources in a request-response manner. Real-time application, which typically require frequent or continuous client-server contact and the capacity to respond quickly to changes, had not been optimised to run on it. The real-time web has been transformed by WebSocket technology, which allows programmers to make dynamic and responsive apps. According to study of Paul Murley, Zane Ma, Joshua Mason, Michael Bailey, Amin Kharraz [18] *WebSocket* is a relatively new technology that has gained a lot of popularity since it enables low-latency, two-way communication between web clients and servers. Rui Lu and Zhenyu Chen [27] suggests that WebSocket is the ideal option if full-duplex, bidirectional communication with low latency is a top goal and cross-browser support is less crucial. However, SignalR is a better choice because it offers WebSocket-like features with fallback options for browsers that do not support them. In addition, WebSocket frameworks like SignalR offer approaches for scaling and optimisation that enable the handling of numerous concurrent users. Eilon Lipton [19] study claims that for creating real-time web application with need to update clients instantly, SignalR is a compelling tool. The SignalR library offers a straight forward application programming interface (API) for integrating real-time functionality into web applications and supports a few transport protocols, including WebSockets and long-polling.

3   HTTP vs WebSocket and SignalR

A network protocol is a description of collection of rules or instructions that directs how a network should behave. This involves the design and structure of messages, their transmission and reception in chronological sequence, and the decisions made in response to the data they contain [16]. This section discusses three network protocols/library and their corresponding attributes and fitness in context of relal time web applications.

3.1   Data Transfer

Data transport protocols primarily used in web applications are HTTP, WebSocket, and SignalR, however each of these protocols vary in terms of functionality and use cases. The common protocol used for sending data over the internet is HTTP is a request-response protocol. On web servers, content is stored. Web servers are also known as HTTP servers since they support the HTTP protocol. These HTTP servers maintain the data of the Internet and make it available to HTTP clients upon request HTTP uses TCP connection to send and receive data as it is stateless where each request is independent of all previous and upcoming requests [10]. The *Web Socket* protocol allows for real-time, bidirectional communication between client and server. Data can be transferred in both directions without the need for repeated requests as it establish a persistent connection between the client and server. WebSocket is especially useful for applications based on real-time updates and low latency such as chat programmes, online games etc [11]. The *SignalR* is a library which uses WebSocket whenever possible, which includes functionality like connection management, automatic reconnections, and support for different transport protocols (including WebSockets, Server-Sent Events, and Long Polling). WebSockets and SignalR offer extra functionality for real-time communication, while HTTP is the standard protocol used for data delivery only. Low latency real-time applications benefit from WebSockets, although SignalR offers more functionality and makes real-time application development easier [22].

3.2   Connection

With the use of HTTP protocol, client and server communicate via requests generated by client to server where each request is independent of any prior requests. For each pair of request-response cycle, a connection is established that is terminated after each cycle is finished. This connection approach is appropriate for online applications, including e-commerce websites or static websites, where real-time communication is not necessary [24].

WebSocket applies a stateful connection mechanism. Here the same connection instance is maintained even after data exchange process is complete among client and server. Second, WebSocket is bi-directional where data may be sent and received in real time from client and server simultaneously. Consecutively, WebSocket helps to construct exceptionally responsive and dynamic applications those offer a seamless user experience by utilising WebSocket [1]. *SignalR* offers a high-level abstraction over Web-Sockets that helps developers to concentrate on core functionality implementation of the respective application without having to worry about the nitty-gritty of the communication protocol. It is an effective tool for creating real-time web applications that need sophisticated capabilities and flexibility [25].

3.3 Scalability

The scalability of a web application is the capacity to manage multiple concurrent connections preserving high performance and cost in response to change in application. A HTTP server is capable to process concurrent requests effectively as the most popular application like Google and Facebook process billions of requests per day using HTTP. Beside more concurrent demands, scalability also means to preserve good performance, de pendability, and security while scaling up. HTTP is scalable but for accomplishing scalability user need careful planning and optimizing technique [10]. In comparison, *WebSockets* are persistent that makes them better that HTTP requests. The WebSocket client establishes a new connection and then reuses the same for communication with server. Both the server and client can broadcast and reply to events across the same persistent connection. This is known as duplex connection. A load balancer can be used to open connections, but once established the connection remains with the server until it is terminated or disrupted. The use of vertical and horizontal scaling allows the application to add extra WebSocket servers to manage increased load [11].

However, scalability in SignalR is achieved by – vertical scaling and horizontal scaling. Scaling vertically increased the power (e.g., CPU, RAM) of an existing computer and is also referred to as scaling up. Scaling horizontally increased the number of machines on the network to share the processing load is a process known as scaling out. The advantage of SignalR over WebSocket is that SignalR enables high scalability and fault tolerance by distributing messages over numerous servers. Performance can be enhanced by, for instance, distributing data over several servers to avoid obstruction [5].

3.4 Communication

HTTP is clearly a request-response. The client generates a request whereas the corresponding server waits until it receives a complete request and sends a reply only after that. The client and server cannot send message at the same time as http is a half-duplex in nature [23]. Whereas *WebSocket* utilises full duplex communication by default. Web-Sockets is an option when a full duplex communication is taking place [26]. Depending on the transport mode it uses, SignalR can communicate in full or half duplex.

3.5 Methods and Events

Methods are used in HTTP such as *GET*, POST, *PUT*, and DELETE for exchange of information [21]. A web page or picture file can be retrieved from a server using the GET protocol. For example, when submitting a form or uploading a file, POST is used to transmit data to a server for processing. Similarly, PUT is the method for updating already-existing data like updating a file or a database record and DELETE is used to delete data from a server, including database records and files. HTTP completely functions based on polling mechanism [10].

In contrast to HTTP, WebSocket is an event driven protocol. Both client and server listen to events to handle data change, error occurrence and connection status. WebSocket also has two methods send() and close () for the purpose to transmit data and close the connection respectively. Other than methods, WebSocket also has four events named as *open*, *message*, *error* and *close* with callbacks *nooepn*, *onmessage*, *onerror* and respectively [11].

SignalR uses methods *OnReconnected*(), OnDisconnected(), OnConnected() When a *onclose* transport con nection is automatically re-established after becoming lost [11], the Hub's OnReconnected event handler fires. After the end of a SignalR connection, the OnDisconnected event handler is called. The framework offers a collection of events that can be started at various points in a SignalR connection's lifespan [20].

3.6 Error Handling

Even though all three protocols handle errors, the precise methods and procedures employed does vary depend ing on connection models and protocols lifecycles. HTTP relies on error codes and response messages. HTTP error codes occurs when web page is no longer accessible (404 not found) or if there is a server issue (500 inter nal error). Although, error codes of HTTP are not a part of web pages, alternatively, they are server response indicating how the request was handled [10].

On the other side WebSocket is event driven and uses error events and close codes When there is an unex pected error or problem (like when some data couldn't be transmitted), the error event is fired. An error event is always swiftly followed by a close event since errors force the WebSocket connection to terminate [11]. Where as *SignalR* offers extra error handling tools built on top of WebSocket such as automatic reconnection, connection management [12].

3.7 Reliability

HTTP is not the ideal choice for real-time applications as it demands a constant communication. Though TCP abstraction is the web socket. In other words, TCP is the protocol that underlies the Web Socket protocol (Web socket messages are transferred over TCP). TCP is used on top of the HTTP protocol [11]. Implementations of WebSocket have their own frames or packets, which may include extra assurances such message sorting based on correlation ids, so http can be ideal choice rather than http. SignalR offers further characteristics that helps to make real-time applications more reliable and resilient as it supports multiple transport protocols and group management which makes Acknowledgement (ACK) more powerful and flexible (Fig. 1).

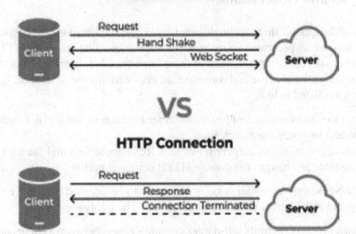

Fig. 1. HTTP vs WebSocket Connection

Sample Heading (Forth Level). The contribution should contain no more than four levels of headings. The following Table 1. Gives a summary of all heading levels.

Table 1. HTTP vs WebSocket vs SignalR

Parameters	HTTP	WebSocket	SignalR
Data transfer	Small amount of data Occasional updates	Frequent updates Data transferred quickly	More flexible than WebSocket Low latency
Connection	Stateless Unidirectional	Stateful Bi-directional	Stateful
Scalability	Less effective when traffic increase	More scalable than HTTP	Alternate transport protocol Distributing data
communication	Half-duplex	Full duplex	Use both full-duplex and half-duplex
Performance	Each request and re sponse are independent. Poor performance	Supports data compression Better performance compared to HTTP	Multi-platform compatibility Automatically select optimum transport protocol
Error Handling	Relies on error codes and response messages	Uses error events and close codes	Automatic reconnec tion
Reliable	Not ideal for real time application	It is ideal choice for real time application	Make ACK more powerful

4 Analysis and Observations

This study discusses all three technologies i.e., HTTP, WebSocket and SignalR. The discussion and details presented above are evidence to claim that that when HTTP, WebSocket and SignalR are compared, each of these three has advantages and disadvantages primarily based on the use case and context of the application to be developed. Important observations are listed below.

a. HTTP is a trustworthy protocol for client-server interaction, but it is not intended for real-time and two-way communication.
b. For a web application that largely relies on CRUD operations and the user doesn't need to respond to changes right away, HTTP is a good option.

The WebSocket, on the other hand is important to be adopted to achieve scalable, low-latency and reliable real-time web applications. The lightweight

c. WebSocket protocol, permits full-duplex communication between a client and server, making it perfect for real-time applications like chat programmes and online games.
d. SignalR builds on top of WebSocket and provides additional features such as au tomatic reconnection, message grouping [12].

Ultimately, the application-specific requirements determine the protocol and library to be used. While making the technology decision, developers must carefully weigh the trade-offs between performance, dependability, and ease of implementation. Real-time communication technologies like WebSocket and SignalR are trust worthy, but their dependability varies depending on the use case and how the technology is implemented by the developer.

5 Conclusion

Realtime web applications are important to industry and business. Each company wants web pages updating in real time for better user experience and no delay in in formation. This paper does an exhaustive and comparative discussion on three web application technologies i.e., HTTP, WebSocket, SignalR. The HTTP is basic proto col, exhaustively used for web application development and usage. However, HTTP fails to support real time updates due to request-response and polling mechanism for refresh. Alternatively, WebSocket and SignalR are event driven and duplex connection-based techniques that ensures a all-time connectivity and refresh on client time in real time.

References

1. Pimentel, V., Nickerson, B.G.: Communicating and displaying real-time data with WebSocket. IEEE Internet Comput. **16**(4), 45–53 (2012). https://doi.org/10.1109/MIC.2012.64
2. Łasocha, W., Badurowicz, M.: Comparison of WebSocket and HTTP protocol performance. J. Comput. Sci. Inst. **19**, 67–74 (2021). https://doi.org/10.35784/jcsi.2452
3. Link: https://www.rfc-editor.org/rfc/rfc6455.html

4. Rajak, C.K., Soni, U., Biswas, B., Shrivastava, A.K.: Real-time web based Timing display Application for Test Range Applications. In: 2021 2nd International Conference on Range Technology (ICORT), Chandipur, Balasore, India, pp. 1–6 (2021). https://doi.org/10.1109/ICORT52730.2021.9581663. Ingebrigtsen, Einar. SignalR Blueprints. Packt Publishing Ltd (2015)

5. Murley, P., et al.: WebSocket adoption and the landscape of the real-time web. In: Proceedings of the Web Conference 2021 (2021)

6. Link: https://developer.mozilla.org/en US/docs/Web/HTTP

7. https://developer.mozilla.org/en-US/docs/Web/API/WebSockets_API

8. https://learn.microsoft.com/en-us/aspnet/signalr/overview/getting%20started/introduction-to-signalr

9. http://www.staroceans.org/e-book/O'Reilly%20-%20HTTP%20-%20The%20Definitive%20Guide.pdf

10. https://pages.ably.com/hubfs/the-websocket-handbook.pdf

11. https://learn.microsoft.com/en-us/aspnet/signalr/overview/guide-to-the%20api/handling-connection-lifetime-events

12. https://developer.mozilla.org/en-US/docs/Web/API/WebSocket

13. https://developer.mozilla.org/en-US/docs/Glossary/TCP

14. https://learn.microsoft.com/en%20us/aspnet/core/signalr/introduction?WT.mc_id=dotnet-35129-website&view=aspnetcore-7.0

15. Fall, K.R., Richard Stevens, W.: TCP/IP illustrated, vol. 1: The protocols. Addison-Wesley (2011)

16. Pimentel, V., Nickerson, B.G.: Communicating and displaying real-time data with WebSocket. IEEE Internet Comput. **16**(4), 45–53 (2012). https://doi.org/10.1109/MIC.2012.64

17. Murley, P., Ma, Z., Mason, J., Bailey, M., Kharraz, A.: WebSock et adoption and the landscape of the real-time web. In: Proceedings of the Web Conference 2021, pp. 1192–1203 (2021)

18. https://weblogs.asp.net/leftslipper/asp-net-mvc-design-philosophy

19. Aguilar, J.M.: SignalR Programming in Microsoft ASP. NET. Vol. 40. Mi crosoft Press (2014)

20. Castellani, A.P., Gheda, M., Bui, N., Rossi, M., Zorzi, M.: Web services for the internet of things through CoAP and EXI. In: 2011 IEEE International Conference on Communications Workshops (ICC), pp. 1–6. IEEE (2011).

21. https://learn.microsoft.com/en-us/azure/azure-signalr/signalr-concept%20performance

22. https://developer.mozilla.org/en-US/docs/Web/HTTP/Overview

23. Mogul, J.C.: The case for persistent-connection HTTP. ACM SIGCOMM Comput. Commun. Rev. **25**(4), 299–313 (1995)

24. Nayyeri, K., White, D.: Pro ASP. NET SignalR: Real-Time Communi cation in. NET with SignalR 2.1. Apress (2014)

25. Wang, V., Salim, F., Moskovits, P.: The definitive guide to HTML5 WebSocket, vol. 1. Apress, New York (2013)

26. Lu, R., Chen, Z.: Evaluation of WebSocket, Socket.IO, and SignalR for Real-Time Web Communication

An IoT Based Early Alert System to Monitor and Reduce Electrical Energy Consumption at Home in Smart Cities

Vinoth Chakkaravarthy[✉], N. Anbaran Arivukoe, A. Aravinda Krishnan,
N. S. Santhsoh Sivan, and J. Tavamani Rajadurai

Velammal College of Engineering and Technology, Madurai, Tamil Nādu, India
vcvcsri@gmail.com

Abstract. The proposed IoT-based energy monitoring system is a powerful tool for managing and monitoring energy consumption in homes and small businesses. It is designed to measure the voltage and power consumed by appliances connected to it, providing users with real-time energy consumption data and a historical record of energy usage over time. Additionally, the system is equipped with features that enable remote access via a software application, allowing users to monitor their energy consumption and track their daily usage patterns to identify opportunities for energy savings. One of the most significant advantages of this system is its ability to turn off the appliances connected to it. Users can set up the system to turn off appliances when they are not in use, reducing energy consumption and lowering their electricity bills. The software application provides users with a comprehensive view of their energy consumption, allowing them to identify the appliances that consume the most energy and develop strategies to conserve energy. The application's data storage capabilities enable the safe and secure storage of energy consumption data, which can be accessed at any time. The appliance is a sophisticated energy monitoring system that operates by measuring the power and voltage consumption of appliances. It can be easily integrated into a home or small business network, and its data can be accessed remotely through the software application. With its powerful features, including real-time energy consumption monitoring, historical data tracking, and the ability to turn off connected appliances, this IoT-based energy monitoring system is an ideal choice for those looking to conserve energy and lower their energy bills.

Keywords: IoT-based energy monitoring system · voltage · power · remote access · energy consumption data · historical record · conserve energy · data storage · turn off appliances · real-time monitoring

1 Introduction

In recent years, energy management has become a critical issue in homes and small businesses due to the ever-increasing demand for energy and the need to conserve resources. With the advent of the Internet of Things (IoT), a new generation of energy monitoring systems has emerged, enabling consumers to monitor their energy consumption

© The Author(s), under exclusive license to Springer Nature Switzerland AG 2023
S. Kadry and R. Prasath (Eds.): MIKE 2023, LNAI 13924, pp. 136–147, 2023.
https://doi.org/10.1007/978-3-031-44084-7_14

more efficiently and take steps to reduce their energy usage. The proposed IoT-based energy monitoring system aims to provide users with real-time monitoring of energy consumption and control over connected appliances, offering a simple and effective way to conserve energy [1].

The idea of monitoring energy consumption and controlling appliances remotely is not new. Several studies have been conducted on energy monitoring and conservation, and a variety of devices have been developed to help users manage their energy usage [2]. For instance, a study by Han et al. (2012) proposed a system that allows users to monitor energy consumption in real-time and control appliances remotely. The system used sensors and a cloud-based platform to provide users with real-time energy consumption data and offer suggestions for energy conservation [3]. The system consisted of sensors and smart plugs that could be used to turn off appliances when they were not in use, helping users reduce their energy consumption [4]. Similarly, a study by Tian et al. (2013) developed an energy monitoring system that could measure and analyze energy consumption in homes and small businesses.

The proposed IoT-based energy monitoring system builds on the work of previous studies and offers several innovative features that make it an ideal solution for homes and small businesses [5]. First, the system can measure voltage and power consumption accurately, providing users with a precise measurement of their energy usage. Second, the system is integrated with a software application that allows users to access real-time energy consumption data and historical records of energy usage [6]. This feature enables users to monitor their energy usage and identify opportunities for energy conservation. Third, the system has the ability to turn off connected appliances remotely, providing users with a convenient way to conserve energy and reduce their electricity bills [7].

Several studies have shown that energy monitoring systems can be effective in reducing energy consumption [8–10]. For example, a study by Jin et al. (2015) found that energy monitoring systems can help users reduce their energy consumption by up to 15%. Similarly, a study by Garg and Singh (2015) showed that energy monitoring systems can help users identify energy wastage and develop strategies for energy conservation. These studies highlight the importance of energy monitoring systems in promoting energy efficiency and reducing energy consumption.

The proposed IoT-based energy monitoring system offers several advantages over traditional energy monitoring systems. First, it provides real-time energy consumption data, enabling users to monitor their energy usage and identify areas for improvement. Second, the system can be controlled remotely, allowing users to turn off appliances and reduce energy consumption with ease. Third, the system is highly accurate and reliable, ensuring that users receive precise measurements of their energy usage. These advantages make the proposed system an ideal solution for homes and small businesses looking to conserve energy and reduce their electricity bills.

In conclusion, the proposed IoT-based energy monitoring system is a significant innovation that can help consumers manage their energy consumption more effectively. The system's innovative features, including real-time energy consumption monitoring, remote control of appliances, and accurate measurements of energy usage, make it an ideal solution for homes and small businesses. The system builds on the work of previous studies on energy monitoring and conservation and offers several advantages over traditional energy monitoring systems. The next section will discuss the methodology used in developing the proposed system.

2 System Description

The power supply of 220 V is given to the step down transformer. It converts the high volt to low volt and the power supply of 12 V is given to the rectifier. Rectifier has 2 regulators, 7812 and 7805. 7812 is the regulator that is used for the power supply of 12 V for the relay. 7805 is used to provide power supply for the controller, WiFi Module and LCD Display. 2 Current sensors and 2 voltage sensors are connected with it. Voltage sensor is done by connecting the rectifier with the Potential transformer. Analog input is given to the Arduino board. The values are updated to the cloud and that can be accessed by software using the channel number. The values can be viewed on the software. Once the units reaches the threshold. With the relay the power is turned off.

2.1 Hardware Components

2.1.1 Microcontroller Arduino Uno

Arduino Uno is a popular microcontroller board that is widely used in energy monitoring devices for data acquisition, processing, and control. In energy monitoring systems, Arduino Uno is used as the main control unit that interfaces with different sensors and modules to collect data, analyze it, and provide control signals to different components.

By using Arduino Uno, energy monitoring devices can be customized and programmed to meet specific requirements, enabling the implementation of advanced monitoring and control algorithms. Arduino Uno is equipped with various input/output pins that can be used to interface with different sensors and modules, such as current and voltage sensors, WiFi modules, LCD displays, and relays (Fig. 1).

Fig. 1. Microcontroller Arduino Uno

2.1.2 Step Down Transformer

Step-down transformers can also be useful in energy monitoring devices, which measure and monitor the amount of electrical energy used by households, businesses, or other facilities. These devices typically use current transformers (CTs) or voltage transformers (VTs), which are types of step-down transformers, to reduce the voltage or current level of the AC power signal before it is measured. The current transformer reduces the current level of the AC signal, making it easier to measure accurately. Similarly, a voltage transformer can be used to measure the voltage level of the AC signal by reducing it to a lower, more manageable level.

Fig. 2. Step Down Transformer

2.1.3 Current Sensor

Current sensors, also known as current transducers, are important components in energy monitoring devices used for measuring and monitoring electrical energy consumption. These devices measure the electrical current flowing through a conductor and convert it into an electrical signal that can be measured and analyzed. In energy monitoring devices, current sensors are typically used with a microcontroller or a similar electronic device that can process the signals from the sensor and calculate the amount of electrical energy being consumed by the load. By measuring the current flowing through a power line, current sensors can provide real-time information about the power consumption of devices, machines, or entire systems, which is essential for optimizing energy usage, identifying power-hungry appliances, or detecting energy waste and leaks. Current sensors are widely used in energy monitoring systems for various applications, such as monitoring the power consumption of HVAC systems, refrigeration systems, motors, pumps, and lighting systems in industrial, commercial, and residential settings (Fig. 3).

Fig. 3. Current Sensor

2.1.4 Voltage Sensor

Voltage sensors are important components in energy monitoring devices used for measuring and monitoring electrical energy consumption. These devices measure the electrical voltage level of a circuit and convert it into an electrical signal that can be measured and analyzed. In energy monitoring devices, voltage sensors are typically used in conjunction with current sensors to determine the amount of electrical energy being consumed by a load. By measuring both the current and voltage levels of a circuit, the power consumption of the load can be calculated in real-time. This information can be used to optimize energy usage, identify power-hungry appliances, or detect energy waste and leaks. Voltage sensors are essential components in energy monitoring devices that enable the accurate and precise measurement of electrical energy consumption, facilitating energy management and conservation efforts in various industries and applications.

2.1.5 Relay

Relays are important components in energy monitoring devices used for controlling the flow of electrical power to loads or circuits. In energy monitoring systems, relays are used to switch power on and off to the loads being monitored. Relays work by using an electromagnetic coil to open or close a set of contacts. When the coil is energized, the contacts close and power flows to the load. When the coil is de-energized, the contacts open, and power is disconnected from the load. In energy monitoring systems, relays are typically used to control the power supply to specific loads or circuits. For example, a relay can be used to switch off the power supply to a specific appliance during off-peak hours when energy consumption is high. This can help to reduce energy consumption and save costs. Relays can also be used to control the power supply to entire buildings or facilities (Fig. 4).

Fig. 4. Relay

3 Proposed System

- Hardware Components: The proposed system will include a set of hardware components such as sensors, microcontrollers, and relays to measure the voltage and power consumed by the appliances that are connected to the system. These components will be connected to a central hub that will communicate with the software application.
- Software Application: The system will include a software application that will be accessible from a mobile device or computer.
- Cloud Storage: The data collected by the sensors will be stored on a cloud-based platform. The cloud storage will allow for easy access to the data from anywhere in the world and ensure the data is not lost in case of a system failure.
- Data Analytics: The proposed system will incorporate data analytics to provide insights into energy consumption patterns. The data analytics will help users identify which appliances consume the most energy and at what time of day, enabling them to make more informed decisions about their energy consumption.
- Integration with Smart Home Devices: The proposed system can be integrated with smart home devices to provide a more seamless user experience.

3.1 Methodology

1. Identify the objectives and requirements:

   - Define the goals of your early alert system, such as reducing energy consumption, identifying abnormal usage patterns, or optimizing energy efficiency.
   - Determine the specific requirements, including the types of sensors, communication protocols, and alert mechanisms you want to incorporate.

2. Sensor selection:

   - Choose appropriate sensors to collect data on energy consumption. This can include smart energy meters, current sensors, temperature sensors, or occupancy sensors.
   - Ensure that the selected sensors are compatible with IoT platforms and can provide accurate and real-time data.

3. IoT platform and connectivity:

  – Select an IoT platform or framework that supports data collection, storage, and analysis. Examples include AWS IoT, Google Cloud IoT, or Microsoft Azure IoT.
  – Determine the connectivity options for your IoT devices, such as Wi-Fi, Bluetooth, or Zigbee, based on the range and scalability requirements.

4. Data collection and transmission:

  – Install and configure the selected sensors to collect energy consumption data at regular intervals.
  – Establish a secure and reliable communication channel between the sensors and the IoT platform for transmitting the data. Ensure encryption and authentication mechanisms are in place to protect the data.

5. Data storage and processing:

  – Set up a data storage system, such as a cloud database, to store the collected energy consumption data.
  – Implement data processing algorithms to analyze the data and identify patterns or anomalies in energy usage. This can involve statistical analysis, machine learning, or rule-based methods.

6 Alert generation and notification:

  – Define the criteria for triggering an alert based on predefined thresholds or abnormal usage patterns.
  – Implement an alert generation mechanism that can generate notifications when unusual energy consumption is detected.
  – Choose appropriate alert mechanisms, such as email notifications, SMS alerts, or mobile app notifications, to inform users about energy consumption anomalies.

7. User interface and visualization:

  – Develop a user-friendly interface, either a web or mobile application, to visualize energy consumption data and provide insights to users.
  – Display real-time energy consumption information, historical trends, and alerts in an easily understandable format, such as graphs or charts.

8. Testing and validation:

  – Conduct extensive testing of the entire system to ensure its reliability, accuracy, and responsiveness.
  – Validate the system's effectiveness in identifying abnormal energy consumption and generating timely alerts.
  – Gather user feedback and iterate on the system to improve its performance and usability.

9. Deployment and maintenance:

- Deploy the system in the target environment, such as residential homes or apartments.
- Monitor the system regularly to ensure it continues to function properly and provide accurate alerts.
- Address any maintenance or upgrade needs, such as replacing faulty sensors or updating the software, to maintain the system's efficiency over time.

4   Circuit Implementation

See Fig. 5.

Fig. 5. Circuit of the Proposed System

5   Energy Consumption Graph

See Fig. 6.

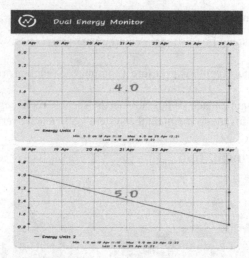

Fig. 6. Energy Consumption Graph

6 Future Enhancements

Enhanced Data Analytics: The current system collects data on energy consumption but there is scope for further analysis of the data to provide insights into energy consumption patterns, such as identifying which appliances consume the most energy and at what time of day. This could help users make more informed decisions about their energy consumption and identify areas for potential energy savings.

Artificial Intelligence: AI could be integrated into the system to provide predictive analysis of energy consumption patterns, enabling users to make more informed decisions about how they use their energy. For example, the system could predict when energy usage is likely to be high and suggest ways to reduce consumption or reschedule energy-intensive tasks.

Renewable Energy Integration: With the increasing adoption of renewable energy sources, such as solar panels, wind turbines and hydroelectric systems, IoT-based energy monitoring systems can be enhanced to measure the energy generated from these sources as well as the energy consumed. This would enable users to optimize their use of renewable energy and potentially sell excess energy back to the grid.

Smart Grid Integration: With the advent of smart grid technology, IoT-based energy monitoring systems can be integrated with the smart grid to enable more efficient energy consumption. For example, the system could receive signals from the grid indicating when energy prices are low, enabling users to schedule energy-intensive tasks accordingly.

Voice-Activated Controls: Integrating voice-activated controls into the system would provide a more convenient way for users to control their appliances. Users could simply issue voice commands to turn appliances on or off or adjust their settings.

Augmented Reality: By integrating augmented reality into the system, users could get a visual representation of their energy consumption in real-time. For example, they could use their smartphone or tablet to view a 3D model of their home and see which appliances are consuming the most energy.

Blockchain Technology: Integrating blockchain technology into the system could provide a more secure and transparent way of managing energy transactions. For example, users could earn rewards for reducing their energy consumption during peak demand periods, and these rewards could be stored on a blockchain-based platform.

7 Advantages

Energy Savings: One of the primary advantages of an IoT-based energy monitoring system is the potential for energy savings. By monitoring and controlling the energy consumption of connected appliances, users can optimize their energy usage and reduce their overall energy bills.

Increased Awareness: IoT-based energy monitoring systems provide users with real-time information on their energy consumption, which can increase their awareness of how much energy they are using and where it is being used. This can help users make more informed decisions about their energy consumption and potentially reduce their energy usage.

Remote Control: IoT-based energy monitoring systems allow users to control their connected appliances remotely, which can be particularly useful for households with multiple occupants or for users who are away from home for extended periods.

Improved Efficiency: By optimizing energy consumption patterns, IoT-based energy monitoring systems can improve the overall efficiency of energy usage. This can lead to reduced energy waste and potentially lower greenhouse gas emissions.

Data Analytics: IoT-based energy monitoring systems provide users with detailed information on their energy consumption patterns. This information can be used to identify inefficiencies and areas for improvement, enabling users to make informed decisions about their energy consumption.

Integration with Smart Home Devices: IoT-based energy monitoring systems can be integrated with smart home devices such as thermostats, lighting, and security systems, providing users with a more seamless user experience.

Cost-Effective: IoT-based energy monitoring systems are generally cost-effective compared to traditional energy monitoring systems. Additionally, the potential for energy savings can help users recoup the initial investment in the system over time.

Environmental Benefits: By reducing energy waste and potentially lowering greenhouse gas emissions, IoT-based energy monitoring systems can provide environmental benefits and contribute to a more sustainable future.

8 Limitations

- While an IoT-based energy monitoring system has many benefits, it also has some limitations that should be considered. Here are some potential limitations:
- Cost: IoT-based energy monitoring systems can be expensive to install and maintain. This could limit their adoption among households and small businesses that cannot afford the upfront costs.

- Connectivity: The reliability of the system depends on the strength of the internet connection. Poor connectivity can affect the accuracy of the data collected and the ability to control appliances remotely.
- Compatibility: The system may not be compatible with all types of appliances. Older appliances may not be able to connect to the system, or may require additional hardware to do so.
- Security: IoT devices are vulnerable to cyber-attacks, and an energy monitoring system is no exception. Hackers could potentially gain access to the system and control appliances remotely or steal sensitive data.
- Data privacy: Collecting data on energy consumption raises concerns about data privacy. Users may not want their energy usage data to be shared with third-party companies or government agencies.
- Reliability: The accuracy of the system's data collection depends on the accuracy of the sensors used. If the sensors are not properly calibrated or are faulty, the data collected may not be accurate.
- Limited Control: While the system allows users to control their appliances remotely, it may not provide the same level of control as manual control. Users may not be able to adjust the settings of their appliances as precisely as they would if they were physically present.
- Overall, while an IoT-based energy monitoring system has many advantages, it is important to consider these potential limitations and address them appropriately.

9 Conclusion

In conclusion, the IoT-based energy monitoring appliance offers an efficient and effective solution to monitor and manage energy consumption in homes and businesses. By providing real-time energy consumption data and historical records of energy usage, the system enables users to make informed decisions about their energy usage and identify opportunities to save energy and reduce costs.

The ability to remotely turn off connected appliances to conserve energy also offers added convenience and control. The system provides accurate measurements of energy usage for each appliance, allowing users to understand the energy consumption of individual appliances and make informed decisions about their usage.

Moreover, the system is easily accessible through a user-friendly software application that provides a comprehensive overview of energy consumption, making it easy for users to understand their energy usage and identify areas for improvement. By storing historical records of energy usage, the system also allows users to track their energy consumption over time and identify patterns and trends.

Overall, the IoT-based energy monitoring appliance represents a significant step forward in energy management technology and has the potential to revolutionize the way we monitor and manage energy consumption in homes and businesses. The system provides a cost-effective and easy-to-use solution for energy management that has numerous benefits for both individuals and society as a whole, including reduced energy consumption, cost savings, and a more sustainable future.

References

1. Lam, K.P., Gao, S., Yang, S., Li, Y., Li, X., Wang, D.: An Internet of Things-enabled energy management system for demand response in smart homes. Energies **12**(7), 1289 (2019)
2. Kalaivani, M., Rajeswari, M., Elango, C.: IoT-based energy management system for smart home automation. Int. J. Eng. Technol. **7**(2.6), 42–46 (2018)
3. Malini, A., Venkatesh, N., Sundarakantham, K., Mercyshalinie, S: Mobile application testing on smart devices using MTAAS framework in cloud. In: International Conference on Computing and Communication Technologies (2014). https://doi.org/10.1109/ICCCT2.2014.706 6751
4. Wang, Y., Liu, L., Liu, Z., Guo, S.: An intelligent and energy-efficient home appliance control system based on IoT. J. Ambient. Intell. Humaniz. Comput. **10**(10), 3971–3981 (2019)
5. Hsu, C.C., Huang, Y.C.: Design and implementation of an IoT-based energy management system. IEEE Access **8**, 29056–29066 (2020)
6. Park, J.J., Lim, J.J., Chung, T.Y.: IoT-based energy management system for improving energy efficiency in residential buildings. J. Clean. Prod. **181**, 322–331 (2018)
7. Li, X., Li, Y., Li, L.: An IoT-based energy monitoring system for smart homes. Int. J. Comput. Intell. Syst. **13**(1), 498–506 (2020)
8. Nguyen, H.V., Nguyen, H.T., Nguyen, Q.N., Nguyen, N.T., Bui, T.D.: An IoT-based energy management system for households. Int. J. Adv. Comput. Sci. Appl. **11**(5), 402–409 (2020)
9. Kumar, N., Kumar, A.: IoT-based energy monitoring and control system using MQTT protocol. Int. J. Adv. Trends Comput. Sci. Eng. **9**(3.3), 3363–3367 (2020)
10. Huang, S.S., Tsai, M.T., Liu, C.M.: An IoT-based energy management system for green buildings. Sustainability **11**(5), 1365 (2019)
11. Zhang, J., Liao, W.: IoT-based real-time energy monitoring and management system. J. Comput. Theor. Nanosci. **15**(1), 15–21 (2018)
12. Li, L., Li, Y., Li, X., Yang, S., Wang, D.: An IoT-enabled energy management system for demand response in residential buildings. IEEE Trans. Industr. Inf. **15**(6), 3431–3441 (2019)
13. Shaban, A.M., Fathy, H.K., Zayed, T.S., Salah, M.B.: IoT-based energy management system for smart buildings. J. Build. Eng. **29**, 101175 (2020)
14. Koyuncu, H., Karaca, E., Dogdu, E.: Development of an IoT-based energy monitoring system for smart homes. SN Appl. Sci. **3**(5), 1–12 (2021)
15. Islam, M.R., Islam, M.N., Faruque, M.A.: An IoT-based energy management system for sustainable smart homes. J. Electr. Syst. Inform. Technol. **6**(2), 147–160 (2019)

Securing the MANET by Detecting the Flooding Attacks Using Hybrid CNN-Bi-LSTM-RF Model

B. Deena Divya Nayomi[1], L. Venkata Jayanth[2(✉)], A. Vinay[1,2], P. Subba Rao[1,2], and L. Shashi Vardhan[1,2]

[1] Department of CSE, G. Pullaiah College of Engineering and Technology, Kurnool, India
deena.divya20@gmail.com, subbaraopelluri23@gmail.com
[2] G. Pullaiah College of Engineering and Technology, Kurnool, India
jayanthpranay890@gmail.com

Abstract. Security in Mobile Ad Hoc Networks (MANETs) is complicated by attacks such request route flooding, which are simple to launch but hard to defend against. An attack can be launched by a rogue node by delivering a flood of route request (RREQ) packets or other worthless data packets to non-existent destinations. As the network's resources have been exhausted trying to handle this deluge of RREQ packets, it has been rendered incapable of performing its usual routing function. The majority of the available literature on identifying such a flooding assault use a threshold based on the rate of RREQ generation attributable to a certain node. These algorithms are effective to a point, but they have a high misdetection rate and hinder the efficiency of the network. Using a (CNN), a Bidirectional Long Short and the (RF) for classification, this study suggests a novel technique for detecting flooding threats. The method uses each node's route discovery history to recognise shared traits and routines among members of the same class, allowing it to determine whether or not a given node is malicious. The effectiveness of the projected method is measured by associating the results of NS2 simulations run under normal and RREQ attack scenarios with respect to attack detection rate, packet delivery rate, and routing load. Simulation findings demonstrate that the proposed model can identify over 99% of RREQ flooding assaults across all scenarios with route discovery, and outperforms state-of-the-art methods for RREQ flooding attacks in terms of packet delivery ratio and routing burden.

Keywords: Route request · Mobile Ad Hoc Networks · Flooding attack · Convolutional Neural Network · Random Forest

1 Introduction

A MANET is network that does not rely on a predetermined topology to facilitate communication activities. Temporary network formed by wireless nodes that rely on multihop communications since no underlying infrastructure is present [1]. Self-organization and decentralised control are hallmarks of MANETs, making it possible for individual

S. Kadry and R. Prasath (Eds.): MIKE 2023, LNAI 13924, pp. 148–160, 2023.
https://doi.org/10.1007/978-3-031-44084-7_15

nodes to work together to achieve the network's goals and ensure reliable communication. Issues with routing, security, access control, dependability, and power consumption are only some of MANET's difficulties [2]. To overcome these obstacles, a secure routing protocol is used in order to identify rogue nodes and isolate them from the communiqué network so that network performance may be improved. In MANETs, data announcement must be protected at all times. The majority of DoS attacks [3, 4] originate from security breaches caused by packet flooding on the network, which uses up more resources and leads to congestion. By using a trustworthy route, you may lessen the chances of interacting with malicious nodes. For MANETs to be a secure, low-infrastructure communication channel, trust management is essential [5]. The on-demand routing architecture employed by AODV relies heavily on routing protocols to determine which paths to take. Security parameters are included in the route reply packet and the discovery packet by changing them [6].

Without intermediary devices like routers or base stations, communication between MANET nodes is possible. Transmissions such as single-hop and multi-hop are used in MANETs to facilitate communication between mobile nodes via intermediary nodes that function as routers to transmit and relay messages [7, 8]. MANETs' instantaneous deployment and lack of need for a preexisting infrastructure make them a strong contender for use in a wide variety of contexts, including but not incomplete to: military and police communications; fire and rescue; inter-vehicle networks; emergency and disaster recovery; personal area networks; healthcare; and educational and medical settings. The cooperative and dispersed nature of a MANET's routing design makes it more susceptible to denial-of-service assaults [11]. DoS attacks are launched to stop their intended recipients from making use of the system's resources and, by extension, its services. Mobile nodes in MANETs are particularly vulnerable to intrusion, and once infiltrated, they may be used to achieve DoS assaults. (DoS) attacks occur when several nodes across a network all launch simultaneous DoS attacks. DDoS bouts are more hazardous and harder to counteract in real time [12].

There are two broad types of denial-of-service attack: vulnerabilities (in which a known flaw in the target process is exploited) and floods (in which a large number of service requests or fake traffic are generated). Hateful nodes will launch a large number of packets with route RREQ for IP addresses including destinations, draining the battery capabilities of intermediate nodes and increasing their energy consumption, in a flooding DoS attack [14]. In MANETs, the discovery of a route is often initiated packets containing the RREQ protocol; nevertheless, the flooding of such RREQ fake packets without any adherence for regulating rate in a network can have a profoundly degrading effect on system throughput [15].

The following are the most important results from this research.

Create an accurate attack-detection system using an ensemble CNN-BiLSTM-RF construction trained using ML and DL replicas.

Provide a module for data preprocessing that looks at the temporal features of the dataset.

To emphasise the importance of the proposed research, a comparison will be made between the suggested ensemble CNN-BiLSTM-RF construction and the more traditional learning and DL replicas.

The residue of the paper is organised as shadows: Sect. 2 summarises the relevant papers, and Sect. 3 provides a brief account of the suggested model. Sections 4 and 5 show the results of the validation analysis and draw a conclusion…

2 Related Works

Using security localization and an improved multilayer network, Gebremariam et al. [16] presented detection and localization against numerous assaults (MLPANN). The suggested approach detects and localises WSN DoS attacks using a combination of localization and machine learning techniques. Simulation in MATLAB is used to construct the suggested system, while the IBM SPSS toolbox and Python are used to handle the data. Using a multilayer perceptron artificial neural network, we can identify 10 different types of assaults, such as denial-of-service (DoS) attacks, by dividing the dataset into training and testing sets. Results from applying the proposed system to benchmark datasets show that it significantly outperforms the state-of-the-art with an average detection accuracy of 100%, 99.65%, for different types of DoS assaults, respectively. The proposed approach achieves better results than state-of-the-art alternatives in all measures of localization performance (accuracy, f-score, precision, and recall). Lastly, simulations are run to evaluate the security performance of the proposed strategy for identifying and pinpointing malicious nodes. This technique yields a low localization error approximation of the unknown node's position. The results of the simulations demonstrate the efficacy of the proposed system in detecting and securely localising malicious assaults in wireless sensor networks that are scalable and hierarchically distributed. With this method, we were able to reduce the worst-case localization error to 0.49 % and improve the average to 99.51 %.

A hybrid deep learning strategy, devised by Elubeyd and Yiltas-Kaplan et al. [17], integrates elements from three distinct deep learning algorithms. Both theoretical analysis and empirical testing showed that this method obtained very high accuracy (99.81% and 99.88%, respectively) on two separate datasets. Specifically for software-defined networks, this study represents a major advancement in the field of network security. DoS/DDoS attacks may be avoided and SDN security can be improved with the help of the suggested technique. Since SDNs play a key role in today's network architecture, safeguarding them from assaults is essential to preserving the reliability and accessibility of such resources. Overall, the work proves that a hybrid deep learning technique can effectively spot DoS/DDoS assaults in SDNs, and it points the way for further investigation into this topic.

Stacked auto-encoder based technique for MANET was presented by Meddeb et al. [18] to improve intrusion detection systems (Stacked AE-IDS) is a method for minimising correlation using neural networks in Machine Learning (ML). It employs numerous processing layers in an effort to model important aspects at a high level and to obtain a suitable representation feature from Data Correlation. When the input and output dimensions are the same, the Stacked AE-IDS approach attempts to recreate the input while minimising the correlation between the two. We suggest a two-stage process for implementing Deep Learning in IDS. Classification is performed using a Deep Neural network (DNN) using the auto-output encoder's as training data (DNN-IDS). It leverages the most likely assaults to disrupt routing services in Mobile Network and zeroes down on DoS attacks within labelled datasets available for intrusion detection.

Tekleselassie et al. [19] provide a new way to project information about assaults on wireless networks onto a grid-like data structure, which can then be used to train the EfficientNet CNN model. Determine how the attribute values should be arranged in a matrix before it can be recorded as an image. The goal is to create an accurate and lightweight IDS module that can be implemented in IoT networks by merging the most significant subset of features with EfficientNet. Use the AWID dataset (which contains information on Wi-Fi assaults) to analyse the suggested model. Obtain a 99.91% F1 score of 0.11% for optimal performance. This indicates that the suggested model successfully leveraged the spatial information in tabular data to preserve detection correctness, and that its results were equivalent to those of previous statistical machine learning models. The false positive rate is kept at roughly 0.11 %. So, the suggested model was compared to three existing machine learning models, and it was shown to be competitive with their performance. Where it is assumed that the spatial info must be taken into account by mapping out the tabular data on a grid.

In this research, we suggest (CLPDM-SSA) using the sparrow search algorithm introduced by Venkatasubramanian et al. [20] and by [21, 22]. This proposal employs a cluster-based meta-heuristic detector to single out a malicious node in a real-world data gathering system hit by a packet drop (PD) assault. Results: Throughput,, and false positive rate are utilised to validate the deployment of NS-2. The results show significant improvements once SSA intelligence was integrated. The method analyses the central processing unit and memory to identify false positives of suspected malicious nodes.

A novel Intrusion Detection System (IDS) is presented to increase network performance by identifying DDoS assaults in wireless networks by Nalayini and Katiravan [23], and by [24,25]. In order to pick the features that contribute the most to improving classification accuracy, we present a novel approach Feature Optimization Method (SFSH-FOM). In this paper, We provide a new deep learning method, Fuzzy Temporal Features integrated Convolutional Neural Network, for accurate classification (FT-CNN). To further enhance performance, we also provide an unique cross-layer feature fusion technique in this study, which makes use of FT-CNN and LSTM. Using assessment criteria like detection IDS was put to the test on benchmark datasets including KDD'99, NSL-KDD, and DDoS; the findings demonstrate that the proposed IDS is more effective than competing methods.

2.1 Research Gaps

There have been a lot of studies showed in this area, but the threshold value has seldom been taken into account. The study also proposes a deep learning model to determine the threshold value.

While many studies have been conducted on filtering-based systems, trust, and game theory, there are still obstacles to overcome before an effective solution can be designed.

Thirdly, there is a lack of study on how to protect AODV-based mobile ad hoc networks from attacks like hello flooding. We are unaware of any method for protecting Mobile Adhoc Networks from hello flood attacks.

3 Proposed System

3.1 Problem Statement

The security of MANET is crucial for revealing and avoiding the many assaults that pose a risk to MANET. An example of a DoS attack that poses a danger to network operations is the Flooding attack, which sends out bogus packets in an effort to slow down or halt legitimate data transmissions between nodes. To discover the shortest way between network nodes, the original AODV is an on-demand routing protocol; however, it does not have any means of detecting or avoiding the Flooding attack.

3.2 Our Contribution

To that end, we investigated the nature of the Flooding assault and its impact on the network, and we improved the deep learning model's ability to withstand this type of attack. By empowering nodes in the network with the ability to decide whether the request is received from an attacker node or from a normal node, our proposed model helps to avoid false judgement on nodes by putting them in a suspicious list before judging them, thereby reducing the negative effects of fake request packets on the network. Lastly, we attempted to make the limit value..

In this research, we provide a new statistically-based method for preventing flooding attacks on Mobile Adhoc Networks. The suggested approach makes use of the distribution as a statistical justification for navigating to the node that is causing network disarray due to an excess of RREQs. Spreading in a Mobile Ad hoc Network is computed by determining the standard deviation of RREQs answered by nodes with different characteristics. The detection and prevention of Mobile Adhoc Networks operating under the AODV protocol are greatly aided by this method. The statistical cutoff value is the foundation of this technique. This cutoff is based on the dispersion of the RREQs generated by several nodes in the MANET relative to the mean. If there are 'n' nodes in the MANET, and that for each I in the range.

$$Mean\ of\ Route\ Requests(MRREQ) = \sum_{i=1}^{n} \frac{x_i}{n} \tag{1}$$

After the mean has been determined for all RREQs in the Mobile Adhoc Network, the next step is to determine their variance. The dispersion of all nodes' route requests from $x_1, x_2, x_3, x_4, \ldots, x_n$ is calculated as

$$Variance^2 = \frac{\sum_{i=1}^{n}(x_i - MRREQ)^2}{n-1} \tag{2}$$

$$Variance = \sqrt[2]{\frac{\sum_{i=1}^{n}(x_i - MRREQ)^2}{n-1}} \tag{3}$$

After different threshold values have been computed, flood attackers and malicious nodes in the MANET can be located with more ease. STV is what you'll get when you plug in your mean and variance numbers.

$$STV = 2 * \sum_{i=1}^{n} \frac{x_i}{n} * \frac{\sum_{i=1}^{n} \frac{x_i}{n}}{\sqrt[2]{\frac{\sum_{i=1}^{n}(x_i - MRREQ)^2}{n-1}} + 1} \quad (4)$$

The STV is the cutoff point used to identify the MANET's bad actor. The suggested DL model for determining STV's worth is described in detail below. The total number of RREQs generated by the n unique nodes in the mobile ad hoc network is denoted by the variable x, where x ranges from 1 to xn. Now we can see if x i > STV holds true for every xi with I ranging from 1 to n. Node $'i'$ is sending fake RREQs in the mobile ad hoc network to decrease performance if the cost of x i > STV is genuine. After the rogue node has been identified, a packet will be sent out through the mobile ad hoc network to cut it off. This process of sending RREQs to many recipients is carried out by every node in a mobile ad hoc network. In a mobile ad hoc network, malevolent nodes are well isolated. For the forthcoming statistical and threshold-based approach, this algorithm is intended.

Step 1: Start
Step 2: Determine how many RREQs each network node sent and keep track of those numbers in variables. as $x1, x2, x3, x4, \ldots\ldots\ldots\ldots\ldots, xn$ by increasing the source node pawn as xi $++$
Step 3: Find out the mean of the network using Eq. (1).
Step 4: If you want to see how much variation there is in the RREQs sent by different network nodes, you may do it by using Eq. (2–3).
Step 5: Compute Statistical Threshold Value (STV) using Eq. (4).
Step 6: For slightly node x_i where $i = 1, 2, 3, \ldots\ldots\ldots\ldots, n$.
If $x_i > STV$ then change to step 7 else go to step 8.
Step 7: The RREQs from node I are being dropped since it has been identified as a malicious source attempting to flood the network..
Step 8: End.

This technique does a scan of the whole network in order to identify any potential attacker nodes. This approach of isolating malicious node is more effective than other arithmetical and threshold-based strategies flooding attacker harmful nodes in MANET because the value of variance is determined based on the deviation of RREQs made by each node in the network.

3.3 Proposed CNN-BiLSTM-RF Architecture

The suggested method consists of three main parts. In the input layer of deep learning networks, we feed them 111 pairs of Click fraud data. It consists of a single-dimensional (CNN) layer, which together permit sample-based discretization of parameters for feature recognition, speed up training, and avoid over-fitting. The Batch Normalization layer follows the Maxpooling layer to enable parameter normalisation across intermediate layers and save prolonged training durations. There are 64 filters in the 1-D CNN layer, and the activation function is Relu. The kernel size is 2. Maximum pooling length of 2 is used in the Maxpooling layer. The BiLSTM layer receives its input features from this map. The 128 memory blocks of a BiLSTM are used to acquire expertise in the time domain characteristics. After a Maxpooling layer with pooling length 2, and before a Batch Normalization layer, the BiLSTM layer is placed. Afterwards, the input will be Flattened in preparation for the subsequent Dense layers. Filters 128 and 64 are used to add two thick layers. The activation function Relu has been applied to both dense layers. Between the two thick layers, a 0.5-dropout layer is utilised. Even if Max Pooling is used in between each layer in the model, the Dropout Layer is still there to prevent Over Fitting. This is typically the case since combining CNN with BiLSTM increases the likelihood of over-fitting and hence leads to subpar results on the testing set. At last, the attributes are used as input into RF to distinguish between genuine and fake clicks.

In addition, a hybrid model is taught to function using f(X,y) Where X,y [S 1,...,S K]. In order to reduce training and validation error, the Adam optimizer is used to adjust the weights at the end of each training session based on the training and validation loss and accuracy. Accuracy throughout both training and validation is also recorded for each K-set. In order to train the RF model, the f method is used to the output of the trained hybrid CNN-BiLSTM model, which is then utilised to extract hidden features from input data (EF,ytrain). Following feature extraction and RF training, the hybrid DL model is used to extract features for unseen data through f(ENF; ytest) to yield ypred. Accuracy, Precision, Recall, F1 Score, and Area Under the Curve (AUC) are only few of the assessment criteria used to assess performance after prediction results are achieved..

4 Results and Discussion

4.1 Simulation Environment

This subsection provides an illustration of the planned DL-performance AODV's evaluation. The simulation is run for 50 randomly placed nodes in a 1000m x 1000m region using a random way point model. In order to pinpoint the source of the network breach, a simulation duration of 900 s is used, and each scenario is run 10 times. The additional variables and typical values used in simulations are shown in Table 1. The effectiveness of the proposed DL-AODV is demonstrated against an adversarial node density variation attack. Moreover, classic AODV and trust-based AODV systems are compared to the proposed DL-AODV.

The effectiveness of the new DL-AODV is measured by the next set of routing KPIs:.

- **Routing overhead:** Ratio of control/routing packets used to total packets is the metric...

$$Routing overhead = \frac{Number of routing packets}{Number of data packets + Number of control packets} \quad (5)$$

- The term "Packet Lost" refers to the sum total of all packets that were lost while running the simulation.

$$Packet Lost = Number of packets sent - Number of packets received \quad (6)$$

Table 1. Network Imitation parameters

Typical Value	Parameter
802.11p	MAC Protocol
10,20,30,40,50	Node densities
~ 250m	Communication Range(m)
CBR	Traffic Source
0–20 m/s	Max Speed
512 (bytes)	Packet size

- Throughput: The percentage of received data during the allotted simulation period is calculated. A more secure and efficient network is one with a high throughput value.

$$Throughput = \frac{\sum Successfully\ received\ data\ packets\ at\ destination}{Simulation\ time * 1024} \qquad (7)$$

- Average round-trip waiting time: The latency of a network is measured in milliseconds, or the time it takes a data packet to get from one point to another (Fig. 1).

$$E2Edelay = \frac{\sum_{i=1}^{n}(ReceivingTime - TransmittedTime)}{TotalNumberofconnections} \qquad (8)$$

- Reliability: The rate at which data packets are received is compared to the rate at which they are sent. This number is always between zero and one (Table 2).

$$R = \frac{sumofdatapocketsrecived}{sumofpocketstransmitted} \qquad (9)$$

Table 2. Comparative Assessment of PDR

No. of nodes	Packet delivery ratio (%)			
	RF	CNN	Bi-LSTM	CNN-Bi-LSTM-RF
10	89.57	94.87	87.56	99.80
20	90.78	93.93	88.98	99.80
30	88.69	93.54	87.75	99.56
40	88.94	94	85.83	98.36
50	87.56	95.5	84.67	96.67

When the node is 30, the existing models achieved nearly 87% to 93%, where ensemble model achieved 99% of PDR. When the nodes are high, the PDR is low for every technique. For instance, RF achieved 87.56%, CNN achieved 95.5%, Bi-LSTM achieved 84.67% and proposed model achieved 96.67% of PDR (Table 3 and Fig. 2).

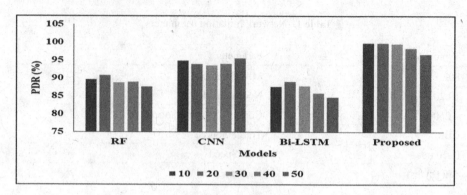

Fig. 1. PDR Analysis

Table 3. Comparative Assessment of throughput

No. of nodes	Throughput (bps)			
	RF	CNN	Bi-LSTM	CNN-Bi-LSTM-RF
10	67	250	95	700
20	38	148	70	520
30	45	125	74	425
40	40	178	69	345
50	35	185	73	240

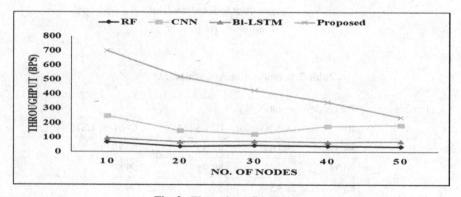

Fig. 2. Throughput Comparison

When comparing with all models, RF and Bi-LSTM achieved low performance on throughput analysis. For instance, when the node is 40, RF has 40bps, CNN has 178bps, Bi-LSTM achieved 69bps and proposed model achieved 345bps. The reason is that the RF works based on tree methodology, which needs to store all nodes in the memory. It consumes high computation for RF and provides poor performance (Table 4 and Fig. 3).

Table 4. Assessment of Energy consumption

No. of nodes	Energy Consumption (J)			
	RF	CNN	Bi-LSTM	CNN-Bi-LSTM-RF
10	10	10	11	6
20	16	17	15	9
30	19	21	18	12
40	25	27	23	16
50	33	34	26	19

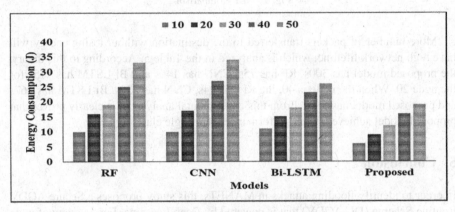

Fig. 3. Analysis of Energy Consumption

When the node is 10, CNN achieved 10J, RF achieved 10J, Bi-LSTM achieved 11J and proposed model achieved 6J. Less energy consumption means better performance of the model, therefore, the proposed model achieved 19J, CNN achieved 34J, RF achieved 33J and Bi-LSTM achieved 26J for the node 50 (Fig. 4).

Table 5. Evaluation of Network life time

No. of nodes	Network Lifetime (s)			
	RF	CNN	Bi-LSTM	CNN-Bi-LSTM-RF
10	60	119	98	150
20	75	140	115	300
30	86	186	140	420
40	94	230	167	530
50	120	265	198	580

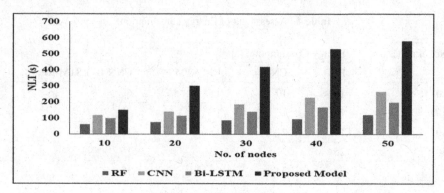

Fig. 4. NLT Comparison

More number of packets transferred to the destination without losing energy will have high network lifetime, which is analyzed in the Table 5. According to this theory, the proposed model has 300s, RF has 75s, CNN has 140s and Bi-LSTM has 115s for the node 20. When the node is 40, the RF has 94s, CNN has 230s, Bi-LSTM has 167s and proposed model has 530s. From this experimental analysis, it is clearly proves that proposed model achieved better performance than single classifiers.

5 Conclusion

In order to identify flooding attacks in MANETs, this study proposes a Secure AODV Routing Scheme (DL-AODV) that is powered by Deep Learning. For a variety of node densities, the effectiveness of DL-AODV was compared to that of standard AODV and trust-based AODV. Furthermore, the suggested DL-AODV significantly boosts the secure communication by enhancing the accuracy and throughput of intrusion detection with CNN and Bi-LSTM classifier. In addition, in comparison to previous DL based AODV protocols, the suggested method increases network burden. Just 50 nodes are included in this analysis of the planned DL-AODV, which has a top speed of 20m/s. As the node density and speeds in complete urban settings are very dynamic, DL-AODV is best suited for information transmission in semi urban environments.

References

1. Sbai, O., El boukhari, M.: September. Data flooding intrusion detection system for manets using deep learning approach. In: Proceedings of the 13th International Conference on Intelligent Systems: Theories and Applications, pp. 1–5 (2020)
2. Vatambeti, R., Sanshi, S., Krishna, D.P.: An efficient clustering approach for optimized path selection and route maintenance in mobile ad hoc networks. J Ambient Intell Human Comput **14**, 305–319 (2023). https://doi.org/10.1007/s12652-021-03298-3
3. Nandi, M., Anusha, K.: An optimized and hybrid energy aware routing model for effective detection of flooding attacks in a manet environment. Wireless Personal Communications, pp.1–19 (2021)

4. Archana, H.P., Khushi, C., Nandini, P., Sivaraman, E., Honnavalli, P.: Cloud-based Network Intrusion Detection System using Deep Learning. ArabWIC 2021: The 7th Annual International Conference on Arab Women in Computing in Conjunction with the 2nd Forum of Women in Research, Sharjah, UAE (2021). https://doi.org/10.1145/3485557.3485562

5. Banerjee, B., Neogy, S.: A brief overview of security attacks and protocols in MANET. In: 2021 IEEE 18th India Council International Conference (INDICON), pp. 1–6. IEEE (2021, December)

6. Kalime, S., Sagar, K.: A review: secure routing protocols for mobile adhoc networks (MANETs). Journal of Critical Reviews 7, 8385–8393 (2021)

7. Kothai, G., et al.: A new hybrid deep learning algorithm for prediction of wide traffic congestion in smart cities. Wireless Communications and Mobile Computing, vol. 2021, Article ID 5583874, pp. 13 (2021). https://doi.org/10.1155/2021/5583874

8. Abdelhaq, M., et al.: The resistance of routing protocols against DDOS attack in MANET. Int. J. Electr. Comp. Eng. (2088–8708) 10(5) (2020)

9. Fiade, A., Triadi, A.Y., Sulhi, A., Masruroh, S.U., Handayani, V., Suseno, H.B.: Performance analysis of black hole attack and flooding attack AODV routing protocol on VANET (vehicular ad-hoc network). In: 2020 8th International conference on cyber and IT service management (CITSM), pp. 1–5. IEEE (2020, October)

10. Divya, N.S., Bobba, V., Vatambeti, R.: An adaptive cluster based vehicular routing protocol for secure communication. Wireless Pers Commun 127, 1717–1736 (2022). https://doi.org/10.1007/s11277-021-08717-4

11. Srinivas, T.A.S., Manivannan, S.S.: Prevention of hello flood attack in IoT using combination of deep learning with improved rider optimization algorithm. Comput. Commun. 163, 162–175 (2020)

12. Mahajan, R., Zafar, S.: DDoS attacks impact on data transfer in IOT-MANET-based E-Healthcare for Tackling COVID-19. In: Data Analytics and Management: Proceedings of ICDAM, pp. 301–309. Springer Singapore (2021)

13. Nishanth, N., Mujeeb, A.: Modeling and detection of flooding-based denial-of-service attack in wireless ad hoc network using Bayesian inference. IEEE Syst. J. 15(1), 17–26 (2020)

14. Kurian, S., Ramasamy, L.: Securing service discovery from denial of service attack in mobile ad hoc network (MANET). Int. J. Comp. Netw. Appli. 8(5), 619–633 (2021)

15. Gebremariam, G.G., Panda, J., Indu, S.: Localization and Detection of multiple attacks in wireless sensor networks using artificial neural network. Wireless Communications and Mobile Computing (2023)

16. Elubeyd, H., Yiltas-Kaplan, D.: Hybrid deep learning approach for automatic Dos/DDoS attacks detection in software-defined networks. Appl. Sci. 13(6), 3828 (2023)

17. Meddeb, R., Jemili, F., Triki, B., Korbaa, O.: A Deep Learning based Intrusion Detection Approach for MANET (2022)

18. Tekleselassie, H.: Two-dimensional projection based wireless intrusion classification using lightweight EfficientNet. J. Cyber Sec. Mobi. 601–620 (2022)

19. Kishen Ajay Kumar, V., et al.: Dynamic Wavelength Scheduling by Multiobjectives in OBS Networks. Journal of Mathematics vol. 2022, Article ID 3806018, 10 (2022). https://doi.org/10.1155/2022/3806018

20. Ramana, K., et al.: Leaf disease classification in smart agriculture using deep neural network architecture and IoT. J. Circ. Sys. Comp. 31(15), 2240004 (2022). https://doi.org/10.1142/S0218126622400047

21. Dwaram, J.R., Madapuri, R.K.: Crop yield forecasting by long short-term memory network with Adam optimizer and Huber loss function in Andhra Pradesh, India

22. Ramana, K., et al.: A vision transformer approach for traffic congestion prediction in urban areas. IEEE Trans. Intell. Transp. Syst. **24**(4), 3922–3934 (2023). https://doi.org/10.1109/TITS.2022.3233801. April

23. Nalayini, C.M., Katiravan, J.: A New IDS for Detecting DDoS Attacks in Wireless Networks using Spotted Hyena Optimization and Fuzzy Temporal CNN. Journal of Internet Technology **24**(1), 23–34 (2023)

The Smart Coverage Path Planner for Autonomous Drones Using TSP and Tree Selection

M. Sundarrajan[1], Akshya Jothi[1], D. Prabakar[1],
and Seifedine Kadry[2,3,4]([✉])

[1] SRM Institute of Science and Technology, Chennai, India
akshyaj@srmist.edu.in
[2] Department of Applied Data Science, Noroff University College, Kristiansand,
Norway
skadry@gmail.com
[3] Artificial Intelligence Research Center (AIRC), Ajman University,
Ajman 346, United Arab Emirates
[4] Department of Electrical and Computer Engineering,
Lebanese American University, Byblos, Lebanon

Abstract. The problem of efficiently covering an area with a limited number of sensors or cameras is an important research topic in various fields such as surveillance, monitoring, and inspection. This problem can be posed as a Traveling Salesman Problem (TSP), where the optimal path needs to be found to visit all the coverage points and return to the starting point. However, the computational complexity of solving TSP increases exponentially with the number of coverage points, making it impractical for large-scale applications. This paper proposes a novel approach to solving the coverage problem as a TSP by incorporating tree selection algorithms to improve efficiency and accuracy. The proposed approach represents the study area as a graph, with each node representing a coverage point and the edges representing the distance between them. To reduce computational complexity, we use various tree selection algorithms to select the best possible starting point for the TSP algorithm. To evaluate the effectiveness of the proposed approach, we conduct simulations using different scenarios with varying numbers of coverage points. The simulation results demonstrate that the proposed approach significantly improves the quality of coverage while reducing computational complexity compared to traditional TSP algorithms. Moreover, the proposed approach can be applied in a wide range of real-world applications, such as surveillance, monitoring, or inspection, where efficient and accurate coverage of an area is essential.

Keywords: Path Planning · Drones · Tree Selection Algorithms ·
Unmanned Aerial Vehicles · Area Coverage · Travelling Salesperson
Problem

© The Author(s), under exclusive license to Springer Nature Switzerland AG 2023
S. Kadry and R. Prasath (Eds.): MIKE 2023, LNAI 13924, pp. 161–172, 2023.
https://doi.org/10.1007/978-3-031-44084-7_16

1   Introduction

Unmanned aerial vehicles (UAVs), often called drones, are aircraft that do not have a human pilot on board. They may be designed to fly autonomously according to instructions stored on board, or they can be controlled remotely by a human operator on the ground. Little quadcopters that fit in the palm of your hand and massive fixed-wing aircraft used in the military and industry are just two examples of the sizes and types of drones available. Path planning is an essential component when it comes to the safe and effective operation of drones. It's the process of determining the most effective path for the drone to take in light of its capabilities, the surrounding environment, and the mission at hand. Drones benefit from path planning because it allows them to safely navigate around obstacles, maximize their flight duration, and arrive at their destinations without delay. Drones' safety is greatly improved by using route planning software. Path planning guarantees that drones can safely fly through hazardous settings by avoiding obstacles like trees, buildings, and other structures. This lessens the potential for drone and environmental damage caused by collisions. Drones may use route planning to safely fly across disaster zones and other hostile settings where people shouldn't go. There may be more than one feasible route for a drone to take from its origin to its final destination in any deployment scenario. Choosing a route that shortens the total tour duration and distance is crucial. There could be more than one path for a drone from its source to its destination when deployed in any application, as shown in Fig. 1 and Fig. 2. The major role lies in choosing the path with less time and tour length. Both the figures have the same number of nodes to be traversed, but both paths are different.

 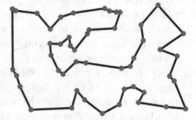

Fig. 1. Nodes = 49 - Path A   **Fig. 2.** Nodes = 49 - Path B

Covering a large area with a small number of sensors or cameras is a common need in many practical applications like surveillance and monitoring. In a TSP formulation, the nodes (sensors or cameras) are cities, and the goal is to find the shortest route from the starting point to each node and back again. The suggested method begins by representing the research region as a graph, with each node representing a coverage point and the edges representing the distance between them. The Traveling Salesman Problem (TSP) technique is then used to determine the best route to every node and then back to the origin. The best feasible jump-off point for the TSP method is chosen using various tree selection techniques, which boosts the approach's efficiency and accuracy. Each algorithm has its own benefits and drawbacks, and the right one may be chosen for every given task. The suggested method employs tree selection methods to both lessen the computing burden of the TSP issue and increase the quality of sensor or camera coverage. Several real-world situations call for effective and efficient covering of a particular region, and this method may be put to use in such cases. Using tree selection methods to increase efficiency and accuracy, the suggested methodology offers a unique and efficient way to address the coverage issue as a TSP. The method has the potential to be used in several practical contexts, as it can both increase coverage quality and decrease computing complexity.

2   Literature Review

Many existing strategies are being developed by writers worldwide to discover the ideal route for an Intelligent UAV. Noor et al. summarises the uses of unmanned aerial vehicles (UAVs)/drones using remote sensing technologies in populated regions [1]. Heidari et al. provide a comprehensive analysis of machine learning's uses in IoD systems, which can be found in [2]. Security, privacy, and standardization are just a few unanswered questions addressed in this discussion of recent IoD system implementations. Daud et al. has broadly studied drones' potential uses in crisis management. Damage assessment, searching for survivors, and managing supplies are just some of the many ways drones have been utilized in disaster management, all of which are briefly discussed by the writers [3]. To categorize flood-affected regions in UAV-captured aerial photos, Akshya and Priyadarsini analyze several machine-learning algorithms. They talk about the difficulties of flood mapping and how unmanned aerial vehicles (UAVs) may help with flood monitoring and control [4]. Path planning for numerous UAVs is addressed in [5], where Shafiq et al. suggest a max-min ant colony optimization strategy. The authors use a minimum and maximum distance criterion to fine-tune the route planning of many UAVs.

To better design routes for unmanned aerial vehicles (UAVs), Zhang et al. take into account restrictions such as height, distance, and angle, and the authors optimize UAV route planning [6]. Chowdhury and De offer a UAV path-planning technique for a 3D dynamic environment inspired by Reverse Glowworm Swarm Optimization [7]. In [8], Graph Convolution Network-based optimum route planning for intelligent UAVs (GCNs). The authors use topological structure and semantic information about objects to improve UAV route planning. For the symmetric traveling salesman problem, Cinar et al. (2020) offer a discrete tree-seed approach (TSP) [9]. For the drone version of the traveling salesman problem, Nguyen [10] suggests using a Monte Carlo Tree Search method. The authors maximize the TSP by balancing exploration and exploitation, two objectives that are made more difficult by the drone's mobility. In [11], Song et al. offers a Graph Neural Network (GNN) strategy for TSP that uses geographical information.

The TSP is optimized by considering the location of all nodes and edges [12]. For the Traveling Salesman Problem (TSP) and the Steiner Tree Problem, Ganesh et al., by focusing on algorithm robustness and scalability, the authors optimize solutions for the TSP and Steiner tree issues. In [13], Bogyrbayeva et al. suggests using drones to solve the Traveling Salesman Problem (TSP) using a deep reinforcement learning technique. In the context of drone and robot-supported packet stations, the Multiple Traveling Salesman Problem (mTSP) is discussed in [14]. The goal of the mTSP is to determine the best routes for a group of salespeople who must each visit a different set of destinations.

In [15], the TSP is discussed with a drone that can refuel at specific locations. The classic Traveling Salesman Problem (TSP) tasks a salesperson to make a single trip to many places before returning to his origin. Yet, the addition of a drone to the situation allows for the transportation of some of the commodities while also allowing for their recharging in certain areas. The application of computational geometric approaches for UAV (Unmanned Aerial Vehicle) route planning is discussed in [16]. In order to plot a course for the UAV that is smooth and free of obstacles, the study suggests a technique based on the Delaunay triangulation and Voronoi diagram. Results from comparing the proposed technique to other well-known route planning algorithms reveal that it provides smoother pathways and more effectively avoids obstacles. The research finds that the suggested strategy is effective for UAV route planning in dynamic situations at real-time speeds.

3   Solving TSP for Path Planning of Drones

Drone path planning is selecting the most efficient flight route or trajectory for a drone to complete a mission. Drone operations rely heavily on careful path planning to guarantee the drone can fly safely and efficiently, avoiding obstacles while using the least amount of power possible. Sensing the surroundings, mapping the environment, and planning the route are the three primary processes in the path planning process. The drone's cameras, LiDAR, and GPS, are used to map and collect data about the surrounding region. With this map, the best route for the drone may be mapped out. The algorithms calculate the best route for the drone based on its capabilities and the features of its surroundings. The next part will focus on the procedures taken to formulate the issue as a TSP and locate the best possible solution for the drone.

3.1   Tree Selection Algorithms for Solving TSP

Tree selection methods are often used to obtain an optimum solution to the TSP (Traveling Salesman Problem) quickly and efficiently. By simplifying the TSP problem using tree selection methods, we may discover solutions more quickly and with less effort. Nevertheless, the number of nodes and the distribution of distances between nodes in the TSP instance may influence the decision of which method to use. The right tree selection method is crucial for finding the best answer to any given TSP problem. This subsection will discuss the tree selection algorithms we have utilized.

1. Kruskal's Algorithm: The well-known Kruskal's approach is used for computing the MST of a weighted undirected graph. The method achieves its goals by sorting the graph's edges in non-decreasing order of their weight and then incrementally adding those edges to the MST without introducing any cycles. After a tree covering the complete graph with the lowest feasible total weight has been constructed, the algorithm finishes running. The time complexity for Kruskal's approach is O(E log E), where E is the total number of edges in the graph. Because of this, it is often used in contexts where locating the MST is crucial, such as in the design and optimization of networks. Kruskal's technique can independently identify the MST of each linked component, which is one of its benefits when used to graphs with unconnected components. Because of this, it may be used to solve issues like clustering, where many clusters inside a network must be located and studied. Kruskal's algorithm offers a robust and flexible method for determining the shortest path in a weighted undirected network. Its ease of use and effectiveness make it a go-to if top-notch network optimization is necessary. The tree obtained using Kruskal's algorithm is shown in Fig. 3.

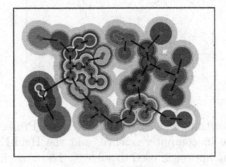

Fig. 3. Kruskal's Algorithm.

2. Prim's Algorithm: To determine the MST of a weighted undirected graph, several researchers turn to Prim's algorithm. The method begins with a single vertex and adds the nearest vertex to the expanding MST at each iteration, checking to make sure that no cycles are produced along the way. Each time through the procedure, the minimum-weighted edge between the current MST and a vertex not yet in the MST is chosen. In the graph, Prim's algorithm has a temporal complexity of O(E log V) when E and V are the numbers of edges and vertices, respectively. Its efficiency and applicability to large-scale applications like network design and optimization justify its adoption. Prim's algorithm is advantageous since it can be coded and used in reality using a priority queue data structure. The program further ensures that the MST is correct and optimum every time. Prim's algorithm for finding the least spanning tree of a weighted undirected graph is robust and commonly used. Network design, cluster analysis,

and optimization are just a few fields that benefit from this method's ease of use, speed, and precision.

3. Boruvka's Algorithm:

For UAV (Unmanned Aerial Vehicle) route planning in a network of nodes and edges, a frequent choice is Boruvka's method, a graph-based technique. The approach generates a minimal network spanning tree (MST) by repeatedly connecting nodes using the edge with the lowest weight. UAV route planning uses a graph with nodes representing waypoints and edges representing flight time or distance between them. The MST produced by Boruvka's algorithm provides a collection of interconnected components, which may be used to plot a course for the UAV. The approach is effective for large-scale route planning issues since it only considers the local minimum-weight edge at each node. The tree obtained using Kruskal's algorithm is shown in Fig. 4.

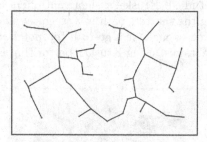

Fig. 4. Boruvka's Algorithm.

4. Held-Karp Algorithm: The Traveling Salesman Problem (TSP) is a classic optimization issue in computer science, and the Held-Karp algorithm is a dynamic programming approach used to solve it. The Traveling Salesman Problem (TSP) may be seen as the challenge of determining the quickest route between a given set of waypoints (locations the UAV must reach to complete its job). To solve the TSP, the Held-Karp algorithm creates a table of optimum solutions for subproblems, each of which entails finding the shortest route that begins at a given waypoint and stops at a certain number of the remaining waypoints. The Held-Karp method finds the best answer by repeatedly solving these smaller problems and filling the table. By iteratively solving these subproblems and building up the table, the Held-Karp algorithm can find the optimal solution for the entire problem. The Held-Karp algorithm has a time complexity of $O(n\hat{2} * 2\hat{n})$, where n is the number of waypoints, which makes it computationally feasible for small to medium-sized instances of the TSP. The Held-Karp algorithm can be expressed using the following recurrence equation:

$$C(S, i) = minC(S - i, j) + d(j, i) \tag{1}$$

for all j in S, where S is a subset of the set of waypoints 1, 2, ..., n, i is a particular waypoint in S, and d(j, i) is the distance between waypoints j and i.

The tree obtained while making use of Kruskal's algorithm is shown in Fig. 5.

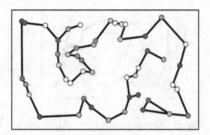

Fig. 5. Held-Karp Algorithm.

Overall, the graph's nature and the task's needs dictate the method used. Graphs with few connections benefit most from Kruskal's method, whereas those with numerous connections benefit most from Prim's and Borvka's algorithms. Borukva was selected since its implementation requires less time overall.

3.2 Solving TSP Using Tree Selection Algorithm

Finding the shortest path between a group of cities and back to the origin is the goal of the Traveling Salesman Problem (TSP), a classic optimization issue. Exact, heuristic, and approximation algorithms are viable options for addressing the TSP. The optimum solution to the TSP may be found with certainty using exact methods like the brute force approach and the branch and bound algorithm. Unfortunately, because of their exponential time complexity, they are only useful for minor problems. For more complex examples of the TSP, heuristic techniques like the closest neighbor algorithm and the 2-opt algorithm are preferable to accurate algorithms because of their speed and efficiency. Heuristic algorithms are effective because they repeatedly enhance a solution until a local optimum is reached. Yet the best answer isn't always found by heuristic algorithms. To obtain a solution near the optimum, approximation techniques like the Christofides algorithm and the Lin-Kernighan heuristic are developed. Approximation techniques may handle more significant instances of the TSP and are often quicker than precise algorithms, but they are not guaranteed always to provide the optimal solution.

The closest neighbor approach is a common method for resolving optimization issues like the traveling salesman problem (TSP) because of its simplicity and ease of implementation. The method selects a random beginning place and then chooses the nearest unvisited spot until every possible location has been explored. While the technique is straightforward, it may not always provide the best results and may get mired in local optima. An example path obtained while computing for 49nodes is depicted in Fig. 6.

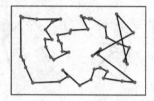

Fig. 6. Nearest Neighbor Algorithm.

The Lin-Kernighan algorithm is a more sophisticated algorithm widely used to solve TSP and other optimization problems. This approach relies on heuristics to repeatedly enhance an existing solution. The algorithm finds a better answer by beginning with a good one and then making a series of adjustments. These actions are based on heuristics that aim to optimize the solution by exchanging edges and making other modifications. An example path obtained while computing for 49 nodes using the Lin-Kernighan algorithm is depicted in Fig. 7.

Fig. 7. Lin-Kernighan Algorithm.

The Christofides algorithm is another popular algorithm for solving TSP and other optimization problems. This procedure is a heuristic, and the result is guaranteed to be within 1.5 times as close to the ideal answer as possible. The technique builds an Eulerian tour of a least-spanning graph tree to solve the problem. A Hamiltonian cycle is created by modifying the tour in some way. An example path obtained while computing for 49 nodes using the Christofides algorithm is depicted in Fig. 8.

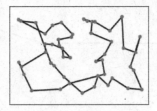

Fig. 8. Christofides algorithm.

4   Experimental Results

QGIS and Concorde TSP Solver were used to collect experimental data. We first tried to figure out which tree selection technique worked best for the chosen random nodes. Table 1 summarises the relevant data under the assumption of a network topology in which the number of nodes is 49. In Fig. 9, we see the effects of changing the total number of nodes on the network and the complexity of the corresponding algorithms. The data indicated that the Held-Karp algorithm consistently generated longer tours than the other three methods. Boruvka's algorithm consistently achieves the smallest runtime while achieving the same tour length as Kruskals's and Prim's algorithms. In light of this, we have used Boruvka's algorithm to carry out additional TSP issue-solving, as it can lead to the best solution while reducing the total computing time.

Table 1. Comparison between various Tree Selection Algorithms.

S.No	Algorithm	Tour Length	Computation Time
1	Kruskal's	355293	0.98
2	Prim's	355293	0.72
3	Boruvka's	355293	0.68
4	Held-Karp	423826	1.42

The minimal spanning tree (MST) of a graph is a tree that connects all of the nodes in the network with the fewest possible edges, and Boruvka's algorithm may help you discover it. Using Boruvka's algorithm, the MST can be located, and then the TSP can be solved in a certain fashion. The Traveling Salesman Problem (TSP) is an example of a combinatorial optimization problem that aims to determine the shortest path that visits each vertex in a graph precisely once and then loops back to the initial vertex.

Many algorithms have been used in our efforts to crack the TSP. Because of its simplicity and effectiveness in solving the Traveling Salesman Problem, the Nearest Neighbor algorithm is utilized (TSP). The program selects a city at random as a starting point and then chooses the closest unvisited city until

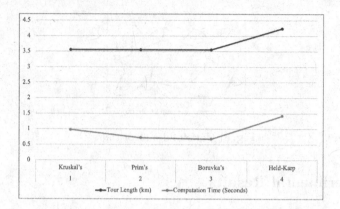

Fig. 9. Tour Length and Computation Time Analysis.

all cities have been visited. The Greedy Algorithm is the next candidate. When applied to the Traveling Salesman Problem, the Greedy algorithm provides a heuristic solution (TSP). Once all cities have been visited, including a final return to the beginning city, the algorithm operates by picking the edge with the lowest cost at each step and adding it to the tour. There are a variety of algorithms to pick from when attempting to solve an optimization issue. Neighbor-finding, Lin-Kernighan, and Christofides algorithms are some of the more popular ones. In Fig. 10, the best route is analyzed in detail, with the number of nodes changed to show how the complexity of the algorithms changes. The data indicated that the Held-Karp algorithm consistently generated longer tours than the other three methods. Time complexity analysis of all algorithms is shown in Fig. 11 (Table 2).

Table 2. Comparing Algorithms for Solving TSP.

S.No	Algorithm	Tour Length	Computation Time
1	Nearest neighbor	500855	1.12
2	Greedy	509984	1.54
3	Nearest Insertion	481857	1.62
4	Christofides Tour	506673	1.45
5	Lin-Kernighan tour	440259	1.36
6	Chained Lin-Kernighan Tour	425824	1.38

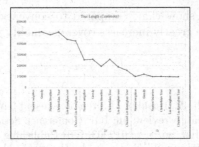

Fig. 10. Tour Length Analysis.

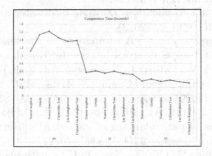

Fig. 11. Computation Time Analysis.

5   Conclusion

Drone path planning is a hot issue in robotics and surveillance from above. Covering a large region with a finite set of sensors or cameras is a significant challenge that may be posed as a transportation scheduling problem (TSP). Unfortunately, large-scale applications are not feasible because of the exponential growth in processing complexity required to solve TSP. This research proposes a solution to this problem by using tree selection algorithms, which have been shown to increase productivity and precision. The method uses a graph representation of the research region, and simulations demonstrate that it greatly improves coverage quality while decreasing computing costs compared to standard TSP methods. Solving TSP, however, becomes prohibitively computationally expensive for large-scale applications as the number of covering points grows exponentially. In this research, we offer a new method for addressing the coverage issue by casting it as a TSP and using tree selection algorithms to boost speed and precision. The suggested method visualizes the research region as a network, where each node represents a coverage point and the links between them are represented by distances. Selecting the optimal initial state for the TSP method is computationally expensive, thus we use a number of tree selection strategies to narrow the search space. In comparison to standard TSP algorithms, the Boruvka method and Lin-kernigham TSP solver are shown to dramatically enhance coverage while decreasing computational cost. We simulate many situations with variable numbers of coverage points to assess the performance of the suggested method.

The simulation results show that, compared to conventional TSP algorithms, the suggested technique greatly increases coverage quality while decreasing computing costs.

References

1. Noor, N.M., Abdullah, A., Hashim, M.: Remote sensing UAV/drones and its applications for urban areas: a review. In: IOP Conference Series: Earth and Environmental Science, vol. 169, no. 1, p. 012003. IOP Publishing (2018)
2. Heidari, A., Jafari Navimipour, N., Unal, M., Zhang, G.: Machine learning applications in internet-of-drones: systematic review, recent deployments, and open issues. ACM Comput. Surv. **55**(12), 1–45 (2023)
3. Daud, S.M.S.M., et al.: Applications of drone in disaster management: a scoping review. Sci. Just. **62**(1), 30–42 (2022)
4. Akshya, J., Priyadarsini, P.L.K.: A comparison of machine learning approaches for classifying flood-hit areas in aerial images. In: Khanna, A., Gupta, D., Bhattacharyya, S., Snasel, V., Platos, J., Hassanien, A.E. (eds.) International Conference on Innovative Computing and Communications. AISC, vol. 1087, pp. 407–415. Springer, Singapore (2020). https://doi.org/10.1007/978-981-15-1286-5_34
5. Shafiq, M., Ali, Z.A., Israr, A., Alkhammash, E.H., Hadjouni, M., Jussila, J.J.: Convergence analysis of path planning of multi-UAVs using max-min ant colony optimization approach. Sensors **22**(14), 5395 (2022)
6. Zhang, R., Li, S., Ding, Y., Qin, X., Xia, Q.: UAV path planning algorithm based on improved Harris Hawks optimization. Sensors **22**(14), 5232 (2022)
7. Chowdhury, A., De, D.: RGSO-UAV: reverse glowworm swarm optimization inspired UAV path-planning in a 3D dynamic environment. Ad Hoc Netw. **140**, 103068 (2023)
8. Jothi, A., Priyadarsini, P.L.K.: Optimal path planning for intelligent UAVs using graph convolution networks. Intell. Autom. Soft Comput. **31**(3), 1577–1591 (2022)
9. Cinar, A.C., Korkmaz, S., Kiran, M.S.: A discrete tree-seed algorithm for solving symmetric traveling salesman problem. Eng. Sci. Technol. Int. J. **23**(4), 879–890 (2020)
10. Nguyen, M.A., Sano, K., Tran, V.T.: A Monte Carlo tree search for traveling salesman problem with drone. Asian Transp. Stud. **6**, 100028 (2020)
11. Song, Y., Bliek, L., Zhang, Y.: Revisit the algorithm selection problem for TSP with spatial information enhanced graph neural networks. arXiv preprint arXiv:2302.04035 (2023)
12. Ganesh, A., Maggs, B.M., Panigrahi, D.: Robust algorithms for TSP and Steiner tree. ACM Trans. Algorithms **19**(2), 1–37 (2023)
13. Bogyrbayeva, A., Yoon, T., Ko, H., Lim, S., Yun, H., Kwon, C.: A deep reinforcement learning approach for solving the traveling salesman problem with drone. Transp. Res. Part C: Emerg. Technol. **148**, 103981 (2023)
14. Kloster, K., Moeini, M., Vigo, D., Wendt, O.: The multiple traveling salesman problem in presence of drone-and robot-supported packet stations. Eur. J. Oper. Res. **305**(2), 630–643 (2023)
15. Yurek, E.E., Ozmutlu, H.C.: Traveling salesman problem with drone under recharging policy. Comput. Commun. **179**, 35–49 (2021)
16. Priyadarsini, P.L.K., Gadupudi, P., Paladugula, K.: Path planning of intelligent UAVs using computational geometric techniques. In: 2022 International Conference on Electronic Systems and Intelligent Computing (ICESIC), pp. 190–193. IEEE (2022)

UAV Smart Navigation: Combining Delaunay Triangulation and the Bat Algorithm for Enhanced Efficiency

Akshya Jothi[1][iD], M. Sundarrajan[1][iD], R. Gayana[2][iD],
and Seifedine Kadry[3,4,5]([✉])[iD]

[1] SRM Institute of Science and Technology, Chennai, India
akshyaj@srmist.edu.in
[2] K. Ramakrishnan College of Engineering, Tiruchirappalli, India
[3] Department of Applied Data Science, Noroff University College,
Kristiansand, Norway
skadry@gmail.com
[4] Artificial Intelligence Research Center (AIRC), Ajman University,
Ajman 346, United Arab Emirates
[5] Department of Electrical and Computer Engineering,
Lebanese American University, Byblos, Lebanon

Abstract. In the realm of unmanned aerial vehicles (UAVs), path planning is crucial. The objective is to create a safe, practical, and optimum flight route for UAVs to fly across unfamiliar terrain while avoiding obstacles. Because of their high computational complexity and inability to handle changing surroundings, traditional route planning algorithms confront hurdles in real-world applications. As a result, academics have created many sophisticated ways to handle UAV route planning challenges. This research provides a unique technique for intelligent UAV route planning. The program first divides the research region using the Delaunay Triangulation method. After obtaining the partitions, the coverage points are located, and the issue is framed as the Traveling Salesperson Problem (TSP). The Bat Optimization Algorithm (BOA) is used to find an optimum route for the TSP. The suggested technique seeks to find the best route for UAVs in dynamic situations. The suggested approach was verified by comparing it to other existing algorithms. The findings demonstrate that the suggested method can construct an optimum route that meets all requirements while avoiding obstacles. Moreover, the method may swiftly identify a new route as the environment changes. The study findings show that the suggested approach beats the other techniques in total computing time, saving roughly 0.03 s. Moreover, the suggested technique reduces the UAV's route by around 0.02 cm, substantially improving existing algorithms.

Keywords: Path Planning · Intelligent UAVs · Bat Optimization · Delaunay Triangulation · Area Partitioning · Convex Hull

© The Author(s), under exclusive license to Springer Nature Switzerland AG 2023
S. Kadry and R. Prasath (Eds.): MIKE 2023, LNAI 13924, pp. 173–184, 2023.
https://doi.org/10.1007/978-3-031-44084-7_17

1   Introduction

Unmanned Aerial Vehicles (UAVs), commonly known as drones, are aircraft that fly remotely without the presence of a human pilot. UAVs may be outfitted with a broad range of sensors and cameras, making them ideal for anything from military reconnaissance and surveillance to search and rescue missions, disaster relief, and agricultural monitoring. Aerial photography and videography are examples of how they may be utilized for entertainment. UAVs are getting more sophisticated as technology progresses, capable of autonomous flying and difficult tasks such as obstacle avoidance and course planning.

UAVs, or unmanned aerial vehicles, may be used with CPP to optimize flight patterns and boost efficiency. CPP entails determining the quickest and most direct path to a location while considering topography, weather, and barriers into consideration. By employing CPP, UAVs can traverse difficult surroundings and achieve their goals swiftly and securely. UAVs may use CPP in various ways, including real-time mapping and route planning. Swarm intelligence, which includes coordinating the motions of several UAVs to accomplish a shared objective, is another method that UAVs might employ CPP. Swarm intelligence may enhance efficiency and lessen the danger of collision or other incidents by applying algorithms to optimize the flight paths of each UAV.

In this study, we used Bat optimization (BO) to solve the CPP issue for intelligent UAVs. BO is a metaheuristic algorithm inspired on bat echolocation behavior that may be used to address the critical path planning (CPP) issue for unmanned aerial vehicles (UAVs). CPP seeks the shortest and most efficient route between two places while accounting for numerous limitations such as obstructions, topography, and weather conditions. The BO method employs a population of bats that alter their location and velocity by criteria to find the best answer. The bats communicate with one another and modify their behavior depending on the results of prior searches. This enables them to effectively traverse the search space and arrive at a near-optimal answer. To apply BO to the CPP issue for UAVs, the method is updated to consider the problem's special restrictions, such as the requirement to avoid obstacles or travel over difficult terrain. The bats are depicted as unmanned aerial vehicles (UAVs), and their locations and speeds are modified depending on environmental data acquired by sensors and cameras on board. The BO method is used to develop a series of optimum UAV pathways, which are then assessed and compared to determine the most efficient and effective route. UAVs can navigate through difficult surroundings and reach their goals swiftly and securely by employing BO to solve the CPP issue, making them a useful tool for a broad variety of applications. BO is a successful method for addressing the CPP issue for unmanned aerial vehicles, providing a viable solution for real-world applications. Figure 1 depicts the general architecture of the suggested technique.

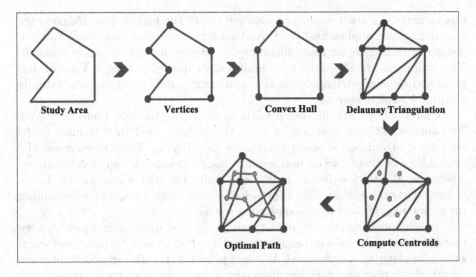

Fig. 1. Architecture Diagram.

2   Literature Review

Many exciting strategies are being developed by writers worldwide to discover the ideal route for an Intelligent UAV. As the utilization of unmanned aerial vehicles (UAVs) grows, so do the applications for which they are used. [1] discusses recent advancements and future potential for using UAVs in the cryosphere. The article shows how, because to their high geographical and temporal resolutions, unmanned aerial vehicles (UAVs) are becoming a useful tool for exploring the cryosphere. Fan, B. et al. in [2] provide an in-depth examination of the development and present applications of unmanned aerial vehicle (UAV) technology. It encompasses everything from the airframe and motor system to the sensors on the UAV. This article also discusses the current and potential applications of unmanned aerial vehicles (UAVs) in agriculture, construction, and surveying. The author of [3] outlines how UAVs may be utilized effectively in various industries such as agriculture, mining, SAR, and environmental monitoring. The author of [4] investigates the viability of utilizing unmanned aerial vehicles (UAVs) and artificial intelligence (AI) in the mining industry. The article discusses the advantages of employing unmanned aerial vehicles (UAVs) with sensors and cameras for geological mapping, mineral exploration, and mine monitoring.

As a first step in finding an ideal route, the flight region must be divided into several divisions so that the CPP algorithms may be easily deployed. McLaughlin et al. proposed in [5] utilizing an immunity-based approach to evaluate whether or not a UAV's sensors and actuators are malfunctioning. Acevedo et al. [6] provide a successful distributed area division technique in their study to support cooperative monitoring applications using many UAVs. Luna et al. propose a rapid multi-UAV path planning technique to get the highest possible coverage in aerial sensing applications [7]. Priyadarsini et al. [8] provide a computational geometric technique for route planning for intelligent UAVs in their research. This approach provides a route planning algorithm employing a Voronoi diagram and Delaunay triangulation that can work in difficult environments while ensuring complete coverage.

Different efforts are also being made to determine the best route for UAVs. The authors offer an overview of how AI has been used to determine paths for groups of unmanned aerial vehicles in [9]. (UAVs). The writers stress the practicality of these tactics and explore their advantages and disadvantages. Akshya et al [10] describe a graph-based route planning algorithm for UAVs utilized for area coverage. This approach enhances route planning by segmenting the area into grids and using graph theory ideas.

Certain route-planning algorithms employ evolutionary algorithms in very particular ways. Priyadarsini et al. demonstrate how to apply evolutionary algorithms for efficient and effective UAV route planning [11]. The approach enhances overall route planning using evolutionary algorithms to discover the shortest path across each part. Gul et al. [12] propose a meta-heuristic optimization approach to solve the issues of multi-objective route planning for autonomous guided robots. The technique combines particle swarm optimization, grey wolf optimization, and evolutionary programming to optimize route planning while reducing parameters like energy consumption, path length, and collisions. Song et al. [13] provide an improved particle swarm optimization approach for smooth route planning of mobile robots using a continuous high-degree Bezier curve. Yahia et al. [14] use meta-heuristic tools to explore path-planning optimization for UAVs in detail. Malyshev et al. propose using evolutionary optimization approaches to synthesize trajectory planning algorithms [15]. The method uses evolutionary algorithms to fine-tune the parameters of the planning algorithm, resulting in more effective trajectory planning. Overall, the articles in this collection provide a broad range of techniques to enhance UAV route planning via AI, optimization algorithms, and segmentation methods. These strategies are intended to improve the utility, performance, and security of UAVs in various situations.

3   Solving Coverage Path Planning Problem

When it comes to deploying UAVs in real-time applications, one of the most crucial restrictions is the solution of CPP. It is critical for unmanned aerial vehicles (UAVs) to find the shortest route for numerous reasons. For starters, it enables the UAV to finish its task more quickly, saving time and energy. This is particularly crucial in applications like search and rescue, surveillance, and inspection when the UAV must cover a big area quickly. Second, determining the shortest route reduces the likelihood of an accident or collision with a barrier, such as a building or a tree. The UAV may avoid needless diversions or obstructions by choosing the most direct path, lowering the risk of damage or harm. Finally, selecting the shortest route may increase the UAV's battery life, a significant factor for battery-powered UAVs. The UAV can preserve energy and remain in the air for extended periods by selecting the most efficient path, expanding its range, and minimizing the need for regular recharging. The following sub-sections will go through the numerous procedures we took in this research to solve the CPP issue successfully.

3.1   Area Partitioning - Delaunay Triangulation

We explored building a polygon with random coordinates as vertices as a starting step to discover an optimum route by solving the CPP problem. This polygon will be used as the region of investigation for the suggested algorithms. The procedure is picking random coordinates inside the flight space and linking them to form a polygon with the randomly chosen locations as vertices. Figure 2 depicts the randomly chosen vertices. This polygon acts as the UAV's flight path's border, and the UAV will fly inside it while avoiding any obstacles or no-fly zones. This strategy may be effective when the flying environment is complicated and creating a precise flight plan is challenging.

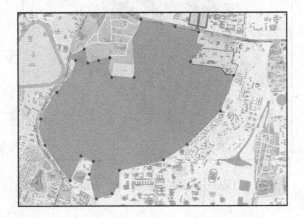

Fig. 2. Selection of Vertices.

After selecting random coordinates as vertices, we constructed a convex hull using the Graham Scan technique to improve route planning for UAVs further. The Graham Scan technique is a well-known approach for calculating the convex hull of a collection of points in a plane [26]. By using the Graham Scan technique to build a convex hull, the final form is a polygon with a minimum perimeter that encompasses all of the randomly picked vertices, as illustrated in Fig. 3. This is particularly advantageous when the vertices are not uniformly distributed, since it creates a smoother and more efficient route for the UAV to follow. After constructing a convex hull, the next step we have adopted for area partitioning is the application of Delaunay triangulation [27], as seen in Fig. 4.

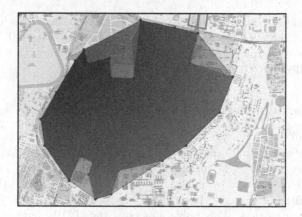

Fig. 3. Convex Hull - Study Area.

Fig. 4. Delaunay Triangulation - Study Area.

The circumcircle of each triangle in the Delaunay triangulation has no additional points in its interior, which is one of its important features. This attribute may be stated numerically using the triangle circumcircle equation:

$$(x - m)^2 + (y - n)^2 = rad^2 \qquad (1)$$

where (x, y) are the coordinates of a point, (m, n) are the coordinates of the triangle's circumcenter, and rad is the circumcircle's radius. Delaunay triangulation, in the context of UAV route planning, may assist in discovering possible areas of interest or risks inside the flight space by creating a precise network of triangles linking the randomly generated vertices. This is particularly important when the flying environment is complicated or has many obstacles, as it may assist in identifying regions where the UAV should avoid or pay special attention.

3.2   Computation of Coverage Points

The next stage is to find the coverage spots after partitioning the region. As the covering point in the suggested method, we calculated the centroids for each triangle. Figure 5 shows the computed centroids for the partitioned region.

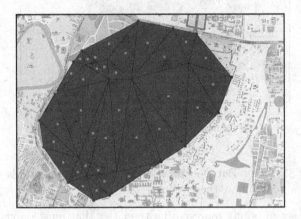

Fig. 5. Centroids - Delaunay Triangulation.

The centroid of a triangle is the point at where the triangle's three medians connect, and it may be computed using the formula below:

$$Cent = (M + N + P)/3 \qquad (2)$$

where Cent is the centroid and M, N, and P are the coordinates of the triangle's vertices. The centroids may be utilized as possible UAV waypoints. After we have the centroids, we can rename the issue the Traveling Salesperson Problem (TSP), which can be solved using any evolutionary technique. We used the Bat optimization technique in our suggested study.

3.3   Solving TSP - Bat Optimization

The Traveling Salesman Problem (TSP) entails determining the shortest route that visits a set of cities and returns to the originating location. In our proposed work, we assume the coverage point to be a city, and the UAV must visit every calculated centroid. Bat optimization is a metaheuristic optimization technique based on bat echolocation activity. The calculated route using BOA is shown in Fig. 6.

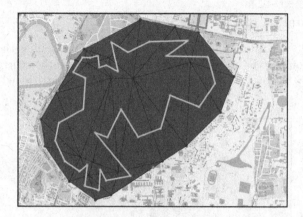

Fig. 6. Convex Hull - Study Area.

To solve TSP using Bat optimization through the centroids of Delaunay triangulation, we can follow these steps:

1. Generate a set of points representing the cities that must be visited.
2. Define the objective function to be the total distance traveled during the tour, which is the sum of the distances between consecutive centroids in the tour.
3. Initialize a population of bat solutions, each representing a potential city tour.
4. For each bat solution, calculate its fitness value (i.e., the total distance traveled during the tour) using the objective function.
5. While the stopping criteria are not met, perform the following steps:
   (a) Generate a new solution by adjusting the bat's position based on its current position and velocity.
   (b) Evaluate the fitness of the new solution.
   (c) Update the position and velocity of the bat based on the current and new solutions.
   (d) If the new solution has a better fitness value than the current solution, replace the current solution with the new solution.
6. Return the best solution found as the solution to the TSP.

We may limit the number of cities to examine and simplify the issue by utilizing the centroids of the Delaunay triangulation as prospective cities to visit. Moreover, bat optimization is well-suited to handling combinatorial optimization problems like TSP, and it can often discover high-quality solutions in a reasonable period.

4   Experimental Results

The experimental results were obtained from QGIS. In the first stage, polygon geometries were taken and their values, such as area and perimeter, were calculated. Table 1 summarizes some of them.

Table 1. Polygon Geometries.

ID	Area	Perimeter
1	0.000231	0.078581
2	0.000354	0.084567
3	0.000476	0.092546
4	0.000598	0.097834
5	0.000673	0.099361

After obtaining the convex hull for the vertices chosen, the area and perimeter of the polygon were computed again, and various ID examples are presented in Table 2. According to both tables, even if the overall area to be covered increases when using the convex hull approach, the perimeter that the UAV must travel is lowered while simultaneously lowering the number of vertices. This might allow the UAVs to travel as little as possible inside the research region.

Table 2. Convex Hull Geometries.

ID	Area	Perimeter
1	0.000246	0.078531
2	0.000367	0.084524
3	0.000489	0.092529
4	0.000612	0.097820
5	0.000682	0.099317

After the geometries are determined, we must apply the Delaunay Triangulation to the acquired area and compute the centroids for each division. The centroids serve as coverage points, which are then utilized to pose the issue as a TSP. We calculated the distance matrix for the centroids before solving the TSP, and a few of them are shown in Table 3. After getting the centroids, we used the Bat optimization technique to solve TSP and find the best route. Figure 7 depicts the optimum route achieved by using the suggested approach.

Table 3. Centroids - Distance Matrix.

InputID	TargetID	Distance
22	9	2165.882
22	2	1566.086
22	7	1862.168
22	5	1694.044
22	19	180.407
22	19	275.138
22	20	271.170
19	16	438.149
19	8	1894.559
19	22	325.767

The ideal route is also compared to other known techniques such as Ant Colony and Particle Swarm optimization. Table 4 summarizes the findings for one particular polygon. We can derive the tour length and duration of travel by UAV from the table for the region partitioned and the method employed. The same procedure is followed with four more polygons, and the entire comparison analysis graph is displayed in Fig. 8.

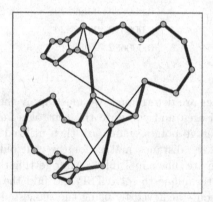

Fig. 7. Optimal Path - Bat Optimization Algorithm.

Table 4. Comparison of paths obtained using various algorithms.

Algorithm	Path Length	Execution time
Ant Colony Optimization	0.532	0.68
Particle Swarm Optimization	0.533	0.72
Bat Optimization	0.530	0.65

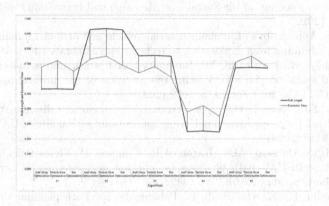

Fig. 8. Optimal paths obtained for various polygons.

5   Conclusion

Unmanned aerial vehicle (UAV) path planning is a significant job in the area of robotics and autonomous systems. It entails creating a safe, practical, and optimum flight route for unmanned aerial vehicles (UAVs) to fly through new terrain while avoiding obstacles. The research provides a unique route-planning technique for intelligent unmanned aerial vehicles (UAVs). The method uses Delaunay Triangulation, the Traveling Salesperson Problem (TSP), and the Bat Optimization Algorithm (BOA) to determine an ideal flight route for unmanned aerial vehicles (UAVs) in dynamic situations while avoiding obstacles. The performance of the proposed method was verified by comparing it to other existing algorithms, and the findings demonstrate that it beats other algorithms in terms of total computing time and route length traveled by the UAV. Moreover, the suggested method can swiftly discover a new route as the environment changes. Because of the large savings in computing time and route length followed by the UAV, the technique provides a possible solution to path planning challenges for intelligent UAVs. According to the analysis findings, the suggested approach outperforms other current algorithms in terms of calculation time by roughly 0.03 s. Moreover, as compared to existing methods, the suggested technique considerably improves the route traveled by the UAV by reducing it by around 0.02 cm. Further study may be conducted to investigate its applicability in other connected sectors and improve its performance.

References

1. Gaffey, C., Bhardwaj, A.: Applications of unmanned aerial vehicles in cryosphere: latest advances and prospects. Remote Sens. **12**(6), 948 (2020)
2. Fan, B., Li, Y., Zhang, R., Fu, Q.: Review on the technological development and application of UAV systems. Chin. J. Electron. **29**(2), 199–207 (2020)
3. Muchiri, G.N., Kimathi, S.: A review of applications and potential applications of UAV. In: Proceedings of the Sustainable Research and Innovation Conference, pp. 280–283 (2022)
4. Choi, Y.: Applications of unmanned aerial vehicle and artificial intelligence technologies in mining from exploration to reclamation. Minerals **13**(3), 382 (2023)
5. McLaughlin, R., Perhinschi, M.: Immunity-based sensor and actuator abnormal condition evaluation on a UAV using the partition of the universe approach. In: AIAA SCITECH 2023 Forum, p. 1662 (2023)
6. Acevedo, J.J., Maza, I., Ollero, A., Arrue, B.C.: An efficient distributed area division method for cooperative monitoring applications with multiple UAVs. Sensors **20**(12), 3448 (2020)
7. Luna, M.A., Ale Isaac, M.S., Ragab, A.R., Campoy, P., Flores Peña, P., Molina, M.: Fast multi-UAV path planning for optimal area coverage in aerial sensing applications. Sensors **22**(6), 2297 (2022)
8. Priyadarsini, P.L.K., Gadupudi, P., Paladugula, K.: Path planning of intelligent UAVs using computational geometric techniques. In: 2022 International Conference on Electronic Systems and Intelligent Computing (ICESIC), pp. 190–193. IEEE (2022)
9. Puente-Castro, A., Rivero, D., Pazos, A., Fernandez-Blanco, E.: A review of artificial intelligence applied to path planning in UAV swarms. Neural Comput. Appl. 1–18 (2022)
10. Akshya, J., Priyadarsini, P.L.K.: Graph-based path planning for intelligent UAVs in area coverage applications. J. Intell. Fuzzy Syst. **39**(6), 8191–8203 (2020)
11. Priyadarsini, P.L.K.: Area partitioning by intelligent UAVs for effective path planning using Evolutionary algorithms. In: 2021 International Conference on Computer Communication and Informatics (ICCCI), pp. 1–6. IEEE (2021)
12. Gul, F., Rahiman, W., Alhady, S.S.N., Ali, A., Mir, I., Jalil, A.: Meta-heuristic approach for solving multi-objective path planning for autonomous guided robot using PSO-GWO optimization algorithm with evolutionary programming. J. Ambient. Intell. Humaniz. Comput. **12**, 7873–7890 (2021)
13. Song, B., Wang, Z., Zou, L.: An improved PSO algorithm for smooth path planning of mobile robots using continuous high-degree Bezier curve. Appl. Soft Comput. **100**, 106960 (2021)
14. Yahia, H.S., Mohammed, A.S.: Path planning optimization in unmanned aerial vehicles using meta-heuristic algorithms: a systematic review. Environ. Monit. Assess. **195**(1), 30 (2023)
15. Malyshev, D., Cherkasov, V., Rybak, L., Diveev, A.: Synthesis of trajectory planning algorithms using evolutionary optimization algorithms. In: Olenev, N., Evtushenko, Y., Jaćimović, M., Khachay, M., Malkova, V., Pospelov, I. (eds.) OPTIMA 2022. CCIS, vol. 1739, pp. 153–167. Springer, Cham (2022). https://doi.org/10.1007/978-3-031-22990-9_11

A Comparative Analysis of Data Backup and Network Consistency in Cluster-Base Wireless Sensor Network Protocols

Mujahid Tabassum[1][✉], Tripti Sharma[2], Saju Mohanan[1], and Isah A. Lawal[1]

[1] Noroff School of Technology and Digital Media (Noroff Accelerate), Noroff University College, 4612 Kristiansand, Norway
{mujahid.tabassum,saju.mohanan,isah.lawal}@noroff.no

[2] Department of Information Technology, University of Technology and Applied Sciences, 133 Muscat, Oman
tripti.sharma@utas.edu.om

Abstract. The fast growth of information and communication technologies (ICTs) has had a big impact and changed a lot of businesses. Wireless sensor networks (WSN) are used in many industries for mobility, scalability, reliability, smart monitoring, and management. Utilization of modern equipment in the agriculture industry is essentially required to boost people's income and create a positive impact on their social lives. WSNs comprise various self-designed devices that gather extensive data from the environment. Numerous techniques have been developed recently to enhance WSN productivity in various industries. WSN has become one of the emerging innovations. In wireless sensor networks, the biggest problems are data loss, node failure, and the need to use more energy to make the sensor nodes last longer. To overcome these limitations, in this research, a novel Tri-Head Fuzzy C Multipath Routing (THFCMR) protocol is proposed for WSNs. The proposed THFCMR approach is designed by a tri-head static fuzzy C means clustering with hybrid energy-efficient distributed clustering (HEED) integrated with a hybrid energy-efficient multipath routing protocol (HEEMP) approach. This paper focused on reviewing the cluster base protocol's usability and weaknesses, along with a proposal for a solution to enhance network consistency and data availability in WSNs.

Keywords: Wireless Sensor Network · Data Loss · Data Reliability · Data Storage · Machine Learning

1 Introduction

Wireless Sensor Network (WSN) demand has increased over the last two decades due to its flexibility and convenience. Despite their differences, all sensor networks share a few essential qualities. These sensor nodes are made up of low-cost components that communicate with one another using low-power wireless technology. They can also be outfitted with traditional instruments to keep an eye on the goings-on in the physical

S. Kadry and R. Prasath (Eds.): MIKE 2023, LNAI 13924, pp. 185–192, 2023.
https://doi.org/10.1007/978-3-031-44084-7_18

world (Garg et al., 2021). These sensor units collaborate to accomplish a common goal and can be distributed in any order across the detecting region. Compared to conventional methods, wireless sensing has several advantages, including lower costs for implementation and maintenance, increased versatility, a wider variety of applications, and larger installations.

WSNs collect and send data to another sensor node, the Cluster Head (CH), or Base Station (BS). Thus, each sensor node has a sensing, power, CPU, memory, and transmission unit. Sensor nodes turn analog data into digital data for wireless transmission (Keerthana et al., 2020). Sensor nodes store and analyze data using RAM and flash memory (Sharma et al., 2021). Data availability means all data is sent to BS correctly and in full, with no node failures or losses (Xu et al., 2022). Network consistency and data availability are critical aspects of any wired or wireless network. WSN's most significant challenges are preserving data availability and maintaining network consistency upon node or network failure (Choudhary and Goyal, 2022). WSN has critical characteristics, including less power and limited storage capacity due to its low-cost infrastructure. These constraints create data loss and network failure problems in WSN applications, especially in healthcare and agriculture. Many algorithms have been proposed to tackle these issues. However, these schemes consume more battery power and increase processing capabilities because of multipath communication and multi-device backup services. As a result, overall network life reduces, and the sensor dies faster.

A reliable, efficient WSN solution that provides network consistency and data availability might save energy, resources, and memory space and prolong the network's lifetime. This study develops a clustering multipath routing algorithm to improve WSN productivity by improving network consistency and data dependability. In this research, a novel Tri-Head Fuzzy C Means Clustering Multipath Routing (THFCMR) is designed with a fault-tolerant clustering-based multipath algorithm to enrich data availability by securely retrieving missed data, reducing cluster redundancy in data transformation, traffic, and energy balance, and extending WSN lifespan. The algorithm lowered transmission energy and end-to-end latency. Controlling data redundancy, providing data backup cluster nodes, sustaining network performance after node failure, and providing a coherent and stable communication platform optimises network performance.

2 Problem Statement

In a WSN, data is crucial; when the communication connection is lost between the Sensor Nodes (SN) and CHs or between the CHs and BS, data is lost, and packets are retransmitted. Packet retransmissions consume extra energy, which leads to faster and higher nodes dying within WSN clusters. This phenomenon reduced the overall network lifespan of WSNs. Sensor nodes die due to battery depletion, which also causes node failure and affects the WSN network's performance and reliability. Data availability and accessibility are lost when communication between sensor nodes and CHs, or between CH and BS, breaks down since sensor nodes and CHs are only in charge of delivering data, not storing it (Rehman et al., 2022). As a result, the BS stores transmitted data and makes it accessible to the end user. Network consistency controls data loss and accuracy (Telecomworld, 2022). WSNs are primarily deployed in outdoor environments where they might be subject to harsh terrain and weather constraints such as rain and heavy wind (Rehman et al., 2022). These circumstances create different kinds of faults on the network, such as hardware failure, interference, data loss, packet loss, path loss, congestion, and others (Suma & Harsoor, 2022). When one or more nodes die, or CHs and BS are not working, the network consistency and availability are reduced throughout the network regarding data loss, end-to-end delay, packet delivery, and packet drop ratios. Several studies (Ibrahim et al., 2021) have been done to tackle these issues. However, most of these studies focus on specific issues and have several limitations (Aiswariya et al., 2018). In WSNs, these issues are the norm rather than the exception. However, data loss and node failure are extremely important fields of study (Gnanavel et al., 2022). For example, many sensors are deployed in the paddy field over a large area for agricultural applications. These sensors continuously sense and forward the data to the BS. If any sensor node or CH fails due to unexpected problems, it should not affect the overall network performance regarding data availability. WSN should remain capable of achieving work efficiency and productivity under all circumstances.

3 Literature Review

The problem with data availability in WSNs is caused by the way the hardware is built and how it talks to each other. These limitations include a limited capacity for calculating and storing low power and a limited storage capacity. Currently, various hierarchical routing protocols, including Low Energy Adaptive Clustering Hierarchy (LEACH), Power-efficient Gathering in Sensor Information Systems (PEGASIS), HEED, and others, are available but they do not provide data backup and network consistency on node or network failure (Bhattacharya, 2020; Fanian and Rafsanjani, 2019; Zhu and Wei, 2019).

Angelin and Kiruthika (2019) propose a hybrid energy-efficient distributed clustering to extend network lifetime by dispersing energy usage. The findings showed that the HEED protocol achieves the best energy management, the longest network lifetime, and the best implementation of security due to a better load balancing model implemented by the protocol due to multi-hop routing and better spreading of cluster heads throughout the network. The limitation of this approach was that it did not provide any data backup due to node failure. Hence, considerable data loss occurs that cannot be retrieved further.

A revolutionary trust-aware localized routing and class-based dynamic encryption approach for secured data transfer were presented by Hema Kumar et al., 2021. Before measuring the trusted data forwarding support (TDFS) value, the path to the destination was determined, and criteria like the count of successful transmissions, the history of data transfers, and the counts of repeated transmissions of the same packets were taken into consideration. Only one neighbor was chosen to receive data per measured TDFS. Additionally, they created a blockchain for data transmission security, storing encrypted data in each block, resulting in increased data security. However, more time is required for data transfer, and more untrusted nodes are present in the network, resulting in data loss.

Shende and Sonavane (2020) created the E-AMRP CrowWhale-ETR, which combined the CSA and WOA. To determine the best paths based on the nodes' trust and energy, the CWOA was established. The suggested work thus obtained the highest detection rate, the shortest latency, the highest energy, and the highest throughput. The end-to-end error produced by this strategy was relatively high.

Borkar et al., (2019) presented an adaptive chicken swarm optimization method for the best CH selection. By suggesting innovative IDS with an adaptive SVM classification, malicious sensor nodes were found, which diminishes the failure in the node and thus produces a high delivery rate. But this model produces less security and very low transmission compared to other existing approaches, and the delay during transmission is also high.

Gobinath and Tamilarasi, (2020) described a system that successfully detects the failure node before transferring data in the network. The involvement of low energy nodes is avoided in the early stages thanks to the robust failure detection and congestion aware routing (RFDCAR) protocol. However, route switching could increase electricity costs. As a result of moving to another way, the network path is degraded or congested.

By combining the ideas from the Taylor series and the modified Cat Salp Swarm Algorithm (C-SSA), Vinitha and Rukmini (2022) created the Taylor CSSA. Through effective CH selection and data transfer, multi-hop routing was accomplished. The proposed Taylor C-SSA, which chooses the best hops, has guaranteed the network's energy efficiency. The disadvantage of this model was its very low throughput and fewer nodes were alive, reducing the network's lifetime.

4   Proposed Research Methodology

WSN algorithm must consider data loss, node failure, storing area, computing power, and battery power. Data backup, an average correct data packet transmitted, data loss, end-to-end delay, normalized overheads, number of alive nodes, packet drop ratio, and leftover energy metrics can assess WSN data availability and network consistency. In the first step, a quantitative analysis is performed by reading related research and with some experiments. In this research paper we have simulated experiments of the HEED to understand the research gap, benchmarking, and limitations of existing techniques in term of data loss, node failure, and network lifespan aspects. In the second phase, we have proposed Tri-Head Fuzzy C Means Clustering Multipath Routing (THFCMR). The suggested THFCMR technique uses tri-head static fuzzy c means clustering with a HEED approach and a HEEMP approach.

Table 1 shows the metrics used to measure performance of THFCMR method. These metrics help to evaluate the proposed approach with the existing approach to prove that the proposed approach produces less data loss and node failure and extends the lifespan of the WSN.

4.1   Performance Evaluation and Benchmarking

To observe the WSN protocol's communication pattern and for benchmarking, we have utilized the HEED protocol as a sample and simulated it in MATLAB. This quantitative analysis was performed to understand the limitations regarding data loss and backup, node failure, and network energy consumption. HEED was introduced in 2004 (Ullah,

Table 1. Performance Measurement Metrics.

Evaluation Objectives	Measurement Metrics
Data Backup	Data backup, Average correct data packets transferred, Data loss, End-to-end delay, Normalized overhead
Node Failure	Packet Delivery Ratio, Throughput, Routing Overheads
Network Lifetime	No of Alive nodes, energy consumption, packet drop ratio, residual energy

2020) as a multi-hop WSN clustering technique that provides energy-efficient clustering routing with explicit energy consideration. In the experiment, 100 nodes were uniformly dispersed into a field. The limitation of this approach is that CH chooses at random. The worst-case situation is that cluster head nodes are not uniformly distributed, which will affect data gathering and cause data loss due to node failure. HEED claims that arbitrary CH selection increases transmission overhead. Member nodes attempt to transmit with their CH, while CHs transmit with other CHs or the base station. Reconstructing clusters requires energy from periodic CH rotation or selection. This protocol randomly selects the CH, which might cause communication overhead issues that waste energy and harm quality of service. A network's lifespan is shortened by frequent CH changes during selection, which requires more energy to reconstruct the cluster. It also does not offer any node failure or data backup solutions.

4.2 Proposed Research Methodology

The proposed Tri-head fuzzy c means clustering multipath routing is developed to select the tri-cluster heads and aggregate the data from the normal sensor nodes. The suggested THFCMR technique uses tri-head static fuzzy c means clustering with a HEED coupled with a HEEMP approach. Initially, a group of sensor nodes is grouped using a fuzzy c-means technique. Then, three node Cluster heads (ACH, TCH, and BCH) are selected.

   The main node initiates data transmission between the base stations by a hybrid energy-efficient multi-path routing protocol. In this method, the backup cluster head (BCH) aids in minimizing data loss brought on by node failure. The backup cluster head is designed to hold data in local storage for data retransmission during link failure to obtain reliable data transmission. Finally, routing is important in WSNs. The hybrid energy-efficient multi-path routing protocol used in WSNs creates a channel between the source and sink for data transmission. By determining the best route for data transmission, this method reduces data loss and lengthens the network's life in the paddy field.

Clustering
One of the best ways to understand and organize information into a cluster is to use fuzzy C-Means grouping. As the name implies, the logic of the FCM clustering algorithm is that one node can combine with more than one group.

Backup Cluster Head
Each node transfers its perceived data to the sink during this phase. While the rest of the

nodes use the following routing strategy, Direct_Set nodes transfer their data directly to the sink. Each node has a Neigh_Table that is filled at this point. Neigh_Table contains the ID of the neighbor node (Neigh_ID), the position of the neighbor node (Neigh_Pos), the residual energy of the neighbor node (Neigh_Res), and the cluster head of the neighbor (Neigh_CH). Each cluster member iterates over its Neigh_Table, shortlisting those of its neighbors whose CH IDs in the fourth field of the Neigh_Table match its own. The ProgressNodeGroup (PNG) stores all the selected candidates' nodes between themselves and TCH (by comparing their locations to their locations). The Neighbor Group (NG) of node i comprises nodes m, j, and k. PNG is made up of all NG nodes that are present between node i and the sink (nodes j and k).

Optimal Path

The most frequent fault detection method is monitoring node performance with a backup cluster head. Whenever data is sent from a cluster member to the aggregation cluster head, the backup node monitors the nodes' functionality using the beacon message it receives from the CH. The energy needed to examine the aggregation cluster head performance is low because beacon packs are considerably smaller than data packs. If the aggregation cluster head does not respond after three rounds, the backup cluster head indicates that the aggregation cluster head's error has been found. Following that, the backup cluster head is chosen as a new aggregation cluster head and can pick a backup from NCHs (Non-Cluster Head nodes). The cluster members' data is then sent to the new aggregation cluster head. Additionally, anytime data is communicated from the aggregation cluster head to the transmitting cluster head, a copy of it is maintained in the backup cluster head so that, in the event of a transmitting cluster head failure, the backup cluster head will function as the transmitting cluster head and send the data to the base station. In contrast, a new transmitting cluster head may be chosen from non-cluster head nodes, negating the need to gather the data again.

5 Future Work and Conclusion

WSNs can self-manage, self-configure, self-diagnosis, and self-healing, making them an excellent choice for intelligent agriculture monitoring. WSN's data handling is essential when any node fails due to an unexplained cause. The proposed THFCMR is developed to select the cluster head, aggregate data, and backup data from the normal sensor nodes. In a WSN node failure and data loss are the main issues in the WSN, which creates higher packet drop, delay, and energy consumption during communication. By reducing the amount of data lost during backups, the backup cluster head helps to prevent node failure and data loss. Thus, the proposed approach diminishes the data loss and node failure and extends the WSN network's life span. The proposed THFCMR method's performance is analyzed using throughput, delivery rate, number of nodes alive, drop rate, end-to-end delay, energy consumption, and overhead ratio in the next section. The proposed tri-head fuzzy c means clustering multipath routing may enhance WSN performance in various situations, regulate network consistency, and conduct data aggregation and availability upon any node loss. The WSN network lifespan and performance might be improved by choosing a backup cluster head that controls packet retransmission and node failure.

In further work, we will simulate the proposed algorithm to evaluate it performance and compare with related protocols for benchmarking.

References

Angelin, T. S., Kiruthika, R.: Hybrid energy-efficient distributed clustering approach for WSN using linear data acquisition algorithm (2019)

Aiswariya, S., Rani, V.J., Suseela, S.: Challenges, technologies and components of wireless sensor networks. Int. J. Eng. Res. Technol. (IJERT) **6**, 1–5 (2018)

Borkar, G.M., Patil, L.H., Dalgade, D., Hutke, A.: A novel clustering approach and adaptive SVM classifier for intrusion detection in WSN: a data mining concept. Sustain. Comput.: Inform. Syst. **23**, 120–135 (2019)

Bhattacharya, M.: A survey on importance of routing protocol in WSN. Journal of Contemporary Issues in Business and Government, vol. 26(02). (2020)

Choudhary, M., Goyal, N.: A rendezvous point-based data gathering in underwater wireless sensor networks for monitoring applications. Int. J. Commu. Syst. **35**(6), e5078 (2022)

Fanian, F., Rafsanjani, M.K.: Cluster-based routing protocols in wireless sensor networks: a survey based on methodology. J. Netw. Comput. Appl. **142**, 111–142 (2019)

Garg, R.K., Bhola, J., Soni, S.K.: Healthcare monitoring of mountaineers by low power wireless sensor networks. Inform. Med. Unlocked **27**, 100775 (2021)

Gnanavel, S., et al.: Analysis of fault classifiers to detect the faults and node failures in a wireless sensor network. Electronics **11**(10), 1609 (2022)

Gobinath, T., Tamilarasi, A.: RFDCAR: robust failure node detection and dynamic congestion aware routing with network coding technique for wireless sensor network. Peer-to-Peer Netw. Appl. **13**, 2053–2064 (2020)

Hema Kumar, M., Mohanraj, V., Suresh, Y., Senthilkumar, J., Nagalalli, G.: Trust aware localized routing and class based dynamic block chain encryption scheme for improved security in WSN. J. Ambient. Intell. Humaniz. Comput. **12**, 5287–5295 (2021)

Ibrahim, D.S., Mahdi, A.F., Yas, Q.M.: Challenges and issues for wireless sensor networks: a survey. J. Glob. Sci. Res. **6**(1), 1079–1097 (2021)

Keerthana, K., Aasha Nandhini, S., Radha, S.: Cyber physical systems for healthcare applications using compressive sensing. In: Compressive Sensing in Healthcare, pp. 145–164. Elsevier (2020). https://doi.org/10.1016/B978-0-12-821247-9.00013-5

Telecomworld: Introduction to Data Communication, Telecomworld. http://www.telecomworld101.com/Intro2dcRev2/page26.html. Accessed 01 June 2022

Rehman, A., et al.: A revisit of internet of things technologies for monitoring and control strategies in smart agriculture. Agronomy **12**(1), 127 (2022)

Sharma, N., Kaushik, I., Agarwal, V. K., Bhushan, B., Khamparia, A.: Attacks and security measures in wireless sensor network. In: Intelligent Data Analytics for Terror Threat Prediction: Architectures, Methodologies, Techniques and Applications, pp. 237–268. (2021)

Suma, S., Harsoor, B.: Detection of malicious activity for mobile nodes to avoid congestion in Wireless Sensor Network. In: 2022 IEEE fourth international conference on advances in electronics, computers and communications (ICAECC), pp. 1–6. IEEE (2022, January)

Shende, D.K., Sonavane, S.S.: CrowWhale-ETR: CrowWhale optimization algorithm for energy and trust aware multicast routing in WSN for IoT applications. Wireless Netw. **26**, 4011–4029 (2020)

Tabassum, M., Zen, K.: Evaluation and improvement of data availability in WSNs cluster base routing protocol. J. Telecommun., Electron. Comput. Eng. **9**(2–9), 111–116 (2017)

Vinitha, A., Rukmini, M.S.S.: Secure and energy aware multi-hop routing protocol in WSN using Taylor-based hybrid optimization algorithm. J. King Saud Univ.-Comput. Inform. Sci. **34**(5), 1857–1868 (2022)

Xu, X., Tang, J., Xiang, H.: Data transmission reliability analysis of wireless sensor networks for social network optimization. J. Sens. **2022**, 1–12 (2022)

Zhu, F., Wei, J.: An energy-efficient unequal clustering routing protocol for wireless sensor networks. Int. J. Distrib. Sens. Netw. **15**(9), 1550147719879384 (2019)

Ullah, Z.: A survey on hybrid, energy efficient and distributed (HEED) based energy efficient clustering protocols for wireless sensor networks. Wireless Pers. Commun. **112**(4), 2685–2713 (2020)

Development of IoT-Healthcare Model
for Gastric Cancer from Pathological Images

Mohammad Riyaz Belgaum[1,2] , Shaik Maryam Momina[1](✉) ,
L. Nousheen Farhath[1] , K. Nikhitha[1] , and K. Naga Jyothi[1]

[1] Department of Computer Science and Engineering, G. Pullaiah College of Engineering and
Technology, Kurnool, India
riyaz@gpcet.ac.in, mominaskm@gmail.com
[2] Faculty of Computing and Informatics, Multimedia University, Cyberjaya, Malaysia

Abstract. The diagnosis of stomach cancer automatically in digital pathology
images is a difficult problem. Gastric cancer (GC) detection and pathological
study can be greatly aided by precise region-by-region segmentation. On a technical level, this issue is complicated by the fact that malignant zones might be any
size or shape and have fuzzy boundaries. The research employs a deep learning-
based approach and integrates many bespoke modules to cope with these issues.
The channel refinement model is the attentional actor on the chin channel. While
implementing the feature channel, the learnt channel weight can be used to eliminate unnecessary features. Calibration is essential to improve classification precision. The results of channel recalibration can be improved with the help of a
re-calibration (MSCR) model. The top pooling layer of the network is where
the multiscale attributes are sent. The outcomes of channel recalibration may
be enhanced by using the channel weights found at various scales as input to
the next channel recalibration perfect. Our unique gastric cancer segmentation
dataset, carefully glossed down to the pixel level by medical authorities, is used
for extensive experimental comparisons. The numerical comparisons with other
approaches show that our strategy is superior.

Keywords: Deep learning · Digital pathology image · Gastric cancer detection ·
Multiscale channel · Pooling Layer · Recalibration

1 Introduction

The Industrial Internet of Things (IIoT) is one of the most rapidly expanding net-works
right now that can collect and exchange huge volumes of data utilizing sensors in a
healthcare context [1]. Medical IoT, IoHT, or IoMT refers to the Internet of Things used
in medicine and healthcare and is often referred to as a "expert application" [2]. The
Internet of Medical Things (IoMT) is a framework for interconnecting digital resources
in the healthcare industry. It is used to evaluate the physical properties of sensor nodes
that collect data from the patient's body using smart portable devices. Integrating AI
approaches provides quick and flexible analysis and diagnoses of medical data, while

© The Author(s), under exclusive license to Springer Nature Switzerland AG 2023
S. Kadry and R. Prasath (Eds.): MIKE 2023, LNAI 13924, pp. 193–201, 2023.
https://doi.org/10.1007/978-3-031-44084-7_19

IoMT paves the way for wireless and distant devices to connect securely over the Internet. Network topology, energy transfer, and processing power are just a few of the unknowns that IoT devices must deal with while sending data over the cloud [3].

The 5-year survival rate for people diagnosed with GC at an early stage can surpass 90%. However, the 5-year existence rate drops below 30% [4] because almost half of patients with GC have previously progressed to the advanced phase at the time of diagnosis. Pathologists typically perform this laborious and time-consuming procedure. Diagnostic accuracy is being significantly impacted on a global scale by a serious scarcity of pathologists and a large workload of diagnostics [5, 6]. Thus, a novel method must be developed to rapidly and precisely detect GC from diseased images. To stratify and select those who may benefit from adjuvant treatment, however, physical histological analysis of tumors specimens is currently not reliable enough [7]. The development of concise and reliable approaches to predict overall GC is, therefore, urgently needed in order to aid in the creation of tailored therapy plans and the maximization of their advantages.

Slowly but surely, oncologists have begun to take notice of deep learning. The discipline of cancer has benefited greatly from the advancements made by deep learning, which have been shown to be superior to those made by traditional machine learning methods. For many picture interpretation tasks, the (CNN) has proven to be the superior deep learning method. Several advancements have been achieved with the use of AI to detect tumors and forecast the prognosis of GC based on photographs of the disease in its various stages [8] collected from the IoT enabled microscopes and image sensors. To make the produced AI models more useful to clinical practice, we first identified a number of obstacles that needed to be solved. For example, a substantial sample size from many centres should be obtainable for training and verifying the proposed model to assure the robustness. The generated model needs to be practical for use in clinical practice while also being easy to use by those without an AI background or in regions with a less developed economy [9].

In this research, we built deep learning-based models using the refinement model as a starting point, and we dubbed them the multiscale channel squeeze-and-excitation model. The residual sections of the paper are structured as follows: The correlated texts are presented in Sect. 2. The suggested model and its experimental analysis are described briefly in Sects. 3 and 4 discussing the results. Finally, Sect. 5 presents conclusion and limitations.

2 Related Works

Whole-slide images (WSIs) of human sections are projected by Hu et al. [10] to highlight diagnostically relevant psychoneuroimmunology (PNI) regions. This framework is based on multi-task deep learning. The suggested system completes the task of recognizing PNI while also segmenting the gastric cancer region using a neural detection model and a PNI decision-making module.

Based on the attention mechanism, Guo et al. [11] proposed a compact micro fuzzy pathology detection algorithm; the YOLOv5 is enhanced under compact and micro fuzzy scenarios of cancer cell detection across the board in digital pathology; this algorithm is evaluated on the gastric cancer slice dataset. Test results show an F1 score of 0.616

and a mAP of 0.611 for the deep blur scenario, indicating that it can serve as a decision support for clinical judgement.

Ma et al. [12] offer a deep learning (DL)-based automatic system for diagnosing early gastric cancer (EGC). This work specifically constructs different DL architectures to enable the automatic interpretation of EGC pictures using a novel an-notated dataset obtained from a single-center. The experimental results on the submitted dataset revealed the potential uses of DL in helping of 0.64 for the included classification and segmentation assignments.

Based on ShuffleNetV2, Fu et al. [13] present a multi-dimensional convolutional lightweight network called MCLNet. The computational complexity, memory footprint, and GPU parallelism of the ShuffleNetV2 model are all quite modest. The problem is that ShuffleNetV2 only uses two-dimensional convolutions, so the amount of information recovered from the data is low. To compensate for the absence of 2-D in global feature extraction, employing 1-D increases information transmission between channels and enhances the information.

Using gastric cancer (GC) tissue slides, Lee et al. [14] attempted a fully automated classification of micro satellite instability (MSI) status. The (ROC) curve areas under the curves (AUCs) (TCGA) were 0.893 and 0.902, respectively. The 0.874 AUC achieved by the classifier on the external validation Asian FFPE cohort shows that the classifier trained with TCGA FFPE tissues is effective. It appears that DL has the potential to autonomously learn the most effective features for determining MSI status in GC tissue slides. This research proved that a DL-based MSI classifier could be useful for preliminary case screening.

For studies on gastric tumor image segmentation, Wang et al. [15] propose a stomach cancer lesion dataset. To get multi granular edge information and refinement, an encoder stage-specific boundary extraction refinement module is presented. The next step is to construct a selective fusion module that can be used to combine features from specific phases. The experimental results demonstrate that the projected method outdoes existing approaches on the CVC-Clinic DB and Kvasir-SEG datasets, with an accuracy of 92.3% and a similarity coefficient (DICE) of 86.9%.

3   Proposed System

3.1   Dataset

Our pathology pictures are taken from actual patient records. We have taken 500 pathological images at a resolution of 2048 by 2048 under an optical magnification of 20. Each picture was extracted from a larger slide of gastric tissue showing typical malignant spots in the stomach. Some samples from our dataset [16] are displayed in Fig. 1.

Model parameters were trained using the training set, and the testing set was used to validate the models' predictions. There will be 350 training images and 150 test photographs used. Each image in the training set was cropped from its original 2048 by 2048 resolution to four 1024 by 1024 patches before being fed into the networks. By subtracting the dataset's mean pixel value and dividing by the standard deviation, we have created uniform image patches.

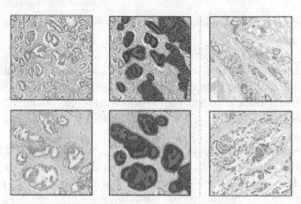

Fig. 1. Gastric cancer segmentation dataset, featuring six sample image-label pairings. (Color figure online)

The areas covered in yellow have cancer in the above Fig. 1. Size differences, hazy borders, and pliable characters are all made very obvious in these pictures. Five hundred abnormal photos with careful annotations make up our dataset.

3.2 Image Preprocessing

Excessive abnormalities from the staining method typically led to noise in the collected images. The following procedure was used during the preprocessing stage.

1. Scaling down all deep learning models have a strict need that all input images have the same size computationally. Hence, the image was scaled down to a size of 224 × 224 pixels in order to shorten the processing time.
2. Noises like additive, random, impulsive, and multiplicative noises removal are crucial. Gaussian noise, pepper noise, speckle noise, and Poisson noise all show up frequently in medical pictures. In this study, a median filter was employed to get rid of the salt and pepper noise over the entire slide image.
3. Third, normalizing stains is a crucial step in the whole slide image (WSI) preprocessing phase of digital pathology. This research made advantage of the well-known Macenko stain normalization technique often applied on histopathology slides.
4. Fourth, data augmentation is a technique for greatly expanding the types and quantities of data used to train models. To boost the amount of data without changing the appearance of the photographs, we rotated them by 90°, 180°, 270°, flipped them horizontally, and flipped them vertically. This resulted in a sixfold increase in the amount of information collected.

A stratified cross-validation technique was used to divide the raw data obtained in-to two equal parts: 80% for training, for testing. The sum of images in each class increases from 35 for Margin Negative to 49 for Margin Positive after 6 augmentation (with 90°, 180°, 270°, horizontal flip), omitting the testing data set, which must be the unique dataset.

Channel Recalibration Model: To teach the network to pay attention only to relevant information, the channel recalibration model assigns a weight to each input channel. The input feature U is compressed in channel order by the channel recalibration model.

$$z_c = F_{sq}(u_c) = \frac{1}{H \times W} \sum_{i=1}^{H} \sum_{j=1}^{W} u_c(i, j) \qquad (1)$$

Once all of the channels in the input feature have been extruded, their respective weights can be calculated by activating the extrusion result using the subsequent formula.

$$s = F_{ex}(z, W) = \sigma(W_2\delta(W_1 z)) \qquad (2)$$

where s is the feature channel's weight; F ex (*,*) is the excitation function; z is the feature's extruded result; (*) is the sigmoid function; (*) is the ReLU function [17], and (*) and (*) are the sigmoid and ReLU functions, respectively. The weights of the first fully associated layer (FC) are W1 and the second fully connected layer (FC) is W2.

Excitation's first fully connected layer takes the sum of feature channels as a function of c and converts it to c/r, where r is the density ratio, with the result that only non-zero values are retained at output following the ReLU function. To maintain parity with the sum of channels in the input feature, the second completely connected layer brings back the number of feature channels to c. The sigmoid function is used to derive the final weight, with a range of 01.0 possible.

$$\ddot{x}_c = F_{scale}(u_c, s_c) = s_c u_c \qquad (3)$$

To summarise the above formula: It is the weight of the cth channel in the input feature, and x_c is the channel's output characteristic following channel feature recalibration. The features of a given channel are multiplied by the channel weight, where F scale is a scaling function. By multiplying the eye of a given channel by the associated channel weight, Eq. (3) implements the recalibration of the feature channel, which in turn suppresses the characteristics that are irrelevant to the classification result, leading to higher precision in the classification.

Proposed Model Description: Channel recalibration models can be better-quality by using multiscale features instead of single-scale ones, and by combining the weights learnt at diverse scales to get the final feature channel weights. CNN with multiscale features are frequently used for target identification and recognition as well as picture semantic segmentation. Using feature information at various scales helps increase the precision of the final result. The input sent to the max pooling layer, where they are scaled differently using a pooling kernel size of 2. The feature channel weights across all scales are combined via maximum-value splicing.

Additive Fusion: In order to do multiscale channel recalibration, the input features are increased by the obtained weights in the order of the respective channels, and the result is the channel weight produced by the additive fusion method using the following:

$$U'_{2way\_add} = (S_{c0} + S_{c1})U_{s0} \qquad (4)$$

In the above formula U'_{2way_add} is used to accomplish multiscale channel recalibration, the channel weight acquired by the preservative fusion technique 2 is the sum of the channel weights for the two feature scales, and this weight is then multiplied by the input topographies in the order of the respective channels. What follows is a procedure:

Maximum Fusion: Maximum fusion, as contrast to additive fusion, chooses individual channels. The weight of the channel is equal to the sum of the highest values on each scale. The current procedure for recalibrating multiscale channels goes as follows:

$$U'_{2way\_max} = (S_{c0} + S_{c1})U_{s0} \tag{5}$$

Multiscale channel recalibration with extreme scale is represented as U'_{2way_max} where max(,) is the maximum function. The channel weight is the sum of the channel weights under each scale, sorted in order of preference.

Splicing Fusion: When fusion technique first maps the channel weights from each scale onto a final scale through a convolution layer, and then combines the results. If the batch picture size and the sum of channels in the input feature is C, then the channel weight size is NC11. Depending on which splicing coordinate axis is chosen, two distinct types of splicing fusion implementations can be identified.

(a) Splicing coordinates, or cat1, are located along the second co-ordinate axis (axis1). The current expression for the process of recalibration of multiscale channels is

$$U'_{2way\_cat1} = F_{conv1}(S_{c\_cat1})U_{s0} \tag{6}$$

where N is the number of input channels, C is the sum of output channels, F conv1 (*) is the mapping function for the convolutional layer conv1, the output of multiscale channel recalibration achieved by splicing and fusing cat1 at 2 scales.

(b) Splicing coordinates (cat2) are located along the third coordinate axis (axis3). The current expression for the process of recalibration of multiscale channels is

$$U'_{2way\_cat2} = F_{conv2}(S_{c\_cat2})U_{s0} \tag{7}$$

where U'_{2way_cat2} is the channel weights gotten in the two scales along the third co-ordinate axis, size N2C21, F conv2 (*) is convolution layer, where 21, and S (c cat2) is the consequence of the recalibration realised by splice and fusing cat2 at two scales. Input and output channel counts are both set to a value of C.

4 Results and Discussion

4.1 Implementation Details

TensorFlow is the foundation of our proposed model. All of the convolutions in a pretrained context have been batch normalised. Although it has been shown that a larger batch size is beneficial for segmentation, a fixed batch size of 8 is selected in order to seek for more effective designs. A regular (SGD) algorithm is used with a weight decay of 2e−4 for back propagation, picking a loss function based on the cross entropy of each pixel across all categories. The authors used the polynomial learning rate policy popularised by DeepLab. The training steps have been multiplied, and the learning rate has correspondingly adjusted.

$$\left(1 - \frac{iter}{maxiter}\right)^{power} \tag{8}$$

where *iter* stands for a training step and *maxiter* for the total number of iterations used during training. Our starting power D is 0:9, and our learning rate is 1e−3. Two NVIDIA TITAN X graphics processing units are responsible for all the matrix calculations. One Intel Core I7-6900k octa-core processor running at 3.2 GHz has completed the remaining calculations. The computed memory size is 64G.

4.2 Validation Analysis of Proposed Model

In this section, the validation analysis are based on 70%–30% and 80%–20%, where Table 1 presents the analysis based on 70% of training data and 30% of testing data. The existing models focused on either their own dataset or simply classify the gastric cancer. Therefore, this research work considered the generic model and implemented with the dataset and results are averaged.

Table 1. Analysis of Projected Perfect for 70%–30%

Methodology	Accuracy	Precision	Recall	F1-Score	Specificity
DBN	85.5	87	86	86	86
CNN	87	91	87	89	87
LSTM	91.5	91.5	91.5	91.5	95
PROPOSED	95.2	95	96	95	96

From the above table it is evitable that the proposed model showed an improvement over all other models in every metric.

Table 2. Analysis of Projected Perfect for 80%–20%

Methodology	Accuracy	Precision	Recall	F1-Score	Specificity
DBN	87.30	90.3	88.56	88.50	88.50
CNN	92.67	92.45	91.67	91.39	90.67
LSTM	93.34	94.78	93.78	94.09	96.27
PROPOSED	96.5	97.0	97.0	97.0	98.7

All models perform well when the model's training ratio is high which is observed from Table 2. Our approach clearly excels over the competing models in key respects. Our approach is more delicate with fine-grained characteristics and sensitive to regions of varying sizes. The models are graphically analysed in Figs. 2, 3, 4 and 5.

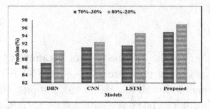

Fig. 2. Precision Analysis

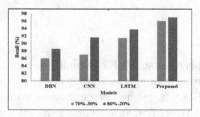

Fig. 3. Recall Analysis

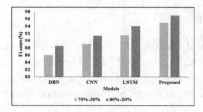

Fig. 4. Comparative analysis on F1-score

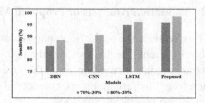

Fig. 5. Analysis of specificity

5   Conclusion and Future Work

For stomach cancer detection, the authors present a deep learning architecture that combines multi-scale modules with targeted convolutional operations. This study explores the classification task of stomach cancer pathology images and proposes a multiscale channel CNN with Res- network construction. The reliability of the feature channel weight learning process could be greatly improved through the fusing of feature weights learnt at various scales. By incorporating properties at several scales, network data can be made more useful. The obtained dataset has been subjected to rigorous comparative analyses, which prove that the proposed method is more precise and well-organized. Until then, further work is required on datasets and methods to advance the integration of deep learning and pathological diagnostics.

Although our approach has yielded promising results, it does have some restrictions which could be considered as future work. Diagnostic challenges will increase in clinical pathological picture analysis. There is still a great deal of in-depth work, from tests to clinical trials, that needs to be done.

References

1. Alansari, Z., Soomro, S., Belgaum, M.R., Shamshirband, S.: The rise of Internet of Things (IoT) in big healthcare data: review and open research issues. In: Saeed, K., Chaki, N.,

Pati, B., Bakshi, S., Mohapatra, D.P. (eds.) Progress in Advanced Computing and Intelligent Engineering. AISC, vol. 564, pp. 675–685. Springer, Singapore (2018). https://doi.org/10.1007/978-981-10-6875-1_66

2. Rayan, R.A., Tsagkaris, C., Papazoglou, A.S., Moysidis, D.V.: The internet of medical things for monitoring health. In: Internet of Things, pp. 213–228. CRC Press (2022)

3. Belgaum, M.R., Soomro, S., Alansari, Z., Musa, S., Alam, M., Su'ud, M.M.: Challenges: bridge between cloud and IoT. In: 2017 4th IEEE International Conference on Engineering Technologies and Applied Sciences (ICETAS), pp. 1–5. IEEE (2017)

4. Sung, H., et al.: Global cancer statistics 2020: GLOBOCAN estimates of incidence and mortality worldwide for 36 cancers in 185 countries. CA: Cancer J. Clin. **71**(3), 209–249 (2021)

5. Morales, S., Engan, K., Naranjo, V.: Artificial intelligence in computational pathology–challenges and future directions. Digit. Sig. Process. **119**, 103196 (2021)

6. Ramana, K., et al.: Early prediction of lung cancers using deep saliency capsule and pre-trained deep learning frameworks. Front. Oncol. **12** (2022)

7. Tie, J., et al.: Circulating tumor DNA analysis guiding adjuvant therapy in stage II colon cancer. New Engl. J. Med. **386**(24), 2261–2272 (2022)

8. Vobugari, N., Raja, V., Sethi, U., Gandhi, K., Raja, K., Surani, S.R.: Advancements in oncology with artificial intelligence—A review article. Cancers **14**(5), 1349 (2022)

9. Rajpurkar, P., Chen, E., Banerjee, O., Topol, E.J.: AI in health and medicine. Nat. Med. **28**(1), 31–38 (2022)

10. Hu, Z., et al.: A multi-task deep learning framework for perineural invasion recognition in gastric cancer whole slide images. Biomed. Sig. Process. Control **79**, 104261 (2023)

11. Guo, Q., et al.: Pathological detection of micro and fuzzy gastric cancer cells based on deep learning. Comput. Math. Methods Med. **2023** (2023)

12. Ma, L., Su, X., Ma, L., Gao, X., Sun, M.: Deep learning for classification and localization of early gastric cancer in endoscopic images. Biomed. Sig. Process. Control **79**, 104200 (2023)

13. Fu, X., Liu, S., Li, C., Sun, J.: MCLNet: an multidimensional convolutional lightweight network for gastric histopathology image classification. Biomed. Sig. Process. Control **80**, 104319 (2023)

14. Lee, S.H., Lee, Y., Jang, H.J.: Deep learning captures selective features for discrimination of microsatellite instability from pathologic tissue slides of gastric cancer. Int. J. Cancer **152**(2), 298–307 (2023)

15. Wang, P., Li, Y., Sun, Y., He, D., Wang, Z.: Multi-scale boundary neural network for gastric tumor segmentation. Vis. Comput. **39**(3), 915–926 (2023)

16. Sun, M., Zhang, G., Dang, H., Qi, X., Zhou, X., Chang, Q.: Accurate gastric cancer segmentation in digital pathology images using deformable convolution and multi-scale embedding networks. IEEE Access **7**, 75530–75541 (2019)

17. Yan, J., Wang, B.: Two and multiple categorization of breast pathological images by transfer learning. In: 2021 6th International Conference on Intelligent Informatics and Biomedical Sciences (ICIIBMS), vol. 6, pp. 84–88. IEEE (2021)

Glaucoma Detection Using the YOLO V5 Algorithm

M. Anusha, S. Devadharshini[✉], Faazelah Mohamed Farook, and G. Ananthi

Department of Electronics and Communications Engineering, Thiagarajar College of Engineering, Madurai 625015, India
devasubramanian1654@gmail.com, gananthi@tce.edu

Abstract. High intraocular pressure causes the eye disease glaucoma, which can eventually result in complete blindness. On the other hand, early detection and treatment of glaucoma can prevent complete blindness in a patient. However, we regularly experience delays as a result of challenging glaucoma screening procedures and a shortage of human resources, which might raise the worldwide vision loss ratio. In the final stage, it is envisaged that a confined region comprising glaucoma lesions and associated classes will develop. To prove the technique's viability, it was put to the test on a challenging dataset, specifically an online retinal fundus image database for glaucoma research (ORIGA). Due to the existing dearth of intelligence and security research on outdoor gantry cranes, a method based on the updated you-only-look-once (YOLO)v5 network for intelligent anti-intrusion detection is proposed. The first step is to offer a broad detection strategy. The YOLOv5 network's goal is to retain speed while achieving the highest detection precision: Add multi-layer receptive fields and fine-grained modules to the backbone network to improve the performance of features. The training of YOLO V5 resulted in an accuracy of 92.5% by the end of the 100th epoch. The high accuracy hence proves that the model was able to detect effectively.

Keywords: YOLO V5 · Glaucoma · Eye pressure · Image processing · Computer vision

1 Introduction

One of the conditions where the optic nerve of the eye is harmed and the primary cause of vision loss is glaucoma, a serious illness that arises from excessive intraocular pressure. Without adequate care and treatment, the whole population that is struggling with glaucoma risk losing their vision. Internally, there are several disorders with similarities to glaucoma. For the early detection of this condition, several studies have been conducted in this area. For accurate identification, the system employed a variety of deep learning algorithms. As previously said, early identification can stop the problem from growing worse and preserve the patient's vision. So, for the detection of this disease, a precise and adequate detecting model is needed. The creation of such a system has undergone several efforts. The method is also shown to identify the patients' glaucoma pattern. The

© The Author(s), under exclusive license to Springer Nature Switzerland AG 2023
S. Kadry and R. Prasath (Eds.): MIKE 2023, LNAI 13924, pp. 202–212, 2023.
https://doi.org/10.1007/978-3-031-44084-7_20

patterns identified in patients will be classified using the given method's CNN approach. The CNN model will be used to distinguish between the trends in the founded data for the identification of glaucoma. To enhance the effectiveness of the provided technique, a dropout mechanism is also used in the offered mechanism. The main goal is to identify the patterns that the healthy human eye and the diseased glaucoma eye share the most.

The signs of glaucoma cannot be found by a single diagnosis based on prior assessment. Regular eye exams may reveal glaucoma symptoms [1], and additional therapy or examinations may be advised. The following list of medical conditions is assessed for glaucoma approval.

- **Tonometry**: This technique measures a patient's ocular pressure.
- **Optical Coherence Tomography**: This test is crucial for glaucoma diagnosis. Retinal nerve fibre layers around the optic nerve are a key indicator of early glaucoma damage, and they are found using this method.
- **Ophthalmoscopy**: This examination looks at the optic nerve [2]. This test is crucial because glaucoma affects the optic nerve severely and is a condition of great importance. Eye drops are used to make the patient's pupil larger so that the optic nerve may be seen more clearly in order to check for indications of disease-related nerve cell loss in the eye [3].
- **Perimetry**: Glaucoma is a condition that initially impairs peripheral vision. In order to identify visual loss, this test is performed. A visual field test is another name for this assessment. A machine that automatically flashes lights in the person's eye's peripheral is used to examine each eye separately.
- **Gonioscopy**: This examination relates to the drainage angle for intraocular fluid. In the eye, the fluid is continually prepared before flowing out at certain angles. This test is used to determine if angle-closure glaucoma, which is defined as high eye pressure produced by blocked angles, or open-angle glaucoma, which is caused by open angles that are not functioning correctly (Fig. 1).

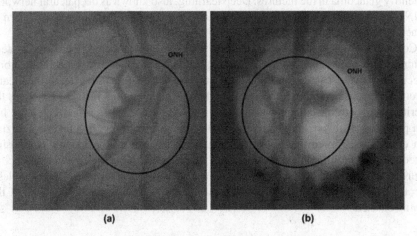

(a) (b)

Fig. 1. Optic nerve head images (a) normal eye (b) glaucomatous eye image.

In this paper, A methodology using YOLO V5 has been proposed.

2 Literature Review

2.1 ORIGA-Light: An Online Retinal Fundus Image Database for Glaucoma Analysis and Research

ORIGA-light, an online repository of clinically accurate retina pictures, is used by researchers as a benchmarking resource to assess image processing algorithms that find and examine key visual indicators strongly associated with glaucoma diagnosis. A constant effort will be made to improve ORIGA-light, both in terms of the quantity of annotated photos and the contributions from the scientific community. The goal of ORIGA-light is to support researchers in developing novel glaucoma mass screening techniques and optimising current image processing technology.

2.2 Automated Glaucoma Diagnosis Using Deep Learning Approach

The strategies that have been researched in this literature so far rely on manually creating various attributes from photos. To identify glaucoma, a technique based on morphological characteristics of the optic disc nerve and an artificial neural network (ANN) classifier was presented. In addition, the authors in presented a glaucoma Risk Index (GRI) approach to detect glaucoma by extracting features and feeding them to a PCA algorithm for dimensionality reduction and a support vector machine (SVM) model to classify the images into either normal or pathological. To identify glaucoma in the photos, a technique based on texture and higher order spectral characteristics with a random forest classifier has been presented. Additionally, using a sequential minimum optimisation classifier, glaucoma was detected in the pictures using wavelet features and a features selection approach. Recently, a method using SVM and Gabor features was presented to identify glaucoma in the pictures. Deep learning, also known as deep neural networks (DNNs), is a current study area because it allows computers to automatically learn to extract extremely complicated properties from input data.

Convolutional neural networks (CNNs) have gained significant popularity as deep learning architectures and have proven successful in various categorization applications. A recent discovery has demonstrated the potential of deep learning in automatically detecting the fovea and optic disc (OD) in digital fundus retinal images. To assess the severity of diabetic maculopathy disease, an effective feature learning technique has been proposed, which exhibits remarkable performance in classifying and recognizing such conditions. Moreover, a novel approach has been implemented for diagnosing glaucoma, relying on contextualizing deep learning techniques to extract distinctive features from the optic disc (OD) after segmenting the region of interest. Nonetheless, several challenges remain, with the most prominent one being the limited size of the training dataset.

3   Proposed Methodology

3.1   Dataset Collection

The publicly accessible database called ORIGA had images of glaucoma-affected eyes' retinal scans to detect and classify glaucoma-affected eyes. The database contains about 650 images, of which 168 images are having glaucoma-affected retina scans, and the remaining are normal human eyes' retinal scans. The dataset had a wide range of variation in aspects of various parameters like noise, color intensity, blurring, etc. Training the model which such a varied dataset shall definitely help in achieving the generalized prognosis of Glaucoma by the deep learning model employed. Apart from ORIGA, REMEDIO high-quality images collected from hospitals were used to test the model and determine its accuracy. The meticulous collection of the database was considered a safer option to ensure the proper detection of glaucoma (Figs. 2 and 3).

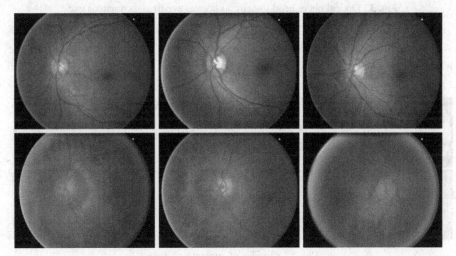

Fig. 2. Dataset comprising of retinal scans of eyes with and without glaucoma

3.2   Annotation of Images

The important aspect with respect to training of the deep learning model, when employing YOLO versions happens to be the proper demonstration of the object to be detected. Here in the case of the chosen problem, its to help the algorithm locate the positions of glaucoma-affected areas from the suspected samples for effective and correct training. The annotation of the dataset hence plays a crucial role, in marking the region of interest. A YAML file was developed after annotation with two pieces of information with respect to the former:

 (i) The class connected to the identified region
(ii) Co-ordinate values of the created bounding box on the glaucomatous area

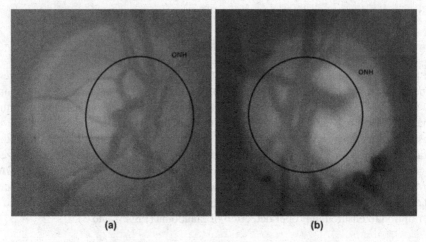

Fig. 3. Optical nerve head images: a) normal eye b) glaucomatous eye

The training file, which is developed from the above YAML file is used to train the network (Fig. 4).

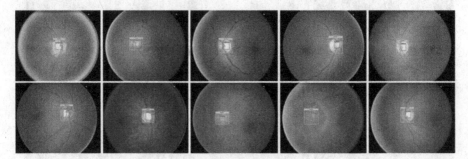

Fig. 4. Samples of annotations done

3.3 Selection of Pre-trained Model

YOLO (You Only Look Once)
The well-known challenge in computer vision happens to be object detection. In order to solve it, for decades, two stages were considered to be accomplished prior to training the model. They are:

(i) Extracting distinct regions of the image using sliding windows of various widths
(ii) Applying the classification problem to identify which class the objects belong.

The major drawbacks associated with these two methods happen to be:

(i) Requirements for a lot of processing to be done

(ii) Training has to be split down into several phases

The result hence happens to be a lack of optimization of speed.

For the first time, in 2015, researcher Joseph Redmon and colleagues unveiled the YOLO method, an object identification system that executes all of the necessary processes to recognize an item using a single neural network. Object recognition is reduced to a single regression problem, with bounding box coordinates and class probabilities replacing picture pixels. At the same time, this unified model predicts numerous bounding boxes as well as class probabilities for objects covered by boxes. For detecting and calculating object coordinates, the YOLO algorithm has obtained exceptional specifications that beat the existing approaches in terms of both speed and accuracy. The YOLO algorithm detects object by their bounding boxes and then localize them by using bounding box coordinates. As a result, the predicted bounding box vectors belong to the output vector y, whereas the ground truth bounding box vectors correspond to the vector label y. Vector label y and anticipated vector y can be indicated, where the purple cell has no item and the bounding box confidence score in the purple cell equals 0, then all remaining parameters will be ignored.

Finally, YOLO cleans all bounding boxes that are empty or contain the same objects as another bounding box using Non-Maximum Suppression. NMS removes all overlap bounding boxes with an intersection over union (IOU) value greater than the threshold value by setting a threshold value.

YOLO V5

The YOLO method underwent five iterations over the next five years, with many of the most cutting-edge ideas coming from the field of computer vision research. Joseph Redmon, the creator of the YOLO algorithm, created and studied the first three versions. After the debut of YOLOv3, he indicated that he will retire from the computer vision industry. Early in 2020, Alexey Bochkovskiy, the Russian developer who created the first three versions of YOLO using Joseph Redmon's Darknet architecture, published YOLO update version 4, also known as YOLOv4, on the official YOLO GitHub account. The YOLO algorithms were modified and refined after the release of YOLOv4 by Glenn Jocher's research group, which used the PyTorch framework before releasing YOLOv5.

Despite not being made by the team who invented the algorithm, YOLOv5 has outperformed all four preceding iterations (Jocher, 2020).

Unlike earlier iterations, YOLOv5 is now created in Python rather than C. This facilitates the installation and integration of IoT devices. In the future, PyTorch is anticipated to get more contributions and have more possibilities for expansion as one of the larger Darknet communities. Comparing the performance of YOLOv4 [4] and YOLOv5 is difficult because they were created using different frameworks and programming languages. However, YOLOv5 has improved on YOLOv4 over time and has begun to receive some attention from the computer vision community in addition to YOLOv4 [5].. The architecture of YOLO V5 comprises of three parts:

(i) Backbone: CSP Darknet
(ii) Neck: PANet
(iii) Head: YOLO Layer

YOLOv5 Backbone

It makes use of CSP Darknet [6] as the backbone for extracting features from images made up of cross-stage partial networks.

YOLOv5 Neck

It generates a feature pyramids [7] network with PANet [8] to conduct feature aggregation and passes it to Head for prediction.

YOLOv5 Head

Object detection layers that generate predictions from anchor boxes.

Apart from this YOLOv5 uses the below choices for training.

Activation and Optimization

Leaky ReLU and sigmoid activation are used in YOLOv5, with SGD and ADAM as optimizer options.

Rectified Linear Unit (ReLU)

Its key advantage is that it does not simultaneously stimulate all of the neurons.

Sigmoid Function

An activation function is used to transmit a weighted sum of inputs through, and the output is used as an input to the following layer.

Stochastic Gradient Descent (SGD)

It's a neural network model training optimization algorithm.

Adaptive Moment Estimation (ADAM)

It's a deep learning model training optimization approach that replaces stochastic gradient descent.

Loss Function

Binary cross-entropy with logits loss is used (Fig. 5).

3.4 Method of Approach

The training and validation dataset is obtained by splitting the already annotated database for the training and evaluation purposes of the model. The model accesses the input via the YAML file, with the number of classes and object names specified in the code. The input images accessed are initially resized into 416x416 dimensions. Before that, they are populated by augmentation, gaussian blurring, etc., the increase the number of images, so that the model is exposed to even more varieties Then, they are separated into minibatches each of size 64. After pre-processing, the images count to 6400 hence the code has to train the model with 1000 batches for each epoch. A total of 100 epochs were performed. After training, the testing was done with 20% of ORIGA DATASET and REMEDIO High quality images (Fig. 6).

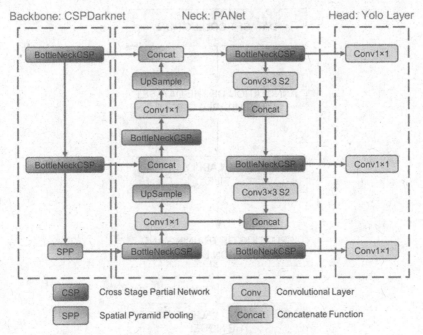

Fig. 5. YOLO V5 architecture

4 Results and Discussions

The suitable training parameters were set to ensure that the algorithm didn't overfit the input dataset and can get the normalized feature detection that could detect glaucoma even in images that the model hasn't been trained with before. The training of YOLO V5 resulted in an accuracy of 92.5% by the end of the 100th epoch. The high accuracy hence proves that the model was able to detect effectively (Fig. 7).

To comparatively prove the efficiency in the employment of YOLO V5 over other deep learning algorithms like YOLO V4 and Convolutional Neural Network, the latter were trained and tested using the same database and resulted in much lower accuracy. Hence it is concluded that the YOLO v5 model happens to be effective for Glaucoma prognosis.

Usage of REMEDIO images which weren't used in training for testing the model helps to eliminate the fact of overfitting associated with the model due to multiple epochs over a limited dataset.

The F1 score could be obtained by the equation:

$$F1\ SCORE = \frac{2 * PRECISION * RECALL}{PRECISION + RECALL} \tag{1}$$

Here precision and recall could be calculated from the confusion matrix by their equations.

$$RECALL = \frac{OVERALL\ TRUE\ POSITIVES}{OVER\ ALL\ TRUE\ POSITIVES\ +\ OVERALL\ FALSE\ NEGATIVES} \tag{2}$$

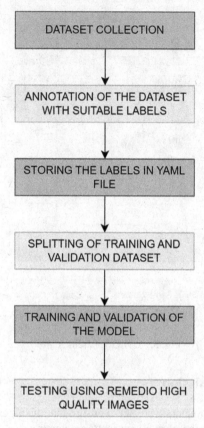

Fig. 6. Method of approach

The recall happens to be 0.911.

$$PRECISION = \frac{OVERALL\ TRUE\ POSITIVES}{OVERALL\ TRUE\ POSITIVES\ +\ OVERALL\ FALSE\ POSITIVES}$$

(3)

The precision happens to be 0.911.

Hence the F1 score happens to be 0.911, which is consistent with the accuracy 92.3%. Hence the model happens to be good in aspects of precision and recall, and hence in the overall generalization.

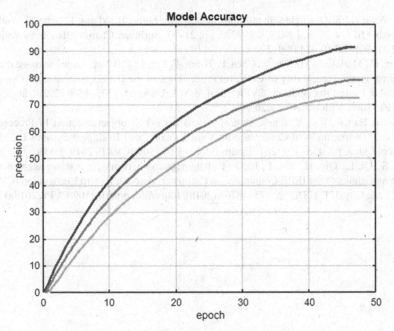

Fig. 7. Accuracy of three different deep learning models

5 Conclusion

The Generalised Image Segmentation Model YOLO V5 method for glaucoma illness prediction was proposed in this study. The Kaggle website's high resolution ORIGA dataset was used for training, testing, and validation. 350 photos of glaucoma comprised the training data set. There were 200 photos of glaucoma in the test data set. The YOLO V5 Deep learning algorithm for glaucoma condition has a 92.3% accuracy rate. YOLO V5 algorithm offers more accuracy as compared to YOLO V3 method for Convolution Neural Network.

References

1. Ovreiu, S., Cristescu, I., Balta, F., Ovreiu, E.: An exploratory study for glaucoma detection using densely connected neural networks. In: Proceedings of 2020 International Conference on e-Health and Bioengineering (EHB), Iasi, Romania (2020). https://doi.org/10.1109/EHB 50910.2020.9280173
2. Abdull, M.M., Chandler, C., Gilbert, C.: Glaucoma, "the silent thief of sight": patients' perspectives and health seeking behaviour in Bauchi, northern Nigeria. BMC Ophthalmol. **16**, 44 (2016). https://doi.org/10.1186/s12886-016-0220-6
3. Quigley, H.A., Broman, A.T.: The number of people with glaucoma worldwide in 2010 and 2020. Br. J. Ophthalmol. **90**, 262–267 (2006). https://doi.org/10.1136/bjo.2005.081224
4. Bochkovskiy, A., Wang, C.Y., Liao, H.Y.M.: YOLOv4: optimal speed and accuracy of object detection (2020). https://arxiv.org/abs/2004.10934. Accessed 10 Aug 2020

5. Liu, W., et al.: SSD: single shot multibox detector. In: Leibe, B., Matas, J., Sebe, N., Welling, M. (eds.) ECCV 2016. LNCS, vol. 9905, pp. 21–37. Springer, Cham (2016). https://doi.org/10.1007/978-3-319-46448-0_2

6. Wang, C., Mark Liao, H., Wu, Y., Chen, P., Hsieh, J., Yeh, I.: CSPNet: a new backbone that can enhance learning capability of CNN. In: Proceedings of the IEEE Conference on Computer Vision & Pattern Recognition (CVPR), Seattle, WA, USA, pp. 1571–1580 (2020). https://doi.org/10.1109/CVPRW50498.2020.00203

7. Zhao, Y., Han, R., Rao, Y.: A new feature pyramid network for object detection. In: Proceedings of the 2019 International Conference on Virtual Reality and Intelligent Systems (ICVRIS), Jishou, China, pp. 428–431 (2019). https://doi.org/10.1109/ICVRIS.2019.00110

8. Liu, S., Qi, L., Qin, H., Shi, J., Jia, J.: Path aggregation network for instance segmentation. In: Proceedings of the IEEE Conference on Computer Vision & Pattern Recognition (CVPR), Salt Lake City, UT, USA, pp. 8759–8768 (2018). https://doi.org/10.1109/CVPR.2018.00913

Deep Learning Ocular Disease Detection System (ODDS)

Priya Thiagarajan(✉) and M. Suguna

Department of Computer Science and Engineering, Thiagarajar College of Engineering,
Madurai, Tamilnadu, India
priyabennet@gmail.com

Abstract. ***Background:*** Prevalent, preventable or treatable causes of blindness include Diabetic Retinopathy (DR), Glaucoma, Cataract and Optic Nerve Head Swelling. Community screening camps are routinely conducted to mainly screen for these diseases/conditions. This is crucial for maintaining good vision and preventing irreversible damage that can lead to blindness. Screening is for early diagnosis, to ensure further investigations and prompt treatment to save vision. The use of intelligent systems to assist the healthcare professionals can help speed up the diagnostic process.

Methodology: Retinal fundus images (RFI) can be used to diagnose several ocular diseases including diabetic retinopathy, glaucoma, cataract and optic nerve head swelling. After preprocessing, the RFI can be classified by trained Neural Networks to detect these diseases. We have curated a dataset with 5600 images in 5 classes and used it to train, validate and test our Ocular Disease Detection System (ODDS). ODDS uses the EfficientNet-B3 model for image classification.

Results: The ODDS performs well with a testing accuracy of 93%. Other performance metrics including precision, recall and F1-scores are also high for all the classes.

Conclusion: The Ocular Disease Detection System (ODDS) will be very useful in community screening programmes and in remote or rural regions. Although the outcomes achieved are highly promising, augmenting the amount of training data from a range of different sources would make the system robust and enhance the practical applicability of the system. The possibility of utilising smartphones equipped with an appropriate lens assembly to record RFI should be explored. This will reduce costs and increase the accessibility of the system.

Keywords: Diabetic retinopathy · Glaucoma · Cataract · Optic nerve head swelling · Retinal fundus · Artificial intelligence · Neural networks · EfficientNet

Novelty:

- Curation of dataset with 5600 images from 5 classes - diabetic retinopathy, glaucoma, cataract, optic nerve head swelling and normal images from various public datasets
- Lighter and more economical neural network model is employed in the proposed system (compared to existing literature)
- High performance metrics achieved

© The Author(s), under exclusive license to Springer Nature Switzerland AG 2023
S. Kadry and R. Prasath (Eds.): MIKE 2023, LNAI 13924, pp. 213–224, 2023.
https://doi.org/10.1007/978-3-031-44084-7_21

1 Introduction

With the rise in population and life expectancy of humans, the healthcare systems of most countries are stretched thin. Doctors and other healthcare workers are overburdened with providing adequate healthcare for the growing population [1]. Intelligent systems based on machine learning algorithms and artificial neural networks are increasingly used in several aspects of healthcare [2], including in disease detection systems. In particular, Artificial Intelligence (AI) image analysis and classification in medical imaging has achieved great success. Intelligent systems have been shown to even outperform human experts in several recent studies [3].

Artificial Neural Networks (ANN) are preferred (over other traditional machine learning algorithms) for image analysis and classification tasks because of their high performance metrics and ability to work with large volumes of complex data [4]. They also have the ability to constantly refine their systems when new data is added for training.

1.1 AI in Ophthalmology

Intelligent systems for image classification are employed in ophthalmology for image analysis of retinal fundus images (RFI) or Optical Coherence Tomography (OCT) images [5, 6]. A few systems have been approved for use in patient care and are currently being used as a triaging or screening tool to assist doctors and other healthcare workers [7]. This project uses classification of RFI for disease detection. A retinal fundus image is a picture of the inner, back surface of the eye, and includes the retina, optic disc and blood vessels [8] (Fig. 1). A fundus camera is used to capture and record retinal fundus images.

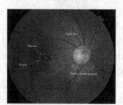

Fig. 1. A normal retinal fundus image

Regular screening of the eye is essential to ensure early detection of diseases. This will help provide suitable treatment for the diseases to reverse the damage or at least to prevent further deterioration and loss of vision. Common eye diseases/disorders generally included in screening programmes include.

- Diabetic Retinopathy
- Cataract
- Glaucoma
- Optic Nerve Head swelling

A brief description of the diseases are given below.

Diabetic Retinopathy

Diabetes mellitus is a chronic disease caused by insufficient blood insulin secretion by the pancreas. This causes elevated levels of blood glucose, leading to several complications including damage to the retina called Diabetic Retinopathy. This causes blocks in the blood vessels of the eye, leading to swelling and leaking of blood or fluids. Symptoms include blurred vision, floaters, and reduced colour perception. In advanced stages of the disease, new blood vessels may develop on the retina, leading to proliferative diabetic retinopathy. This condition can cause severe vision loss and even blindness.

Treatment for diabetic retinopathy includes lifestyle modifications and treatment to maintain healthy blood glucose levels, surgery or laser therapy, depending on the severity of the condition. Regular eye exams are crucial for early detection and treatment. Regular monitoring is also essential after diagnosis.

Glaucoma

Glaucoma occurs when the drainage canals in the eye become blocked. This causes the intraocular pressure to rise in the eye and damage the optic nerve. Glaucoma usually has no symptoms in its early stages, and vision loss can occur gradually over time. It can lead to peripheral vision loss, tunnel vision, and blindness if not diagnosed and treated in the early stages. Treatment options include medications, surgery or laser therapy as dictated by the severity of the condition. The two types of glaucoma are open-angle glaucoma (most common) and angle-closure glaucoma.

Early diagnosis, prompt treatment and regular monitoring are crucial to avoid irreversible damage to vision.

Cataract

Cataract is a vision threatening condition caused by opacification of the lens (or its capsule) of the eye. This prevents the passage of light to the retina, leading to blurred or cloudy vision. Though commonly found in older people, it can occur at any age. Cataracts develop gradually and do not present early symptoms. Left untreated, they can lead to severe or even total vision loss.

Treatment is surgery to remove the opaque lens and replace it with an artificial lens.

Optical Nerve Head Swelling

The optic nerve connects the retina of the eye to the visual centre in the brain. We consider two types of conditions under optic nerve head swelling - papilledema and pseudopapilledema.

Papilledema is caused by the increase in intracranial pressure inside the brain. This causes the optic nerve to become swollen. This is visible in the retinal fundus images as optic nerve head swelling. This may reflect serious and critical changes in the brain, such as brain lesions or infections. Left undiagnosed and untreated, it may progress to severe vision loss or even death.

Several other factors, not related to the brain, can cause the optic nerve head to appear swollen. This includes anatomical variations, congenital conditions and a few other harmless conditions. This is termed as pseudopapilledema.

It is essential to diagnose these conditions early and then evaluate to differentiate between papilledema and pseudopapilledema.

These eye diseases/conditions can go unnoticed for extended periods of time and may not present any symptoms until they have progressed significantly. Each of these conditions has the potential to cause total vision loss if not diagnosed and treated promptly. This makes it essential to have regular eye exams in the community.

Eye camps are conducted regularly to screen for these diseases. It is quite time consuming for the ophthalmologists to review each patient or retinal fundus image. This may delay diagnosis, especially in rural areas. An intelligent system which detects these conditions will be a big boon to the health professionals. It can act as a triaging system - a primary system to classify retinal fundus images to detect these diseases, reducing the workload of the doctors and speeding up the diagnosis for patients.

2 Literature Review

A comprehensive and thorough examination of relevant literature was performed before starting the project. Table 1 outlines the noteworthy and relevant papers, with the AI models employed, and the performance metrics achieved.

The literature survey for intelligent multi disease detection using RFI in ophthalmology, led us to the following conclusions.

- Most research papers use their own data (not available for other researchers to replicate or enhance the models proposed)
- Public access datasets for multiple disease classification in ophthalmology are very few
- Most available datasets also include several rarer eye diseases/ conditions
- This inclusion of rarer conditions leads to highly imbalanced datasets with some classes having many times more data than other classes.
- This in turn leads to poorer performance metrics mainly because of lesser amounts of training data for rarer diseases.

Table 1. AI for Multi-Disease/Disorder Detection (using RFI)

Reference	Diseases Detected	Dataset	AI Model Used	Significant Results
[9]	12 major fundus diseases including diabetic retinopathy, age-related macular degeneration, possible glaucomatous optic neuropathy and papilledema	Dataset created with 56 738 images for training and 8176 images for testing	DL CNN - SeResNext50	Significantly higher sensitivity as compared to human doctors, but lower specificity
[10]	46 ocular conditions in 29 classes	Retinal Fundus Multi-Disease Image Dataset - RFiMD (3200 images)	Multi disease detection pipeline - DCNN pre-trained with ImageNet and transfer learning, Ensemble model	AUROC: 0.95 for disease risk classification 0.7 for Multi label scoring
[11]	39 retinal fundus conditions	249,620 fundus images from heterogeneous sources	2-level hierarchical system with 3 groups of CNN and Mask RCNN	F1 score: 0.923, Sensitivity: 0.978, Specificity: 0.996 and (AUROC): 0.9984
[12]	Glaucoma, Maculopathy, Pathological Myopia, and Retinitis Pigmentosa	Dataset with 250 retinal fundus images	MobileNetV2 and transfer learning	Accuracy: 96.2% Sensitivity: 90.4% Specificity: 97.6%
[13]	Diabetic retinopathy, glaucoma and age related macular degeneration	Dataset with 43055 images collected from 12 public datasets	Ensemble of Inception V3 model (with transfer learning)	Accuracy: 79.2%

3 Methodology

After a comprehensive study of published literature in multi disease screening/detection, the intelligent Ocular Disease Detection System (ODDS) was designed employing neural networks to classify retinal fundus images. The diseases included for classification by the system are prevalent, preventable (by early diagnosis) causes of vision loss. The best model for this task is chosen based on performance metrics. The methodology diagram is given in Fig. 2. The major building blocks of our project are explained with additional information.

3.1 Dataset Curation

The retinal fundus image datasets available with open access for multi disease detection in ophthalmology is very limited. As seen in Sect. 2, most public datasets have several classes including rarer eye conditions which are not suitable for our screening system. They are also imbalanced with some classes having significantly higher number of images than other classes. So we decided to curate our own dataset from other public datasets and used images from

- Indian Diabetic Retinopathy Image Dataset (IDRiD) [14]
- High Resolution Fundus Image Database (HRF) [15]
- Ocular Disease Intelligent Recognition (ODIR-19) [16]
- Digital Retinal Images for Vessel Extraction (DRIVE)
- Machine Learning for Pseudopapilledema Dataset from Open Science Framework (OSF) [17]

Dataset Size: $\simeq$ 5600 images in 5 classes - diabetic retinopathy, glaucoma, cataract, optic nerve head swelling and normal. Preliminary analysis of the data to explore its contents and create visual representations of data samples are done.

3.2 Preprocessing the Data

A pie chart representation of the available data indicates imbalance among the classes (Fig. 5). So 'data trimming' is done by sampling roughly equal numbers of images from each class.

Data augmentation (image flipping and rotation) is also done as part of preprocessing on the training data. This is to help make the model identify different orientations of the image input. The augmentation is done using Keras ImageDataGenerator. The data is split in the ratio of 80:10:10 for training, validating and testing the models.

3.3 The Ocular Disease Detection System (ODDS)

The proposed ODDS is presented in Fig. 3. The captured fundus images can be classified as one of the five classes.

After careful consideration of various models, the Convolutional Neural Network (CNN) was chosen as a base model, as most medical image classification applications uses CNN or a variant of CNN. A detailed study of EfficientNet models revealed enhanced performance in comparison to CNNs, with lesser computational resources. So the ODDS was designed and implemented with EfficientNet-B3.

The Chosen Models: CNN and EfficientNet-B3

At present, CNNs are the most commonly used algorithms for image classification. These networks operate with a set amount of resources, but if additional resources are available, scaling up to improve accuracy can be done by increasing the depth, width, or image resolution. But complex optimization techniques are needed if complex scaling of more than one factor needs to be done.

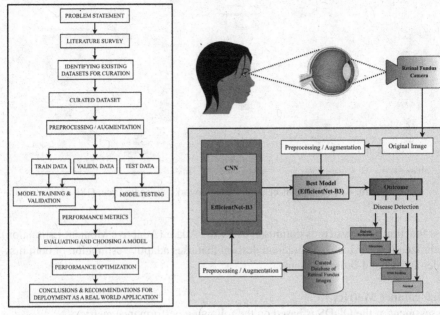

Fig. 2. Project methodology

Fig. 3. The proposed Ocular Disease Detection System (ODDS)

EfficientNet is a family of neural networks (NN) with eight models from B0 to B7. They are designed to achieve peak performance on image classification tasks with minimal number of parameters and optimum computational resources.

EfficientNet models exhibit significantly better performance as their depth, width, and image resolution are uniformly scaled [18]. The different scaling concepts are shown in Fig. 4. With compound scaling, the receptive field of the EfficientNet models is enhanced. This, in turn, enables the capture of finer patterns from larger images.

Previously published research has demonstrated that EfficientNets decrease the number of parameters required for image processing. Compared to other CNN models, they use fewer FLOPS (floating-point operations per second) [18]. Furthermore, studies have shown that even the best CNNs for image processing are surpassed in terms of speed and efficiency by EfficientNets, which are significantly smaller and faster. The models EfficientNet-B0 to B7, use different numbers of parameters and computational complexity. Our chosen model, EfficientNet-B3 gives a good trade off between performance and computational resources [18]. The Keras image classification model that utilises EfficientNet-B3 allows for optional preloading of weights from ImageNet. [19]. The ODDS also uses Imagenet weights. Our system extracts about 11 million features from the images and processes them for the classification. An initial learning rate of 0.001 is used.

Optimization

In deep learning systems, the optimizer is responsible for updating the weights and biases of the model. This is done during each iteration of the training process, to minimise the

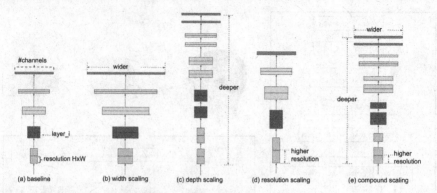

Fig. 4. Scaling concepts in neural networks. Image source: [18]

loss function. Our system is optimised using the Adam (Adaptive Moment Estimation) optimizer. It uses stochastic gradient descent that uses adaptive estimation of both first-order and second-order moments.

Performance Metrics

Assessment of the ODDS is based on the following performance metrics:

- Accuracy (CA)
- Precision (PR)
- Recall (RE)
- F1-score (F1)

The formulae for these are tabulated in Table 2, where T - TRUE, F - FALSE, P - POSITIVE, N - NEGATIVE.

The confusion matrix is also done to study the misdiagnosis patterns.

We have presented the outcomes and performance metrics of our system in the results section.

Table 2. Formulae for performance metrics

$$PR = \frac{TP}{TP+FP}$$

$$RE = \frac{TP}{TP+FN}$$

$$CA = \frac{TP+TN}{TP+TN+FP+FN}$$

$$F_1 = \frac{2TP}{2TP+FP+FN}$$

4 Results

The results of the exploratory data analysis and classification of the curated dataset with a convolutional neural network and with the proposed model using EfficientNet- B3 are given in this section.

4.1 Exploratory Data Analysis

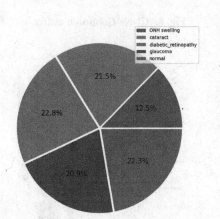

Fig. 5. Distribution of data in different classes of the curated dataset

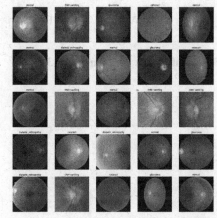

Fig. 6. Data visualisation of the curated dataset

The distribution of image data for the five classes are shown in Fig. 5. It shows a slight imbalance with the optic nerve head swelling images lesser in number than the other classes. This leads to the preprocessing step of trimming the data to maintain approximately similar numbers in each class. This is done by random sampling from the classes with higher numbers of data. A random sample of images from the curated dataset is visualised and presented in Fig. 6.

4.2 Results with CNN

The results of using CNN in our system are presented in Fig. 7 and Fig. 8.

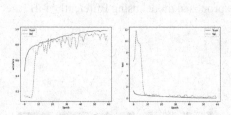

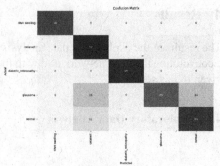

Fig. 7. CNN - Accuracy/Loss vs Epochs

Fig. 8. CNN - Confusion matrix

4.3 Results with EfficientNet-B3

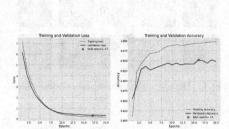

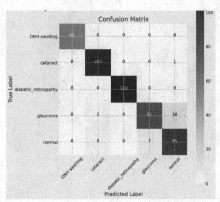

Fig. 9. EfficientNet-B3 - Accuracy/ Loss vs Epochs

Fig. 10. EfficientNet-B3 - Confusion matrix

The outcomes of using EfficientNet-B3 in our ODDS is given in Fig. 9 and Fig. 10. The confusion matrix shows minimal misdiagnosis, mainly between the glaucoma and normal classes. This needs to be looked into to improve performance in future studies. The performance metrics of CNN and EfficientNet-B3 are tabulated in Table 3.

Based on the performance metrics, the proposed ODDS using EfficientNet-B3 model demonstrates a much better performance, when compared to CNN and most existing literature. The metrics also show the chosen model is ideally suited for this disease detection by image classification application. The curated dataset contains images from various sources captured by various devices from different parts of the world. Since our proposed model performs well with our curated dataset, we are certain this model will exhibit similar or improved performance when applied to larger and more diverse datasets as well.

Table 3. ODDS - Performance Metrics of CNN and EfficientNet-B3

MODEL		Precision	Recall	F1 - score
CNN	ONH Swelling	1.00	1.00	1.00
	Diabetic Retinopathy	1.00	1.00	1.00
	Glaucoma	0.97	0.58	0.73
	Cataract	0.73	0.96	0.83
	Normal	0.76	0.88	0.82
	Accuracy		0.87	
EfficientNet-B3	ONH Swelling	1.00	1.00	1.00
	Diabetic Retinopathy	1.00	0.97	0.98
	Glaucoma	0.94	0.82	0.87
	Cataract	0.91	0.98	0.94
	Normal	0.84	0.91	0.87
	Accuracy		0.93	

5 Conclusion

The intelligent Ocular Disease Detection System that we have proposed has demonstrated excellent performance metrics. This represents a significant advancement towards creating an independent, intelligent system that can analyse retinal fundus images to identify instances of diabetic retinopathy, glaucoma, cataract, or optic nerve head swelling. The proposed EfficientNet-B3 algorithm also has the advantage of being much smaller and faster than CNN models and their variants. This system has the potential to be utilised in a variety of contexts for screening and tele health monitoring, and could be instrumental in saving vision in the community.

Further Work: The training dataset can be enlarged to include more images including edge case images (very early or late disease images). Other models including Graph CNN (GCNN) and higher EfficientNet models can be experimented with to find a suitable model to see if the performance can be further improved. A recent development is the use of smartphones (equipped with appropriate lens arrangement) to record images of the retinal fundus [20]. A few studies have shown encouraging results for diabetic retinopathy and glaucoma [21, 22]. This can also be investigated further to simplify the image capture process at significantly lower costs. This will help develop truly mobile fundus capture units for ocular disease detection.

References

1. Benet, D., Pellicer-Valero, O.J.: Artificial intelligence: the unstoppable revolution in ophthalmology. Surv. Ophthalmol. **67**(1), 252–270 (2022)
2. Wang, Z., Keane, P.A., Chiang, M., Cheung, C.Y., Wong, T.Y., Ting, D.S.W.: Artificial intelligence and deep learning in ophthalmology. In: Lidströmer, N., Ashrafian, H. (eds.) Artificial Intelligence in Medicine, pp. 1519–1552. Springer, Cham (2022). https://doi.org/10.1007/978-3-030-64573-1_200
3. Shigueoka, L.S., Jammal, A.A., Medeiros, F.A., Costa, V.P.: Artificial intelligence in ophthalmology. In: Lidströmer, N., Ashrafian, H. (eds.) Artificial Intelligence in Medicine, pp. 1553–1566. Springer, Cham (2022). https://doi.org/10.1007/978-3-030-64573-1_201

4. Malini, A., Priyadharshini, P., Sabeena, S.: An automatic assessment of road condition from aerial imagery using modified VGG architecture in faster-RCNN framework. J. Intell. Fuzzy Syst. **40**(6), 11411–11422 (2021)
5. Anton, N., et al.: Comprehensive review on the use of artificial intelligence in ophthalmology and future research directions. Diagnostics **13**(1), 100 (2022)
6. Zhou, H.: Design of intelligent diagnosis and treatment system for ophthalmic diseases based on deep neural network model. Contrast Media Mol. Imaging 2022 (2022)
7. Vedula, S.S., Tsou, B.C., Sikder, S.: Artificial intelligence in clinical practice is here—now what? JAMA Ophthalmol. **140**(4), 306–307 (2022)
8. Jeong, Y., Hong, Y.J., Han, J.H.: Review of machine learning applications using retinal fundus images. Diagnostics **12**(1), 134 (2022)
9. Li, B., et al.: Development and evaluation of a deep learning model for the detection of multiple fundus diseases based on colour fundus photography. Br. J. Ophthalmol. **106**(8), 1079–1086 (2022)
10. Müller, D., Soto-Rey, I., Kramer, F.: Multi-disease detection in retinal imaging based on ensembling heterogeneous deep learning models. In: German Medical Data Sciences 2021: Digital Medicine: Recognize–Understand–Heal, pp. 23–31. IOS Press (2021)
11. Cen, L.P., et al.: Automatic detection of 39 fundus diseases and conditions in retinal photographs using deep neural networks. Nat. Commun. **12**(1), 1–13 (2021)
12. Guo, C., Yu, M., Li, J.: Prediction of different eye diseases based on fundus photography via deep transfer learning. J. Clin. Med. **10**(23), 5481 (2021)
13. Pandey, P.U., Ballios, B.G., Christakis, P.G., et al.: An ensemble of deep convolutional neural networks is more accurate and reliable than board-certified ophthalmologists at detecting multiple diseases in retinal fundus photographs. Br. J. Ophthalmol. (2023). https://doi.org/10.1136/bjo-2022-322183
14. Prasanna, P., et al.: Indian diabetic retinopathy image dataset (IDRiD). IEEE Dataport (2018). https://dx.doi.org/10.21227/H25W98
15. Budai, A., Bock, R., Maier, A., Hornegger, J., Michelson, G.: Robust vessel segmentation in fundus images. Int. J. Biomed. Imaging 2013 (2013)
16. https://odir2019.grand-challenge.org/dataset/
17. Machine Learning for Pseudopapilledema dataset. https://osf.io/2w5ce/, https://doi.org/10.17605/OS-F.IO/2W5CE
18. Tan, M., Le, Q.: Efficientnet: rethinking model scaling for convolutional neural networks. In: International Conference on Machine Learning, pp. 6105–6114. PMLR (2019)
19. https://keras.io/api/applications/efficientnet/. Accessed 19 Dec 2022
20. Pujari, A., et al.: Clinical role of smartphone fundus imaging in diabetic retinopathy and other neuro-retinal diseases. Curr. Eye Res. **46**(11), 1605–1613 (2021)
21. Gupta, S., Thakur, S., Gupta, A.: Optimized hybrid machine learning approach for smartphone based diabetic retinopathy detection. Multimed. Tools Appl. **81**(10), 14475–14501 (2022)
22. Nakahara, K., et al.: Deep learning-assisted (automatic) diagnosis of glaucoma using a smartphone. Br. J. Ophthalmol. **106**(4), 587–592 (2022)

Efficient Chest X-Ray Investigation Using Firefly Algorithm Optimized Deep and Handcrafted Features

Seifedine Kadry[1]([✉]) [iD], Mohammed Azmi Al-Betar[2], Sahar Yassine[1],
Ramya Mohan[3], Rama Arunmozhi[3], and Venkatesan Rajinikanth[3] [iD]

[1] Department of Applied Data Science, Noroff University College, 4612 Kristiansand, Norway
skadry@gmail.com, sahar.yassine@noroff.no

[2] Artificial Intelligence Research Center (AIRC), College of Engineering and Information
Technology, Ajman University, Ajman, United Arab Emirates
m.albetar@ajman.ac.ae

[3] Department of Computer Science and Engineering, Division of Research and Innovation,
Saveetha School of Engineering, SIMATS, Chennai 602105, India
{ramyanallu,ramaa.sse}@saveetha.com, v.rajinikanth@ieee.org

Abstract. Radiological imaging of the chest (X-ray) is a cost-effective, widely
accepted method of examining the lungs and abnormalities. During this research,
a clinically significant Convolutional-Neural-Network (CNN) framework will be
proposed for the examination of lung abnormalities. The proposed scheme aims
to achieve better detection accuracy from the selected chest X-ray data. The following stages are included in these techniques: (i) CNN segmentation of the
lung section, (ii) Deep-feature extraction utilizing selected CNN schemes, (iii)
Handcrafted-feature extraction, (iv) Optimizing features using Firefly Algorithms,
and (v) Binary classification and cross-validation of fivefold cross-validation. By
implementing the pre-trained VGG-UNet, this framework is capable of extracting
lungs sections from X-ray images. Using this lung segment, handcrafted features
such as Local Binary Patterns (LBP) and Pyramid Histograms of Oriented Gradients (PHOG) are obtained. DFs and HFs are obtained using the FA, and then a
serial concatenation is performed in order to obtain a hybrid feature vector. This
feature vector is used to classify X-ray images into healthy and diseased groups.
For examination in this study, X-ray images of disease classes, such as tuberculosis, COVID19, pneumonia, lung masses, and effusions, are considered. Based on
the experimental results of this study, >98% of disease detection accuracy was
confirmed using the proposed scheme combined with the SoftMax classifier.

Keywords: Classification · CNN scheme · deep-learning · feature selection ·
firefly algorithm · lung infection

S. Kadry and R. Prasath (Eds.): MIKE 2023, LNAI 13924, pp. 225–236, 2023.
https://doi.org/10.1007/978-3-031-44084-7_22

1 Introduction

An essential organ of the human respiratory system is the lung, which exchanges air during inhalation and exhalation. In order to control/cure the infection, it is necessary to conduct appropriate screenings and treatment. Therefore, the disease/infection in the lung causes moderate to severe respiratory issues.

Radiological examinations, such as computed tomography (CT) and chest radiographs (X-rays), are commonly used to determine the presence of lung abnormalities at the clinical level. Due to its simplicity and reputation, the X-ray image is commonly used in hospitals to screen patients for lung infections. A radiologist or pulmonologist examines the X-ray image to determine the extent of the infection.

It has been confirmed by recent literature that lung abnormalities are primarily caused by infectious diseases or acute diseases. Therefore, the examination protocol for one kind of lung disease can be used to analyze the other types. It is, however, time-consuming to follow the traditional clinical procedure to detect lung abnormalities. As a result, many conventional procedures, machine-learning schemes, and deep-learning methods are proposed and implemented by the researchers using X-rays [1, 2].

Because modern computing facilities are readily available, computer scheme-supported disease detection has become widely used in hospitals, reducing the burden of diagnostics. It is also necessary to implement computerized methods to accurately detect the disease during the mass disease viewing procedures due to the large number of images to be examined. In addition, the existing computerized methods are either disease-specific or image-specific, which can only be used for a particular task and may not achieve the same level of accuracy when applied to a different disease or image type.

The development of a distinctive automated disease recognition system is therefore necessary, which can be applied to any medical image, regardless of the condition of the disease. Furthermore, the developed scheme must possess clinical significance and provide better detection accuracy regardless of data dimension (i.e. the image to be examined).

For the purpose of examining the disease/abnormalities within chest X-rays, this research presents a CNN framework. In the proposed scheme, the following stages are considered: (i) Image resizing, (ii) Segmentation of lung segments to eliminate artifacts from the medical image, (iii) handcrafted and deep feature extraction with a selected procedure, (iv) feature selection with a selected algorithm, (v) serial feature concatenation, and (v) validation of the disease detection system.

To reduce the artifact in the X-rays, VGG-UNet-based lung segmentation is implemented [3]. As a result, the visibility of the abnormality/disease is less in X-rays, which affects detection accuracy. In order to extract deep features (DF) using a chosen method, the extracted lung section is analyzed along with handcrafted features (HF), such as Local Binary Patterns (LBPs) with varying weights and Pyramid Histogram of Oriented Gradients (PHOGs) with different bins. To avoid overfitting, a feature reduction process using the Firefly Algorithm with a Lévy flight (LF) search strategy is employed to minimize the extracted features dimension in DF and HF. After reducing the features, serial concatenation is applied to create a hybrid feature vector (DF+HF). Finally, binary classification is applied to categorize the X-ray images, and the results are verified.

A variety of harsh lung abnormality images, including tuberculosis, COVID19, pneumonia, mass, and effusion, is used to evaluate the performance of the framework developed. To confirm the clinical significance of the developed framework, 2000 images (1000 healthy and 1000 disease categories) were examined, of which 80% were used for training, 10% for testing, and 10% for validation, respectively.

In order to verify the performance of the scheme, three approaches are used: (i) individual IDF, (ii) ensemble DF, and (iii) DF+HF. This study examines only the performance of the SoftMax classifier using a 5-fold cross-validation method, and computes a number of performance metrics, including accuracy, precision, sensitivity, specificity, and F1score. Using the proposed methodology, a disease detection accuracy of >98% can be achieved.

The significant contributions of this research include;

- A hybrid feature vector is obtained by reducing features with Firefly
- Disease analysis using individuals and ensembles of DF
- TB, COVID19, pneumonia, masses, and effusion are all detected using this unique disease detection framework.

A context is presented in Sect. 2, the methodology is described in Sect. 3, and the experimental results and conclusions are discussed in Sect. 4.

2 Related Research

X-rays are used to detect lung abnormalities, which are a medical emergency. Table 1 summarizes a few chosen CNN-supported lung abnormality detection methods.

Table 1. Summary of chest X-ray examination procedures

Reference	Implemented method
Bhandary et al. [4]	A deep transfer learning scheme based on lung abnormality using CT and X-ray is discussed
Rahman et al. [5]	CNN segmentation and classification-based detection of TB in chest radiographs are presented
Akcayet al. [6]	X-ray detection based on deep learning is presented here
Gamuchiraiet al. [7]	Implementation of chest X-ray-based TB detection using a deep-learning scheme is presented
Subramanian et al. [8]	A detailed review of deep-learning-based COVID19 is presented
Shilpa et al. [9]	Deep-learning-based lung region segmentation is discussed
Dey et al. [10]	Deep-learning-based detection of pneumonia in X-rays is presented using various classifiers

The methods discussed in Table 1 confirm that the CNN supported schemes helps to provide enhanced disease detection accuracy. Hence, the proposed framework employs CNN scheme based assessment of X-ray with DF, ensemble DF and DF+HF.

3 Methodology

This section demonstrates the procedure executed in the proposed X-ray examination system. The aims of this scheme are to implement a unique framework to evaluate various lung abnormalities, which can be recorded and examined with the help of X-rays.

Figure 1 depicts the proposed X-ray detection framework. This scheme consist three stages;

- Stage 1 helps to screen the patient with the radiological procedure and the collected images are then resized based on the requirement,
- Stage 2 implements VGG-UNet scheme to extract the lung section to be examined. This section removed the artifacts from the X-ray, which helps to achieve better disease detection accuracy
- Stage 3 extracts the DF and HF, implements the feature selection, feature integration and classification task

The results of stage 1 and stage 3 is then examined by the doctor and according to the lung infection rate, a decision regarding the treatment is obtained.

In this scheme, initially the essential X-ray is collected from the patient using the appropriate clinical protocol. The collected image is then resized according to the requirement of the proposed Disease Detection Framework (DDF) and the resized image then processed with VGG-UNet to remove the artifact. The extracted lung section from the test image is then considered to mine the DF and HF using the chosen procedure. This scheme helped to get a DF of dimension $1 \times 1 \times 1000$ and HF of dimension. To reduce the feature vector size, FA based feature selection is then implemented and the reduced features of DF and HF are then considered to construct the hybrid feature vector. This work implements a binary classification with a 5-fold cross-validation.

3.1 Image Database

To verify the clinical significance of the DDF, it is necessary to consider the clinical grade test images. To test the developed DDF, the TB class X-ray images provided by Rahman et al. [5] is considered. This image database consist the necessary test images along with the binary mask. Every image is resized into $512 \times 512 \times 3$ pixels during the implementation of the VGG-UNet. The outcome of the VGG-UNet is the binary image, which is then considered to extract the lung section using a pixel wise multiplication of the original test image and the extracted binary lung section. This procedure is depicted in Fig. 2. Figure 2(a) and (b) presents the test image and binary mask, Fig. 2(c) presented the segmented lung section and Fig. 2(d) presented the extracted lung region (Fig. 2(a) $\times$ Fig. 2(c)). This lung region is then considered to extract the DF and HF using the chosen methods.

After verifying the performance with the TB class X-ray images, other disease class images, such as COVID19 [12, 13], pneumonia [14], lung mass [15], and effusion [15], are considered for the examination. In this work, the VGG-UNet trained with the TB class X-ray images are considered to extract the lung sections from other images. During the experimental investigation, 1000 test images of dimension $224 \times 224 \times 1$ pixels for each class (including the healthy image) is considered in which 80% images are considered

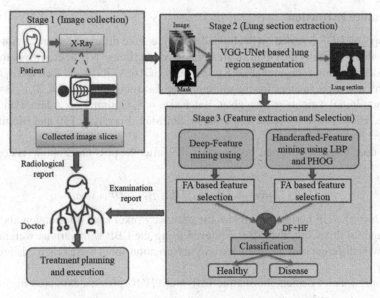

Fig. 1. Developed framework to examine the lung abnormality with X-ray images

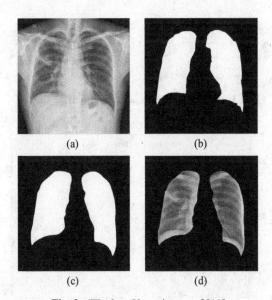

Fig. 2. TB class X-ray image of [11]

to train the DDF, 10% is for testing and 10% images are considered to validate the performance of the DDF. The achieved results are then presented and verified.

3.2 Feature Extraction

The proposed DDF is tested and validated using the three different feature vectors.

DF *mining*: Initially the DF from the segmented lung image is extracted using the pre-trained deep-learning schemes (PDLS), such as VGG16, VGG19 and ResNet18. In this work, the initial parameters for the chosen Deep-Learning-Scheme (DLS) is assigned as follows; Initial weights = ImageNet, Batch size = 8, Optimizer = Adam, Pooling = Average, Hidden layer = Relu, Classifier = SoftMax, Epochs = 150 and Monitoring metrics = Accuracy and loss.

The deep features extracted from each PDLS (after the possible dropout) is depicted in (1);

$$DF_{(1 \times 1 \times 1000)} = DF_{(1,1)}, DF_{(1,2)}, ..., DF_{(1,1000)} \tag{1}$$

HF mining: The HF normally provides the necessary pixel level information about the test image. In this work, the HF is achieved using the LBP with various weights and PHOG with different bins and the necessary information about these scheme is found in [16, 17].

Figures 3 and 4 depicts the sample LBP and PHOG patterns extracted from a test image.

(a) W=1   (b) W=2   (c) W-3   (d) W=4

Fig. 3. The LBP patterns achieved for W = 1 to 4

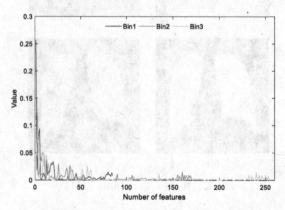

Fig. 4. PHOG features obtained for bin value of 1 to 3

The LBP (for an assigned weights of W = 1 to 4) and PHOG (for an assigned bins of P = 1 to 3) features mined in this work are depicted in Eqs. (2) to (10);

$$LBP1_{(1 \times 1 \times 59)} = W1_{(1,1)}, W1_{(1,2)}, ..., W1_{(1,59)} \tag{2}$$

$$LBP2_{(1\times1\times59)} = W2_{(1,1)}, W2_{(1,2)}, ..., W2_{(1,59)} \tag{3}$$

$$LBP3_{(1\times1\times59)} = W3_{(1,1)}, W3_{(1,2)}, ..., W3_{(1,59)} \tag{4}$$

$$LBP4_{(1\times1\times59)} = W4_{(1,1)}, W4_{(1,2)}, ..., W4_{(1,59)} \tag{5}$$

$$LBP_{(1\times1\times236)} = LBP1 + LBP2 + LBP3 + LBP4 \tag{6}$$

$$PHOG1_{(1\times1\times85)} = P1_{(1,1)}, P1_{(1,2)}, ..., P1_{(1,85)} \tag{7}$$

$$PHOG2_{(1\times1\times170)} = P2_{(1,1)}, P2_{(1,2)}, ..., P2_{(1,170)} \tag{8}$$

$$PHOG3_{(1\times1\times255)} = P2_{(1,1)}, P2_{(1,2)}, ..., P2_{(1,255)} \tag{9}$$

$$PHOG_{(1\times1\times510)} = P1_{(1,85)} + P2_{(1,170)} + P3_{(1,255)} \tag{10}$$

$$HF_{(1\times1\times746)} = LBP_{(1\times1\times236)} + PHOG_{(1\times1\times510)} \tag{11}$$

Ensemble *of DF*: The Ensemble of DF (EoDF) scheme employed in this work is adopted from the recent work of Kundu et al. (2021) [18]. This work verifies that confirms that the EoDF-supported medical image evaluation presents a better result than other methods. The averaging of DF is executed to get the necessary feature vector to achieve better disease detection. The proposed scheme combines the feature vectors of VGG16, VGG19, and ResNet18 to get a single feature vector with dimension $1 \times 1 \times 1000$; which is then considered to classify the X-ray images.

3.3 Feature Optimization with Firefly Algorithm

Feature extraction using heuristic algorithms has become increasingly popular recently to avoid overfitting [19]. A nature-inspired approach, FA, is used to optimize DF, HF, and EoDF [20]. It is extensively used by researchers to find solutions for a variety of optimization problems because of its superiority and optimization accuracy. This FA takes its cues from the social behaviour of fireflies. It initially adopts the Lévy-Flight (LF) strategy to reduce convergence times.

The mathematical expression of FA is depicted below;

Let us consider; there exist two groups of fireflies, like i and j. Due to its attractiveness, the firefly i will shift close to j, and this procedure can be demoted as follows;

$$X_i^{t+1} = X_i^t + \beta_0 e^{-\gamma d_{ij}^t}(X_j^t - X_i^t) + LF \tag{12}$$

where X_i^t = initial position of firefly i, X_j^t = initial position of firefly j, β_0 = attractiveness coefficient, γ = light absorption coefficient and d_{ij}^t = Cartesian Distance (CD) between flies.

The fireflies optimize features based on the CD between normal-class and MS-class features, and they consider features with a maximized CD, and discard features with a minimal CD. The FA parameters are assigned here as follows: 30 flies and 2500 iterations.

The implemented FA based feature selection helps to get the following feature vector values; VGG16 $= 1 \times 1 \times 331$, VGG19 $= 1 \times 1 \times 416$, ResNet18 $= 1 \times 1 \times 329$, HF $= 1 \times 1 \times 288$ and EoDF $= 1 \times 1 \times 374$. The following feature vector is then considered to classify the test images using the SoftMax classifier.

$$Feature_1 {}_{(1 \times 1 \times 619)} = VGG16 + HF \tag{13}$$

$$Feature_2 {}_{(1 \times 1 \times 704)} = VGG19 + HF \tag{14}$$

$$Feature_3 {}_{(1 \times 1 \times 617)} = ResNet18 + HF \tag{15}$$

$$Feature_4 {}_{(1 \times 1 \times 662)} = EoDF + HF \tag{16}$$

3.4 Performance Evaluation

This DD computes the metrics like True-Positive (TP), False-Negative (FN), True-Negative (TN), and False-Positive (FP), and from these values, other metrics presented in Eqs. (16) to (20) are derived [21, 22];

$$Accuracy = AC = \frac{TP + TN}{TP + TN + FP + FN} \tag{17}$$

$$\Pr ecision = PR = \frac{TP}{TP + FP} \tag{18}$$

$$Sensitivity = SE = \frac{TP}{TP + FN} \tag{19}$$

$$Specificity = SP = \frac{TN}{TN + FP} \tag{20}$$

$$F1 - Score = FS = \frac{2TP}{2TP + FN + FP} \tag{21}$$

4 Result and Discussion

An Intel i7 workstation with 20 GB RAM and 4 GB VRAM is used to execute this investigation using Python software. Initially, the VGG-UNet is trained and validated using the TB dataset provided by Rahman et al. [2]. Before executing this scheme, every image is resized into $512 \times 512 \times 3$ pixels and then the image and binary mask is considered to train the VGG-UNet to extract the lung section. The necessary information regarding this segmentation task can be found in [3].

Figure 5 depicts the experimental outcome achieved from the considered scheme for a chosen epoch value of 100. Figure 5(a) presents the training image (image and the mask), Fig. 5(b) and (c) depicts the achieved loss value and the accuracy (>98%) and Fig. 5(d) depicts the extracted lung section in binary form. To demonstrate the results with better visibility, this work considered the "verifies color map".

In order to extract the lung region more accurately, the binary lung section is multiplied by the original test image. An artifact-removed lung section is used to verify the DDF with DF, EoDF, and DF+HF classification performance.

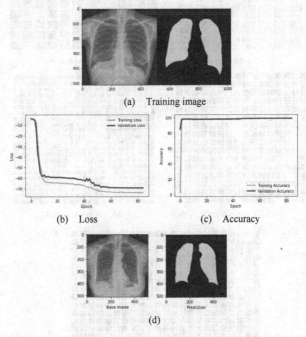

Fig. 5. Lung section extraction from X-ray using VGG-UNet

Classification task on the considered TB database is then executed using the pretrained VGG16 and the achieved result for each convolutional layer is presented as in Fig. 6. Figure 6(a) to (e) shows the various convolutional layer outcomes and it clearly demonstrate that, when the convolutional value increases, the image is transformed into features of finite size.

The classification task is executed using a epoch value of 150 with a SoftMax classifier and the achieved result during this process is depicted in Fig. 7. Figure 7(a) and (b) presents the loss and accuracy with respected to epochs, Fig. 7(c) shows the confusion-matrix initial values, like TP, FN, TN and FP values. Figure 7(d) presents the ROC curve, which presents an Area Under Curve (AUC) value of 0.981. This confirms that the proposed scheme helps to achieve a better TB detection accuracy when a 5-fold cross validation process and the achieved performance metrics of this process is tabulated in Table 2. Alike procedure is repeated with other feature vectors and the achieved results are recorded in Table 2. The result of this table confirms that, compared to the individual DF and EoDF, the DF+HF features, like Feature1 to Feature4 helped to achieve better TB detection accuracy (>98%) and this proves the value of this framework.

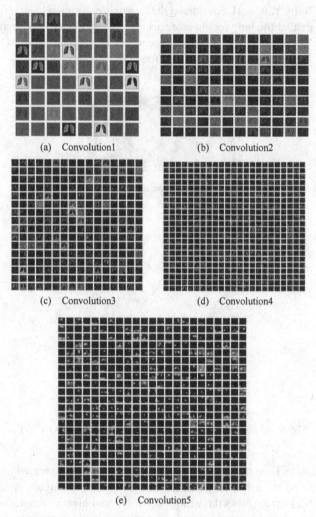

(a)   Convolution1

(b)   Convolution2

(c)   Convolution3

(d)   Convolution4

(e)   Convolution5

Fig. 6. Convolutional layer outcome for VGG16 scheme

The performance of the proposed DDF is then tested using the other X-ray images with diseases, COVID19, pneumonia, mass and effusion and this result also confirms that the proposed scheme helps to achieve a disease detection accuracy of >98% irrespective of the disease conditions. The future scope of this study includes, extending the proposed scheme to examine the disease in other medical imaging modalities, like CT and Magnetic Resonance Imaging (MRI). Further, the merit of the propose technique can be confirmed using the clinically collected medical images.

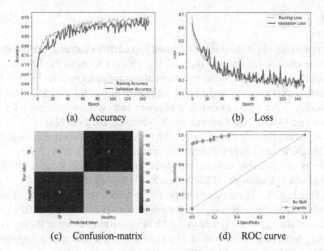

(a) Accuracy

(b) Loss

(c) Confusion-matrix

(d) ROC curve

Fig. 7. Performance metrics achieved with VGG16 when classifier using SoftMax

Table 2. Performance metrics achieved with DDF for various feature vectors

Feature	TP	FN	TN	FP	AC	PR	SE	SP	FS
VGG16	90	9	95	6	92.50	93.75	90.91	94.06	92.31
VGG19	93	6	92	9	92.50	91.18	93.94	91.09	92.54
ResNet18	94	6	91	9	92.50	91.26	94.00	91.00	92.61
EoDF	97	2	98	3	97.50	97.00	97.98	97.03	97.48
Feature1	98	2	99	1	98.50	98.99	98.00	99.00	98.49
Feature2	98	2	98	2	98.00	98.00	98.00	98.00	98.00
Feature3	98	2	99	1	99.00	99.00	99.00	99.00	99.00
Feature4	99	0	99	2	99.00	98.02	100	98.02	99.00

5   Conclusion

X-ray-based detection has been widely used for detection of lung infection because it is simple to use and occurrence is gradually rising. Using X-ray images, this work proposes a DDF for detecting lung infections. Initially, CNN segmentation is used to extract the lung section with better accuracy, and then the extracted lung region is used to mine the necessary DF and HF. Using the FA-based feature selection procedure, we determine the optimal values for DF and HF, and then combine these values serially to create a hybrid feature vector. X-ray images with TB infection are used in this work to perform segmentation and classification, and the result confirms that the proposed scheme leads to better results. A detection accuracy of >98% was achieved with the proposed DDF using X-ray images of COVID19, pneumonia, mass, and effusion.

References

1. Bhattacharyya, A., et al.: A deep learning based approach for automatic detection of COVID-19 cases using chest X-ray images. Biomed. Sig. Process. Control **71**, 103182 (2022)
2. Mahbub, Md.K., et al.: Deep features to detect pulmonary abnormalities in chest X-rays due to infectious diseaseX: Covid-19, pneumonia, and tuberculosis. Inf. Sci. **592**, 389–401 ((2022))
3. Krishnamoorthy, S., et al.: Framework to segment and evaluate multiple sclerosis lesion in MRI slices using VGG-UNet. Comput. Intell. Neurosci. **2022** (2022)
4. Bhandary, A., et al.: Deep-learning framework to detect lung abnormality–a study with chest X-ray and lung CT scan images. Pattern Recogn. Lett. **129**, 271–278 (2020)
5. Rahman, T., et al.: Reliable tuberculosis detection using chest X-ray with deep learning, segmentation and visualization. IEEE Access **8**, 191586–191601 (2020)
6. Akcay, S., Breckon, T.: Towards automatic threat detection: a survey of advances of deep learning within X-ray security imaging. Pattern Recogn. **122**, 108245 (2022)
7. Tavaziva, G., et al.: Chest X-ray analysis with deep learning-based software as a triage test for pulmonary tuberculosis: an individual patient data meta-analysis of diagnostic accuracy. Clin. Infect. Dis. **74**(8), 1390–1400 (2022)
8. Subramanian, N., Elharrouss, O., Al-Maadeed, S., Chowdhury, M.: A review of deep learning-based detection methods for COVID-19. Comput. Biol. Med. 105233 (2022)
9. Gite, S., Mishra, A., Kotecha, K.: Enhanced lung image segmentation using deep learning. Neural Comput. Appl. 1–15 (2022)
10. Dey, N., et al.: Customized VGG19 architecture for pneumonia detection in chest X-rays. Pattern Recogn. Lett. **143**, 67–74 (2021)
11. https://www.kaggle.com/datasets/tawsifurrahman/tuberculosis-tb-chest-xray-dataset
12. https://ieee-dataport.org/open-access/covid-19-and-normal-chest-x-ray
13. https://ieee-dataport.org/documents/covid-19-posteroanterior-chest-x-ray-fused-cpcxr-dataset
14. https://www.kaggle.com/datasets/paultimothymooney/chest-xray-pneumonia
15. https://www.kaggle.com/datasets/nih-chest-xrays/data
16. Gudigar, A., et al.: Global weighted LBP based entropy features for the assessment of pulmonary hypertension. Pattern Recogn. Lett. **125**, 35–41 (2019)
17. Chauhan, A., Chauhan, D., Rout, C.: Role of gist and PHOG features in computer-aided diagnosis of tuberculosis without segmentation. PLoS ONE **9**(11), e112980 (2014)
18. Kundu, R., et al.: Pneumonia detection in chest X-ray images using an ensemble of deep learning models. PLoS ONE **16**(9), e0256630 (2021)
19. Kadry, S., Rajinikanth, V., González Crespo, R., Verdú, E.: Automated detection of age-related macular degeneration using a pre-trained deep-learning scheme. J. Supercomput. **78**(5), 7321–7340 (2022)
20. Yang, X.-S.: Firefly algorithm, stochastic test functions and design optimisation. arXiv preprint arXiv:1003.1409 (2010)
21. Salehinejad, H., et al.: Synthesizing chest X-ray pathology for training deep convolutional neural networks. IEEE Trans. Med. Imaging **38**(5), 1197–1206 (2018)
22. Luo, L., et al.: Deep mining external imperfect data for chest X-ray disease screening. IEEE Trans. Med. Imaging **39**(11), 3 (2020)

Disease Detection and Risk Prediction System Based Web Application Using Machine Learning

Raj Kumar[✉], Ujjawal Singh, Soumya Sahoo, Ipsita Das, and Prashant Kr. Jha

Department of Computer Science and Engineering, C.V. Raman Global University,
Bhubaneswar, Odisha, India
ayerajkumar@gmail.com

Abstract. The purpose of the website We created is to provide consumers with health and disease predictions based on machine learning algorithms. The website is made with the user in mind, offering a straightforward and intuitive interface that makes it simple for users to get about the platform and access their health report and self-health evaluation.

The website's machine learning algorithms are made to assess user information, such as medical history, dietary preferences, and physical attributes, in order to produce a unique health report for each user. The health report will provide details on potential dangers and symptoms of various illnesses, as well as suggestions for safeguards and lifestyle modifications that might help lower such dangers.

The website also has a self-health assessment tool that lets users provide more details about their personal health and way of life. The machine learning algorithms will use this data to produce a more thorough health report that is customised to the user's requirements.

Keywords: Healthcare · Self-Assessment Tool · Diagnose · Machine Learning Algorithm · Web Application for Health and Disease Detection · Decision Tree · Django Backend

1 Introduction

Humans have long placed a high value on their health and well-being. Nonetheless, there has been a huge shift in how people view their health as a result of the development of technology and the rising demand for rapid gratification. Due to their capacity to correctly identify ailments and offer individualised healthcare solutions, machine learning algorithms have recently experienced tremendous growth in popularity in the healthcare sector.

The website we show in this study uses machine learning techniques to forecast ailments and give consumers a detailed health report. By giving users the resources, they need to take charge of their health and make wise decisions about their well-being, our website seeks to empower people. Users can detect potential health issues and take preventive action before it's too late thanks to our self-health assessment tool.

S. Kadry and R. Prasath (Eds.): MIKE 2023, LNAI 13924, pp. 237–249, 2023.
https://doi.org/10.1007/978-3-031-44084-7_23

Our research article intends to offer a thorough description of the website construction process, algorithm selection method, and validation techniques. We think that by offering a straightforward and approachable platform for disease prediction and self-health evaluation, our website has the potential to transform the healthcare sector. One of the primary advantages of our website is that it enables users to take charge of their health. By offering simple access to individualised health reports and self-assessment tools, users are able to recognise potential health hazards and take preventative action before to the onset of serious illnesses. This can lower healthcare expenses and improve the quality of life overall. Our idea is unusual because it combines the most recent advancements in machine learning algorithms with a user-friendly website design to create an accessible and efficient healthcare solution. We think that by using the power of technology, our website has the ability to revolutionise the healthcare business by making it more personalised, convenient, and affordable.

The work done in this research aims to provide a solution for self-assessment and disease prediction using machine learning algorithms. The project has successfully designed and built a web application that uses machine learning models in the backend to process the data for prediction of disease based on trained models. Our web application offers two major services, first one is 'Self-Assessment' for the health examination and another one is 'Diagnosis Section' to test the blood report of patient corresponding to suffering disease to know whether they are in risk or safe.

2 Literature Survey

Heart Health Prediction Using Web Application. It uses machine learning, classification, and data analysis techniques to predict heart conditions based on various parameters such as glucose level, BP, BMI, etc. The model achieved an accuracy of 75-82% using a random forest classifier and is presented as a web application using Streamlit [1].

Heart Disease Prediction Using Machine Learning and Data Mining here the proposed method for detecting heart disease using machine learning algorithms such as K-Nearest Neighbors, Naïve Bayes, Decision Tree, Support Vector Machines, and Random Forest. The models are uploaded to a web server using Flask framework, and users can input their data. The authors conducted experiments and found that K-Nearest Neighbors provides the highest accuracy of 87%. They also discuss previous studies on heart disease detection using machine learning and data mining techniques [2].

Disease Prediction Web App Using Machine Learning. The study evaluates different models such as Support Vector Classifier, Naive Bayes Classifier, and Random Forest Classifier, and uses Django to create a self-assessment system based on the best accuracy model. The objective is to improve healthcare through accurate and early disease detection. The proposed system, called HealthSure, allows users to register and check their symptoms to predict diseases [3].

Heart Disease Prediction Using Machine Learning, The proposal for a web application that aims to predict the occurrence of heart disease and suggest preventative measures. The project will use data mining techniques to extract hidden patterns from datasets and find a suitable machine learning technique for heart disease prediction. The objectives are to develop an easy-to-use platform, require no human intervention, suggest preventative measures. The project aims to improve medical efficiency, reduce costs, and enhance the quality of clinical decisions [4].

Disease Prediction Using Machine Learning and Django and Online Consultation. It is an online platform for users to receive quick medical advice based on their symptoms. The system uses data mining techniques to determine the most likely disease related to a patient's symptoms. Doctors can access patient information and diagnose them online. The system uses the Random Forest algorithm to predict diseases [5].

The authors conducted thorough analysis and normalization of models to understand the need for these predictions. The Voting Classifier of Decision Tree, Sigmoid SVC, and Adaboost achieved the highest accuracy of 88.57% for heart disease, while the voting classifier had an accuracy of 80.95% for diabetes. The study also suggests the potential extension of this methodology to predict a patient's immunity to COVID-19. Future work could involve developing a robust model with automated feature selection to analyze both diseases and their relationship to COVID-19 [6].

The research provided useful disease prediction based on symptoms. A script was developed to extract and format data as needed. Users can input symbols or select options for prediction. Users can also create medical profiles to contribute to the database and improve predictions over time. The analysis revealed similarities between diseases, but the training model would benefit from a larger database [7].

The proposed research had a positive impact on hospitals, catering to dynamic environments and benefiting both patients and organizations. It offers database filtering for monitoring disease outbreaks and ensures data security with individual hospital databases. Unique QR codes help identify patients and reduce wait times. The web app allows patients to access lab reports and medicine notes, eliminating the need for physical copies. Overall, the project improves patient experience and facilitates future consultations [8].

This research implemented a Hospital Management System using Django. The evaluation and assessment by respondents and end-users showed that the system effectively meets their needs and requirements. It was rated positively in terms of acceptability, effectiveness, quality, and productivity in automating hospital management. The conclusion is that the system is efficient, eliminates manual errors, improves hospital operations, enhances patient satisfaction, and overall improves the functioning of the hospital [9].

The research aims to provide a significant solution for icterus sufferers. The proposed rules engine framework is designed to identify parameters and administer appropriate medicines. Future plans involve implementing machine learning to enhance the rules engine's success [10].

3 Proposed System

Our project is a web-based application that is based on a medical and hospital ecosystem. This ecosystem is a comprehensive platform that is designed to provide patients with easy access to a variety of health care services and resources. This platform is a new strategy that has the potential to improve the healthcare business by providing patients with a variety of features that make the process more convenient and effective overall.

The capability of the platform to allow users to do self-evaluations and view health data online is one of the most important characteristics it possesses. Patients will be better able to monitor their own health and recognise any possible issues before they become more serious with the help of this tool. Patients are able to use this function by using the user-friendly interface that the platform provides. This interface makes it simple for patients to navigate and use the platform.

Comparatively with the existing system, we are able to achieve 89% accuracy in self-assessment system and in the other part, the diagnose section where the patient's health report is tested, there in heart disease testing we achieve 90.2% accuracy. Our project aims to build the whole health care environment to provide our users a complete health guide to know the disease - from symptoms to test their report data from blood sample, to know the potential risk of disease.

4 Analyzing Requirement

To build the project, the following requirements needed to be fulfilled:

- *Front-end Development:* The website's user interface was developed using HTML, CSS, and JavaScript.
- *Back-end Development:* Python Django framework was used for the back-end development, which handles user requests, retrieves and stores data, and interfaces with the machine learning algorithm.
- *Machine Learning Algorithm:* An appropriate machine learning algorithm was selected for disease prediction.
- *Self-assessment and Health Report Generation:* The website has a self-assessment module that allows users to input relevant information about their health, lifestyle, and medical history.

The tech stack used to build the project included HTML, CSS, and JavaScript for front-end development, Python Django framework for back-end development, and a machine learning algorithm for disease prediction.

4.1 Front-End Development

The project's front-end development included the design and creation of a user-friendly website interface utilising HTML, CSS, and JavaScript. HTML was utilised to structure the website's content and define text, graphics, and other page elements. Using font, colour schemes, and layout, CSS was utilised to style and make the website visually appealing.

4.2 Back-End Development

The project's backend was developed using the Python Django web framework. Django is a robust and flexible framework that offers a comprehensive solution for web development. It is renowned for its usability, scalability, and safety.

The website's backend was responsible for processing user requests, retrieving and storing data, and communicating with the machine learning algorithm. The backend was developed utilising the Model-View-Controller (MVC) architectural paradigm of Django. This pattern divides the programme into Model, View, and Controller components.

4.3 Machine Learning Algorithm

We are developing symptom checker tools as part of our ongoing work on machine learning, which will be used in the healthcare project we are working on. This will allow patients to perform their own self-assessment check-ups. These tools are intended to assist patients in evaluating their symptoms and providing recommendations regarding the next steps that should be taken.

In this part of the project, machine learning is also being used to develop predictive models in order to assist our healthcare project in predicting the health risks that patients may face. These models are able to identify patients who are at a high risk of developing certain conditions and provide personalised recommendations on how to manage the patients' health by analysing large sets of data pertaining to individual patients.

Decision tree is a method for machine learning that builds a tree-like representation of decisions and their possible outcomes. It is useful for classification and regression issues, and is ideally suited to data with both categorical and continuous characteristics. Decision trees are simple to read, and the resulting model may be represented graphically and explained to non-specialists.

Logistic regression is an approach for supervised learning used to solve classification problems. It models the link between a dependent variable and one or more independent factors and estimates the likelihood that the dependent variable belongs to a specific class. It is frequently used in medical research to estimate a patient's likelihood of developing a particular condition.

Random forest classifier is a technique for ensemble learning that mixes numerous decision trees to enhance accuracy and avoid overfitting. At each split, it randomly selects a subset of the features and trains each tree using a separate random sample of the data. It is frequently employed for high-dimensional datasets with numerous features.

Naïve Bayes algorithm is a probabilistic classification system. Bayes' theorem states that the probability of a hypothesis is updated based on fresh data. The Naïve Bayes assumption is that the features are conditionally independent of the class label, hence the term "Naïve." It excels at text categorization tasks, including spam detection and sentiment analysis.

4.4 Data Management

Our training dataset contains a total of 4920 rows and 133 columns. Out of these columns, 132 represent different symptoms that a person can experience, and the final column represents the prognosis or target disease.

We have a total of 41 diseases in our dataset, and each disease is represented by a set of symptoms that have been mapped to it. Each disease is present in 120 rows of our dataset. This means that for each disease, we have 120 instances of symptom data that can be used to train our machine learning model.

During the training phase, we used the symptom data from these 41 diseases dataset to train our machine learning model. We divided the patient dataset into two files: training.csv and testing.csv. one file was used to train the model, while the other file was used to validate the model's performance during the training process (Table 1).

Table 1. Training Dataset Statistics

Type	Count
rows	4920
columns	133
symptoms	132
disease	41
rows-disease	120

5 Methodology

We've been working on extending the project to create a web application that incorporates a machine lesearning model [11, 12]. The web application provides an interface that allows users to input their symptoms and receive a prediction (ref. Fig. 3) of the most likely disease they might be suffering from, along with the confidence score of having that disease [13–15].

Our project aimed to predict diseases using medical data through a machine learning system employing multiple supervised learning approaches, such as logistic regression, random forest classifier, Naïve Bayes, and decision trees. During the testing phase, we evaluated the performance of these algorithms using k-fold cross-validation, a typical strategy for model selection that involves dividing the data into k subsets, training the model on k-1 subsets, and assessing the model's performance on the final subset. We performed this procedure with k values of 2, 4, 6, 8, and 10 and calculated the ultimate score by averaging the results, represented in Table 2 and Fig. 1.

Based on our evaluation (ref. Table 2), we found that the decision tree algorithm at k-value: 2 was the most effective method for predicting diseases using medical data. This is because decision trees are simple to read and can be displayed and explained to non-experts, making them a valuable tool for medical practitioners and patients. Additionally, decision trees can accommodate both categorical and continuous variables, which are prevalent in medical datasets. We found that the decision tree algorithm accurately classified 89% of the test cases, a high rate of accuracy. In contrast, we observed that some methods, such as logistic regression, random forest classifier, and Naïve Bayes, appeared overfit, with training scores above testing scores. Overfitting happens when a model is overly complicated and has learned the noise in the training data, resulting in inadequate generalization to new data. Therefore, the decision tree approach was the most appropriate for predicting diseases using medical data. So, we choose to build our web application backend using the decision tree model to process and predict the health.

Table 2. Self-Assessment, Testing Score using K-fold cross-validation a different k value.

Algorithm	k = 2	k = 4	k = 6	k = 8	k = 10
logistic regression	100%	100%	100%	100%	100%
decision tree	89%	100%	100%	100%	100%
random forest	100%	100%	100%	100%	100%
naïve bayes	100%	100%	100%	100%	100%

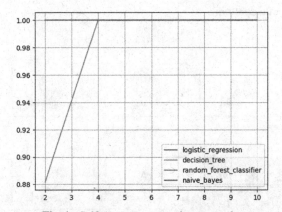

Fig. 1. Self-Assessment, testing score plot

The web application has another service (ref. Fig. 4) for health report generation where the user inputs data from a blood report to determine whether they are at risk or safe [16].

We have performed model preparation on heart disease. To achieve this, four machine learning models were trained on the available data: random forest classifier, K-nearest neighbors (KNN), logistic regression, and naive Bayes. The performance of each model was evaluated using their training and testing scores, and the results are as follows (Table 3):

In general, the higher the accuracy score, the better the model performs. From the results above, we can see that the KNN and logistic regression models performed the best, achieving the highest testing scores of 90.2%. However, between these two models, logistic regression had a slightly better training score, suggesting that it might be more reliable when it comes to predicting new cases [17–21].

Overall, the logistic regression model is likely the best choice for this web application as it demonstrated the highest testing score and a relatively high training score.

Table 3. Report Testing, Heart Disease Accuracy table

Algorithm	Training Score	Testing Score
Random Forest	81%	86.9%
KNN	83.05%	90.2%
Logistic Regression	85.12%	90.2%
Naïve Bayes	84.29%	86.9%

The web application service has a total of six diagnoses (Fig. 2) in addition to heart disease (Fig. 4). These include liver disorder, tuberculosis, kidney disorder, blood sugar test, and diabetes test.

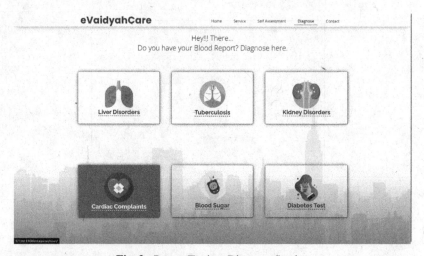

Fig. 2. Report Testing, Diagnose Section

The web application is designed to be user-friendly and accessible to anyone. Users can input their symptoms in a simple and straightforward way, using dropdown menus to select. The web application sends the symptom data to the backend, where our machine learning model processes it and predicts the most likely disease based on the symptom data. The prediction is then displayed to the user along with the confidence score [22–24].

One of the benefits of our platform is that it allows individuals to self-assess (Fig. 3) their health status without requiring a doctor's visit. This can be particularly useful for people who live in remote areas, or who may not have access to medical professionals due to cost or other reasons. Our platform can also help people become more aware of their health status, and encourage them to seek medical attention if necessary.

Suppose you are feeling "breathlessness, chest pain, nausea and back pain" in this case you have to visit our 'self-assessment section' (Fig. 3) to know about the disease that may risks in you.

Fig. 3. Self-Assessment Section

Above in Fig. 3 Our system predicts the potential health disease from the give symptoms. You may risk of 'Heart Attack'. Now you have to visit testing laboratory to fill the following data in 'diagnosis section' (Fig. 4) of heart diagnose, which will examine the report from machine learning trained model to give the health report whether you're in risk or safe of heart attack.

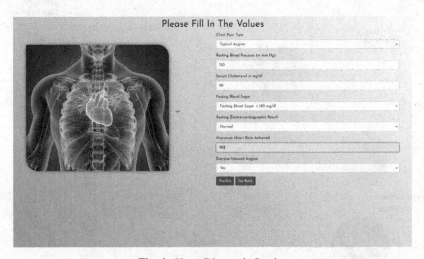

Fig. 4. Heart Diagnosis Section

In summary, our web application provides a user-friendly and accessible platform for health assessment and disease detection, refer Fig. 5 for Data Flow Diagram of our Web Application where we have shown the services that our application offers in convenient way.

6   System Architecture

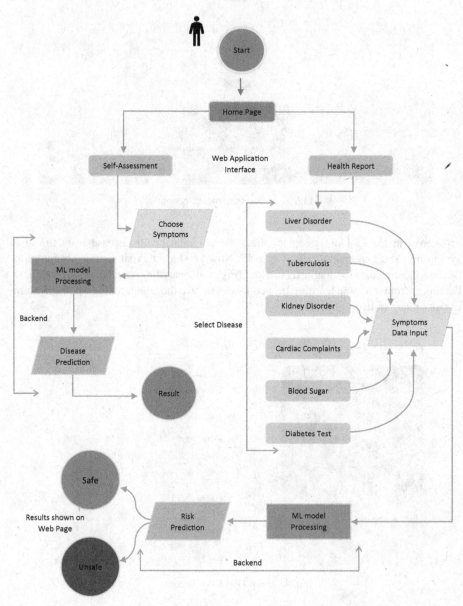

Fig. 5. Data Flow Diagram

7 Future Work

In the future, we plan to develop a new feature for our web application by using recommender system model to calculate the risk percentage of a person's daily or regular food consumption.

This model will be very beneficial for people who want to take control of their health by understanding whether their diet is healthy or unhealthy for them.

To use this model, individuals will need to input their daily diet items and provide information about any existing health conditions they have. Based on this data, the model will provide personalized recommendations on what foods to avoid to help reduce the risk of exacerbating any existing health conditions.

Overall, this model has the potential to significantly improve people's health outcomes by providing tailored and accurate recommendations for their diet.

8 Conclusion

In conclusion, the work done in this research aims to provide a solution for self-assessment and disease prediction using machine learning algorithms. The project has successfully designed and built a web application that uses machine learning models in the backend to process the data for prediction of disease based on trained models.

The dataset used in this project has 133 columns, with 132 of these columns being symptoms that a person experiences, and the last column being the prognosis. These symptoms are mapped to 41 diseases that the project aims to classify. The project has trained its models on a training dataset and tested it on a testing dataset.

Decision tree classifier model was found to give the best accuracy among other models for disease prediction. The project has designed and built a user-friendly web application that allows users to perform self-assessment and generate health reports about their risk or safety from specific diseases.

Overall, this project has successfully achieved its goal of building a machine learning-based solution for disease prediction and self-assessment. It provides an easy-to-use tool that can assist patients in identifying their disease risk, which ultimately leads to early detection and treatment, thereby improving the overall healthcare system.

Acknowledgment. We would like to express our gratitude to all those who have contributed to the successful completion of this research project. First and foremost, we would like to thank our supervisor for providing us with invaluable guidance and support throughout the project.

Once again, we extend our heartfelt thanks to all those who have played a role in making this project a success.

References

1. Kalshetty, J.N., Achyutha Prasad, N., Mirani, D., Kumar, H., Dhingra, H.: Heart health prediction using web application. Int. J. Health Sci. (2022)
2. Srivastava, K., Choubey, D.K.: Heart disease prediction using machine learning and data mining. Int. J. Recent Technol. Eng. (2020)

3. Nawab, M.A., Malpani, T., Kaundal, T., Soni, M.D.: Disease prediction web app using machine learning. Int. Res. J. Mod. Eng. Technol. Sci. (2022)

4. Jagtap, A., Malewadkar, P., Baswat, O., Rambade, H.: Heart disease prediction using machine learning. Int. J. Res. Eng. Sci. Manag. (2019)

5. Tambe, Y., Awhad, S., Keny, A.: Disease prediction using machine learning and Django and online consultation (2022)

6. Dhande, B., Bamble, K., Chavan, S., Maktum, T.: Diabetes & heart disease prediction using machine learning. ITM Web Conf. **44**, 03057 (2022)

7. Srivastava, M., Yadav, V., Singh, S.: Implementation of web application for disease prediction using AI. BOHR Int. J. Data Min. Big Data (2020)

8. Ramees, A.R., Akhil, P.S., Amrutha, M.A., Fathima, J., Elia, N.: Hospital managing QR code web application using Django and Python. Int. J. Creat. Res. Thoughts (IJCRT) (2020)

9. Gawande, M.V., Pisey, P., Shinde, A.A., Ghagre, P., Bhusari, K.: Hospital management system in Django. Int. Res. J. Mod. Eng. Technol. Sci. (2022)

10. Roghit, K.K., Jayachandran, G.: Assessment for hyperbilirubinemia health care using Django. Int. Res. J. Eng. Technol. (IRJET) (2020)

11. Mukherjee, D., Raj, I., Mishra, S.: Song recommendation using mood detection with Xception model. In: Mallick, P.K., Bhoi, A.K., Barsocchi, P., de Albuquerque, V.H.C. (eds.) Cognitive Informatics and Soft Computing. LNNS, vol. 375, pp. 491–501. Springer, Singapore (2022). https://doi.org/10.1007/978-981-16-8763-1_40

12. Abhishek, Tripathy, H.K., Mishra, S.: A succinct analytical study of the usability of encryption methods in healthcare data security. In: Tripathy, B.K., Lingras, P., Kar, A.K., Chowdhary, C.L. (eds.) Next Generation Healthcare Informatics. SCI, vol. 1039, pp. 105–120. Springer, Singapore (2022). https://doi.org/10.1007/978-981-19-2416-3_7

13. Sinha, K., Miranda, A.O., Mishra, S.: Real-time sign language translator. In: Mallick, P.K., Bhoi, A.K., Barsocchi, P., de Albuquerque, V.H.C. (eds.) Cognitive Informatics and Soft Computing. LNNS, vol. 375, pp. 477–489. Springer, Singapore (2022). https://doi.org/10.1007/978-981-16-8763-1_39

14. Mishra, Y., Mishra, S., Mallick, P.K.: A regression approach towards climate forecasting analysis in India. In: Mallick, P.K., Bhoi, A.K., Barsocchi, P., de Albuquerque, V.H.C. (eds.) Cognitive Informatics and Soft Computing. LNNS, vol. 375, pp. 457–465. Springer, Singapore (2022). https://doi.org/10.1007/978-981-16-8763-1_37

15. Patnaik, M., Mishra, S.: Indoor positioning system assisted big data analytics in smart health-care. In: Mishra, S., González-Briones, A., Bhoi, A.K., Mallick, P.K., Corchado, J.M. (eds.) Connected e-Health. SCI, vol. 1021, pp. 393–415. Springer, Cham (2022). https://doi.org/10.1007/978-3-030-97929-4_18

16. Periwal, S., Swain, T., Mishra, S.: Integrated machine learning models for enhanced security of healthcare data. In: Mishra, S., Tripathy, H.K., Mallick, P., Shaalan, K. (eds.) Augmented Intelligence in Healthcare: A Pragmatic and Integrated Analysis. SCI, vol. 1024, pp. 355–369. Springer, Singapore (2022). https://doi.org/10.1007/978-981-19-1076-0_18

17. De, A., Mishra, S.: Augmented intelligence in mental health care: sentiment analysis and emotion detection with health care perspective. In: Mishra, S., Tripathy, H.K., Mallick, P., Shaalan, K. (eds.) Augmented Intelligence in Healthcare: A Pragmatic and Integrated Analysis. SCI, vol. 1024, pp. 205–235. Springer, Singapore (2022). https://doi.org/10.1007/978-981-19-1076-0_12

18. Dutta, P., Mishra, S.: A comprehensive review analysis of Alzheimer's disorder using machine learning approach. In: Mishra, S., Tripathy, H.K., Mallick, P., Shaalan, K. (eds.) Augmented Intelligence in Healthcare: A Pragmatic and Integrated Analysis. SCI, vol. 1024, pp. 63–76. Springer, Singapore (2022). https://doi.org/10.1007/978-981-19-1076-0_4

19. Banerjee, D., Kukreja, V., Hariharan, S., Sharma, V.: Fast and accurate multi-classification of kiwi fruit disease in leaves using deep learning approach. In: 2023 International Conference on Innovative Data Communication Technologies and Application (ICIDCA), Uttarakhand, India, pp. 131–137 (2023). https://doi.org/10.1109/ICIDCA56705.2023.10099755

20. Raj, A., Sharma, V., Shanu, A.K.: Comparative analysis of security and privacy technique for federated learning in IoT based devices. In: 2022 3rd International Conference on Computation, Automation and Knowledge Management (ICCAKM), pp. 1–5. IEEE, November 2022

21. Manikandan, N., Ruby, D., Murali, S., Sharma, V.: Performance analysis of DGA-driven botnets using artificial neural networks. In: 2022 10th International Conference on Reliability, Infocom Technologies and Optimization (Trends and Future Directions) (ICRITO), pp. 1–6. IEEE, October 2022

22. Ali, M.A., Balamurugan, B., Sharma, V.: IoT and blockchain based intelligence security system for human detection using an improved ACO and heap algorithm. In: 2022 2nd International Conference on Advance Computing and Innovative Technologies in Engineering (ICACITE), pp. 1792–1795. IEEE, April 2022

23. Juyal, V., Saggar, R., Pandey, N.: An optimized trusted-cluster–based routing in disruption-tolerant network using experiential learning model. Int. J. Commun. Syst. 33(1), e4196 (2020)

24. Juyal, V., Pandey, N., Saggar, R.: A heuristic light weight security algorithm for resource constrained DTN routing. In: 2016 IEEE International Conference on Computational Intelligence and Computing Research (ICCIC), pp. 1–4. IEEE, December 2016

Weighted Average Ensemble Approach for Pediatric Pneumonia Diagnosis Using Channel Attention Deep CNN Architectures

C. R. Asswin[1], J. Arun Prakash[1], K. S. Dharshan Kumar[1], Avinash Dora[1],
V. Sowmya[1](✉), Meshari Almeshari[2], and Yasser Alzamil[2]

[1] Center for Computational Engineering and Networking (CEN), Amrita School of Engineering,
Coimbatore, Amrita Vishwa Vidyapeetham, Coimbatore, India
v_sowmya@cb.amrita.edu
[2] Department of Diagnostic Radiology, College of Applied Medical Sciences,
University of Ha'il, Ha'il, Saudi Arabia
{m.almeshari,y.alzamil}@uoh.edu.sa

Abstract. Pediatric pneumonia is a serious medical condition in which fluid fills the air sacs of the lungs. While chest X-rays have emerged as a more effective alternative to traditional diagnosis methods, the low radiation levels of X- rays in children have made accurate identification more challenging, leading to human-prone errors. To address this issue, deep learning architectures like Convolutional Neural Networks (CNNs) have been increasingly used for computer-aided diagnosis of chest X-ray images. In this paper, we propose an efficient Channel Attention (ECA) module attached to the end of pre-trained ResNet50 and DenseNet121, VGG19. We also present a weighted average ensemble based on our proposed model's performance. Our approach achieved an accuracy of 95.67%, precision of 94.81%, recall of 98.46%, F1 score of 96.60%, and an AUC curve of 94.74% on the pediatric pneumonia dataset. In conclusion, our proposed architecture holds promise for aiding in real-time pediatric pneumonia diagnosis and potentially improving patient outcomes.

Keywords: Pediatric Pneumonia · Chest X-rays · Computer-Aided Diagnosis (CAD) · Deep learning · Attention · Weighted average ensemble

1 Introduction

In recent decades, many infections and diseases have spread globally. Pneumonia is a lung disease caused by a virus or bacteria that fills the air sacs of the lungs with fluid, resulting in pus-filled pulmonary alveoli. This blockage reduces the lung's capacity to hold oxygen, which can lead to various symptoms such as fatigue and lethargy. The United Nations Children's Fund (UNICEF) has also reported that around 8,00,000 young children which also includes 1,53,000 newborns died due to Pneumonia. Studies suggest that individuals with weaker immune systems are more susceptible to pneumonia thus, geriatric and pediatric populations are the most prominent target of pneumonia

© The Author(s), under exclusive license to Springer Nature Switzerland AG 2023
S. Kadry and R. Prasath (Eds.): MIKE 2023, LNAI 13924, pp. 250–260, 2023.
https://doi.org/10.1007/978-3-031-44084-7_24

with the pediatric population having the highest casualty rate [1]. Numerous diagnostic procedures are employed to diagnose pneumonia in children some of which include measuring blood oxygen levels using pulse oximetry, Complete Blood Count (CBC), and sputum tests. However, abnormal CBC and oxygen levels can be indicative of other lung infections and are not specific to pneumonia. Chest X-ray is used as an affordable alternative for pneumonia detection. But the manual examination of chest X- rays by doctors and radiologists is a long and tedious process with human-prone errors.

These reasons motivate the need for a Computer-Aided Diagnostic (CAD) model, which is accurate with its prompt predictions. CAD serves as a secondary check on the predictions made by physicians and radiologists, thus reducing the possibility of human errors. CAD methods are already in use for different biomedical applications such as the detection of tumors, lesions in medical images, Parkinson's disease [16], hypoxia detection [17], and retinal disease [18]. Deep learning networks like Convolutional Neural Networks (CNN) have simplified the creation of Computer-Aided Diagnosis (CAD) models by automating the process of feature engineering and learning more abstract features, resulting in improved performance [2]. Attention-based convolutional neural networks (CNNs) have shown improved performance, using attention modules to focus on important parts of the input and residual-based attention to prevent gradient loss and improve error propagation. Our proposed solution for pediatric pneumonia diagnosis is simple and replicable. This study discusses the impact of using an efficient channel attention module with weighted average ensemble for pediatric pneumonia diagnosis.

The rest of the paper is divided into the following sections: In Section 2, briefly reviews the literature and summarize the existing limitations. Section 3 introduces the theoretical background on Efficient Channel attention modules, and weighted average ensemble used in the study. Dataset description and proposed methodology is explained in Section 4. The experimental results are analysed and discussed with plots in Section 5. Finally, in Section 6, we conclude our work, summarizing the problem and the limitations of our approach, along with possible future works.

2   Literature Review

The manual feature extraction methods that previously required certain filters have been replaced with deep learning systems. MLP approaches were largely employed in the early studies on pattern classification but with the drawback of being unable to acquire local information. Convolutional Neural Networks (CNNs) were developed to solve this problem. The pediatric pneumonia dataset was introduced by Kermany et al. [3], who also suggested a transfer learning strategy utilizing neural networks. Although transfer learning-based techniques performed better, they still had some limitations such as the loss of spatial information as convolutional layers increased in more complex CNN systems. Gaobo et al. [2] solved this issue by addressing the use of transfer learning and residual connections to stop the loss of spatial information during feature extraction. Kanakaprabha et al. [19] compared the performance of COVID-19 and Pneumonia from X-rays using CNN, which can distinguish between COVID-19 and Pneumonia subtypes, allowing for rapid and accurate diagnosis. Cha et al. [4] found out that using CNN models which have pre- trained weights can be used for training for Pneumonia Chest X-rays classification since it is faster as well as more efficient. Several research papers have discovered that the ensemble model performs better when compared to single pretrained models. Huang et al. [5] compared the performance of various single pretrained models such as EfficientNetV2-B0, EfficientNetV2-B1, EfficientNetV2-B2, and 4 other variations of Efficient Net models and their proposed stacking ensemble model and found out that the ensemble model performed much better than the single pretrained models in terms of classification of multiple chest diseases using chest X-ray images as well as CT images.

Traditional CNNs have limitations in selectively attending to the most informative regions of the input, which can lead to poor performance in scenarios where some parts of the input are more relevant to the task than others. Attention-based CNN models address this limitation by enabling the model to selectively attend to the most important regions of the input image while ignoring the less informative ones. Hu et al. [6] introduced Squeeze-and-Excitation Networks (SENet), which perform channel-wise feature recalibration by leveraging global information. The key idea behind SENet is to learn to selectively emphasize the most important features in each channel of the CNN output. To enhance their representational power, Woo et al. [7] came up with a CBAM module that performs both spatial and channel attention to adaptively recalibrate feature maps. Selective Kernel Networks (SKNet) by Li et al. [3] uses a set of learnable kernels to adaptively select the most informative kernel size for each channel. To preserve the inter-channel dependencies well, Zhang et al. [8] proposed Efficient Channel Attention (ECA) which performs channel- wise feature recalibration by using a lightweight 1D convolutional operation, which can be efficiently integrated into existing CNN architectures. Guo et al. [9] proposed a deep residual neural network model ECA-XNet based on the channel which can capture more rich information and perform much better than many CNN models. Additionally, Fan et al. [10], also proposed a Multi Kernel size Spatial Channel (MKSC) attention network for covid-19 detection from X-ray images which outperformed other attention mechanisms such as SE-Net, ECA-Net, and Spatial Attention. This shows that parallel multi-kernel-size spatial and channel attention modules in MKSC can suppress the shadows and skeletal noises as well as enhance the pathological features of the chest X-ray images.

3 Background

3.1 ECA

The use of attention mechanisms in image classification tasks has demonstrated promising outcomes in terms of enhancing the accuracy and interpretability of deep learning models, as noted by Guo et al. [9]. Various types of attention mechanisms have been proposed for image classification, including soft, hard, self, and non-local attentions. Soft attention mechanisms, such as spatial and channel attention, can help to highlight the most relevant features in an image by assigning different weights to different regions or channels. Hard attention mechanisms, such location-based attention, can be used to selectively crop and focus on specific regions of an image. Self-attention mechanisms, such as the transformer architecture, can capture the relationships between different parts of the image to generate context- aware representations. Non-local attention mechanisms can capture long-range dependencies between different parts of an image, enabling models to better understand the global structure and context of an image as discussed by Zhang et al [11].

Despite the effectiveness of attention mechanisms in image classification, there are still challenges that need to be addressed. These include the scalability and computational efficiency of attention mechanisms. In response to these challenges, Wang et al. [12] proposed the Efficient Channel Attention (ECA) mechanism, which utilizes a lightweight and parallelizable 1D convolution operation to compute attention weights across channels. The ECA mechanism has been shown to outperform existing attention mechanisms in terms of accuracy and efficiency on various image classification benchmarks.

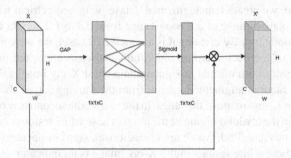

Fig. 1. Diagram of Efficient Channel Attention Module

Figure 1 depicts the efficient channel attention (ECA) module. The ECA module operates on the aggregated features obtained by global average pooling (GAP), which is a common technique used in CNNs for reducing spatial dimensions while retaining important features. The ECA module generates channel weights by performing a fast 1D convolution of size k, where k is adaptively determined via a mapping of channel dimension C. The output of this convolution is then passed through a sigmoid activation function, which scales the weights to a range between 0 and 1. These channel weights represent the importance of each channel in the feature map, allowing the model to selectively focus on the most relevant information for classification.

3.2 Weighted Average Ensemble

Weighted average ensemble is a machine learning technique that combines the predictions of multiple models to improve prediction accuracy. This method assigns weights to individual models based on their performance on a validation set and computes a weighted average of the predictions. Mathematically, the weighted average ensemble can be represented as:

$$y(x) = \frac{\sum_i (w_i * p_i(x))}{\sum_i (w_i)} \tag{1}$$

where $y(x)$ is the weighted average prediction, $p_i(x)$ is the prediction of the ith model for the input x, w_i is the weight assigned to the ith model, and $\sum_i$ represents the sum over all i models in the ensemble.

4 Dataset Description and Experimental Design

In this study, **the Kermany et al.** [3] dataset was used, which is comprised of chest X-ray images from children aged 1 to 5 years old, with a majority of the patients being between 2 and 3 years old. The dataset was collected from the Guangzhou Women and Children's Medical Center and consisted of both anteroposterior and posteroanterior views of the chest. The images were labeled as either normal or pneumonia by experienced radiologists. The use of this dataset allowed the researchers to develop a deep learning algorithm that could accurately detect cases of pneumonia in chest X-ray images.

However, the initial dataset distribution had a class imbalance, with 1583 X-rays labeled as Normal and 4273 X-rays labeled as Pneumonia, as shown in Table 1. To address this issue, we created an augmented dataset with geometrical transformations. This involved a rotation range of 20, zoom range from 0.8 to 1.2, height and width shift of 0.2, and horizontal flip. By creating this augmented dataset, we were able to improve the performance of the deep learning algorithms and ensure that the model was able to accurately classify both normal and pneumonia chest X-ray images. Validation was performed using random augmented images from the training set, as shown in Table 2. Figure 2 shows an X-ray image of the lungs. In the image, the second row of white lesions can be observed in the alveolar region of the lungs. These white lesions correspond to the accumulation of pus and fluid, which are characteristic signs of pneumonia. Therefore, the presence of these white lesions in the X-ray image is an indicator of pneumonia in the patient's lungs.

Table 1. Distribution of the original dataset

Category	Train	Test
Normal	1349	234
Pneumonia	3883	390
Total	5232	624

Table 2. Distribution of the augmented dataset for our study

Category	Train	Test	Validation
Normal	3,884	234	970
Pneumonia	3,883	390	970
Total	7767	624	1940

4.1 Proposed Methodology

The proposed pipeline is inspired from ECA-net to make robust predictions and uses a weighted average ensemble technique for final classification. The model takes in an

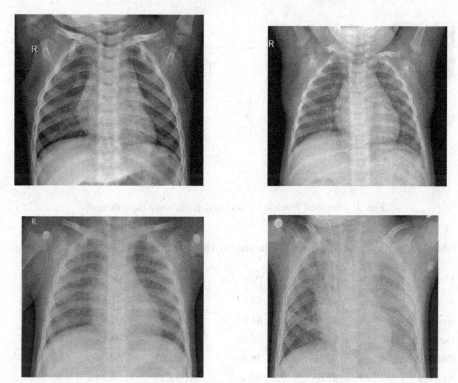

Fig. 2. Samples of Normal X-rays and Pneumonia X-rays from the dataset in the first row and second row respectively.

input image of size 224x224, with each image normalized by the Image data generator and on-the-fly augmentations with geometric transformations.

The features from each of ResNet50, DenseNet121 and VGG19 are passed through the ECA attention block. The resulting attention-aware features are sent as input to a global average pooling layer, followed by the use of 2 dense layers with 512 and 256 neurons respectively. Dropout of 0.4 and 0.2 have been added between each of these dense layers. Finally, the predictions of the individual deep architectures are sent to weighted average ensemble for final classification, which helps to improve the overall accuracy and metrics of the model, as shown in Fig 3.

5   Results and Discussion

The proposed architecture, which utilizes an ensemble of ResNet50-Attention, DenseNet121-Attention, and VGG19- Attention, exhibits superior performance compared to various other baseline deep CNN models when tested on the Kermany dataset [3]. The proposed architecture outperforms several existing deep CNN models, as shown in Table 3.

Table 3 provides a comparison of the performance of various deep CNN architectures in terms of accuracy, precision, recall, F1 score, and AUC. Based on the results, it can

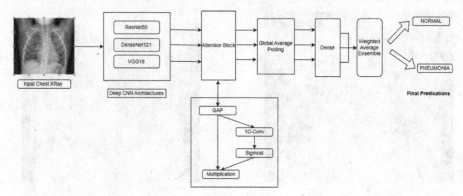

Fig. 3. Proposed pipeline for pediatric pneumonia classification

Table 3. Comparison of performance of the proposed model with several deep CNN architectures

Model	Accuracy	Precision	Recall	F1	AUC
Resnet50	88.62	84.90	99.49	91.62	85.00
Resnet152	86.86	82.63	100.0	90.49	82.48
Densenet121	89.42	85.68	99.74	92.18	85.98
Densenet201	80.76	76.47	100.0	86.66	74.35
VGG19	91.63	90.00	99.23	94.39	90.43
Xception	83.97	79.59	100.0	88.64	78.63
Proposed Pipeline	**95.67**	**94.81**	**98.46**	**96.60**	**94.74**

be inferred that Resnet50, Densenet121, and VGG19 have the highest accuracy among the models tested, with VGG19 having the overall highest recall and F1 score. Overall, these architectures perform well on the given dataset, while others may have slightly lower performance (Table 4).

The proposed deep learning attention models show improved performance compared to their respective baseline models based on evaluation. The ensemble of the predictions from these three models after the weighted average ensemble results in a high accuracy of 95.67% and AUC of 94.74%. These results indicate the potential usefulness of the proposed models for accurate classification tasks in the field of deep learning. Figure 5 displays the confusion matrix of the test data, revealing that the proposed model generated 21 false positives and 6 false negatives. Meanwhile, the validation loss and accuracy plots in Figure 4 indicate that the model was appropriately trained without overfitting.

Figure 6 illustrates the t-SNE plot of the features extracted from the penultimate layer of our proposed models. The plot provides a visualization of the feature representations, revealing a clear formation of clusters with minimal overlaps and outliers. The t-SNE plot serves as evidence of the effectiveness of our proposed models in capturing and distinguishing between relevant features in the pediatric pneumonia dataset.

Table 4. Performance comparison between proposed attention model and baseline deep CNN model.

Model	Accuracy	Precision	Recall	F1	AUC
Resnet50	88.62	84.90	99.49	91.62	85.00
Resnet50-Attn	**94.07**	**94.46**	**96.15**	**95.30**	**93.38**
Densenet121	89.42	85.68	99.74	92.18	85.98
Densenet121-Attn	**94.07**	**94.68**	**95.90**	**95.29**	**93.46**
VGG19	83.97	79.59	100.0	88.64	78.63
VGG19-Attn	**94.55**	**93.41**	**98.21**	**95.75**	**93.33**
Proposed Pipeline	**95.67**	**94.81**	**98.46**	**96.60**	**94.74**

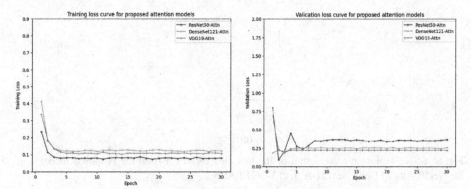

Fig. 4. Training and validation accuracy – loss history of the proposed model

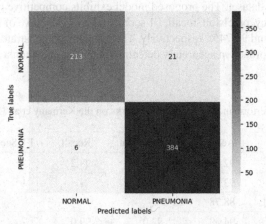

Fig. 5. Confusion matrix for proposed pipeline on the test data

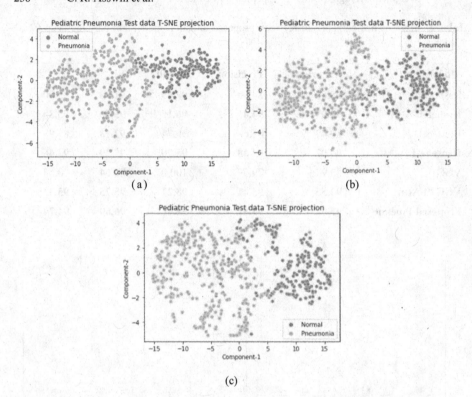

Fig. 6. t-SNE feature representation of the test data extracted from the proposed model – a) ResNet50-Attn b) DenseNet121-Attn c) VGG19-Attn

Table 5 presents a comparison between the proposed model and recent studies on the Kermany et al. [3] dataset. The proposed model exhibits competitive performance with remarkable accuracy, precision, recall, F1 score, and an AUC curve of 95.67%, 94.81%, 98.46%, 96.60%, and 94.74%, respectively. These results demonstrate the effectiveness of the proposed approach in accurately detecting pediatric pneumonia from chest X-ray images.

Table 5. Performance of other recent works on the Kermany et al. [3] dataset

Author	Accuracy	Precision	Recall	F1-Score	AUC
Kermany et al. [3]	92.8	90.1	93.2	-	-
Stephen et al. [13]	93.73	-	-	-	-
Rajpurkar et al. [14]	88.78				
Siddiqi et al. [15]	94.39	92.0	99.0	-	-
Proposed Model	**95.67**	**94.81**	**98.46**	**96.60**	**94.74**

6 Conclusion and Future works

This work proposes a computer-aided pneumonia diagnosis tool using an ensemble of easily replicable channel attention-aware deep CNN architectures. The individual predictions from each proposed model with an external channel attention module are classified into NORMAL and PNEUMONIA using a weighted average ensemble classifier. t-SNE plots are used to analyze the extracted features from the proposed models' penultimate layer. The outliers and overlapping clusters indicate the need for a better model. The proposed pipeline achieves an accuracy of 95.67% and an AUC of 94.74%. The proposed architecture shows competing performance with recent works and its capability to be deployed for real-time use.

Future works can include expanding the work to cover other pneumonia datasets. In addition to that, contrastive learning can be used to build robust models. Finally, various image enhancement techniques and other ensembled models can be implemented for better results. Overall, this study demonstrates the potential for computer-aided tools in the medical field and offers a promising approach to pneumonia diagnosis.

References

1. Ramezani, M., Aemmi, S.Z., Emami Moghadam, Z.:Factors affecting the rate of pediatric pneumonia in developing countries: a review and literature study. Int. J. Pediatrics **3**(6.2), 1173–1181 (2015)
2. Liang, G., Zheng, L.: A transfer learning method with deep residual network for pediatric pneumonia diagnosis. Comput. Methods Programs Biomed. **187**, 104964 (2020)
3. Kermany, D.S., et al.: Identifying medical diagnoses and treatable diseases by image-based deep learning. Cell **172**(5), 1122–1131 (2018)
4. Cha, S.-M., Lee, S.-S., Ko, B.: Attention-based transfer learning for efficient pneumonia detection in chest X-ray images. Appl. Sci. **11**(3), 1242 (2021)
5. Huang, M.-L., Liao, Y.-C.: Stacking ensemble and ECA-EfficientNetV2 convolutional neural networks on classification of multiple chest diseases including COVID-19. Academic Radiology (2022)
6. Hu, J., Shen, L., Sun, G.: Squeeze-and-excitation networks. In: Proceedings of the IEEE Conference on Computer Vision and Pattern Recognition (2018)
7. Woo, S., Park, J., Lee, J.-Y., Kweon, I.S.: CBAM: convolutional block attention module. In: Ferrari, V., Hebert, M., Sminchisescu, C., Weiss, Y. (eds.) Computer Vision – ECCV 2018. ECCV 2018. Lecture Notes in Computer Science, vol. 11211, pp. 3–9. Springer, Cham (2018). https://doi.org/10.1007/978-3-030-01234-2_1
8. Zhang, Z., Wu, Y., Zhang, J., Kwok, J.: Efficient channel attention for deep convolutional neural networks. In: Proceedings of the AAAI Conference on Artificial Intelligence (2021)
9. Guo, Z., et al.: Channel attention residual network for diagnosing pneumonia. In: 2021 4th International Conference on Artificial Intelligence and Big Data (ICAIBD). IEEE (2021)
10. Fan, Y., et al.: COVID-19 detection from X-ray images using multi-kernel-size spatial-channel attention network. Pattern Recognit. **119**, 108055 (2021)
11. Han, Z., et al.: Self-attention generative adversarial networks. In: International Conference on Machine Learning. PMLR (2019)
12. Wang, X., Zhang, X., Qi, J.: Efficient channel attention for deep convolutional neural networks. In: Proceedings of the IEEE/CVF Conference on Computer Vision and Pattern Recognition, pp. 1429–1438 (2020)

13. Okeke, S., et al.: An efficient deep learning approach to pneumonia classification in healthcare. J. Healthc. Eng. 2019 (2019)

14. Pranav, R., et al.: ChexNet: radiologist-level pneumonia detection on chest x-rays with deep learning. arXiv preprint arXiv:1711.05225 (2017)

15. Raheel, S.: Automated pneumonia diagnosis using a customized sequential convolutional neural network. In: Proceedings of the 2019 3rd International Conference on Deep Learning Technologies (2019)

16. Iswarya Kannoth, V., et al.: Parkinson's disease classification from magnetic resonance images (MRI) using deep transfer learned convolutional neural networks. In: 2021 IEEE 18th India Council International Conference (INDICON). IEEE (2021)

17. Vaisali, B., Parvathy, C.R., Hima Vyshnavi, A.M., Tumor Krishnan Namboori, P.K.: Hypoxia Diagnosis using Deep CNN Learning strategy- A theranostic pharmacogenomic approach. Int. J. Prognost. Health Manage. **10**, 7 (2019)

18. Karthikeyan, S., et al.: Detection of multi-class retinal diseases using artificial intelligence: an expeditious learning using deep CNN with minimal data. Biomed. Pharmacol. J. **12**(3), 1577–86 (2019). ProQuest

19. Radha, D.: Analysis of COVID-19 and pneumonia detection in chest X-ray images using deep learning. In: 2021 International Conference on Communication, Control and Information Sciences (ICCISc), vol. 1. IEEE (2021)

Seasonal Disease Based Demand Forecasting for Pharmaceutical Medications Using Random Forest

R. Sakthi Ganesh Dharani, S. V. Lokheshram, and A. Malini(✉) 📷

CSBS Department, Thiagarajar College of Engineering, Madurai, Tamil Nadu, India
{sakthiganeshdharani,lokheshram}@student.tce.edu, amcse@tce.edu

Abstract. In the pharmaceutical industry, accurate demand forecasting is crucial for the efficient management of manufacturing, acquiring, and distribution activities. This paper concentrates on utilising data mining techniques, namely the Decision Tree and Random Forest algorithms, to forecast the demand for pharmaceuticals associated with seasonal ailments that are anticipated to emerge in the upcoming months of 2024. The findings of the analysis demonstrate that the Random Forest algorithm exhibits superior performance in comparison to the Decision Tree algorithm. This is evidenced by the observation of lower mean Root Mean Square Error (RMSE) values of 80.53 and 97.10, respectively, which indicates that Random Forest outperforms Decision Tree. The present work's results offer significant contributions to the pharmaceutical sector by providing valuable insights that facilitate improved planning and resource allocation to address the variable demand patterns that are influenced by seasonal diseases.

Keywords: Distribution · Ailments · RMSE · Resource allocation · Demand patterns

1 Introduction

Healthcare is essential to people's health. Pharmaceuticals, which prevent, treat, and manage illnesses, are vital to healthcare. Accurate medication sales demand forecasting ensures patients have enough drugs. The paper forecasts pharmacological demand for regional conditions. The objective of the research is to determine the most prevalent condition each month, the medication used to treat it, and the anticipated sales of that medication in 2024 using Data Mining Techniques [1].

A decision tree algorithm [2] predicts the monthly demand for disease-related medications using patient history data. In addition to the decision tree model, the study employed a Random Forest technique [3]. The Random Forest method improves accuracy and reduces overfitting by merging several decision trees. Predicting the prevailing condition's monthly drug requirement [4] was the goal.

The work contributes by accurately estimating the demand for pharmaceutical treatments using the Decision Tree and Random Forest algorithms. It identifies the prevalent illnesses in a given month and forecasts the sales of the medication linked to that illness during that month. This aids pharmaceutical companies in making monthly plans for the manufacture of that particular medicine.

The research suggests budget allocation for healthcare and pharmaceutical industries. By forecasting pharmaceutical demand, healthcare practitioners may maximise medicine supply and patient satisfaction. Better inventory management and production planning may save pharmaceutical organisations money and improve customer experience.

2 Literature Review

Demand forecasting literature discusses healthcare research and forecasting. This paper compares healthcare demand forecasting methods, instruments, and models. The literature review will examine healthcare industry trends and predictability issues and suggest promising research areas. This literature review discusses healthcare demand forecasting's best practises, challenges, and future.

Anticipating emergency department admissions implies that congestion, delays, and long waits cause poor patient outcomes like increased morbidity and mortality [5]. Congestion, delays, and long waits may lead to poor patient outcomes. Overcrowding, delays, and long wait times lead to poor patient outcomes.

Text mining estimates bed demand before lab tests [6]. These methods extract text. Prepare enough beds. This meets medical and pharmaceutical standards.

Translational research informs clinical decision-making with temporal data and molecular medicine [7].

MAPE favours the hybrid ARIMA-ANN model [8]. ARIMA-ANN hybrid models are more valuable.

Artificial Neural Networks outperformed traditional demand forecasting methods with a mean percentage error of 4% [9]. This method was less error-prone. Reduced data noise yielded this error rate. Alternative methods average 6% accuracy.

The Croston technique predicts lumpy, unexpected, and slow-moving demand best [10]. The Simple Moving Average (SMA) method is best for accurate and smooth demand forecasting in paediatric intensive care units. Historical patient data is analysed.

Model and method results differ [11]. Thus, the combined model results differ. Coupled model evaluations produce different results. Integrated algorithms that maximise each technique's benefits may have caused this result.

Healthcare demand forecasts depend most on the elderly population [12]. Ageing extends life.

Gradient Boost Machines outperformed decision trees and logistic regression [13]. Gradient Boost Machines won. Despite decision trees and logistic regression's success. This occurred.

Deep Neural Networks outperform Machine Learning in creating and implementing models that predict patient arrivals daily to weekly using calendar and weather data [5]. Calendar and weather data. This holds regardless of model.

Clustering products improves forecast accuracy and simplifies cluster value evaluation [14]. Prediction accuracy improves.

Clustering products increase prediction accuracy and reliability [15]. This forecasting method may improve prediction accuracy.

Tourism demand forecasting [16]. Demand is how much consumers want to buy in a given timeframe. Forecasting methods' pros, cons, and industry applicability are discussed. This industry's managers need forecasts.

The article examines business sales forecasting. Methods are often naïve, extrapolative, and subjective [17]. The research also notes a rise in seasonal adjustments and computer-based forecasts to reduce errors.

In volatile markets, using multiple forecasting methods is better [18]. Combining forecasting sources reduces average error and provides insightful analysis and diagnostics. This method emphasises the benefits of multiple forecasting methods and questions the practise of selecting a single superior forecast.

Random Forest and J48 decision tree classification were compared on 20 diverse datasets [19]. For smaller datasets, J48 outperforms Random Forest. The paper shows these models' pros and cons in various data settings.

This paper estimates 2005–2015 energy demand [20]. Neural networks are efficient but slow. Research hybrid forecasting.

This literature review shows many healthcare demand forecasting methods. According to the survey, demand forecasting ensures healthcare resources are available when needed. The report suggests product clustering and forecasting integration to improve accuracy. Delays and long wait times in the emergency department have been linked to poor patient outcomes, so accurate demand forecasting is essential for patient care. The paper emphasises the need for forecasting models that can handle unpredictable, lumpy, and slow-moving demand.

3 Methodology

3.1 Data Acquisition and Preprocessing

During the data collection phase, accurate information on typical illnesses, associated medications, and monthly prescription sales statistics is collected from databases, patient files, or sources of pharmaceutical sales data. The data is thoroughly cleaned and preprocessed after it has been gathered (Fig. 1).

3.2 Identification of Dominant Disease

The disease dataset consists of 32 diseases and the corresponding number of patients admitted during each month. The most frequently occurring disease in each month is identified.

3.3 Medicine Identification

A drug dataset is used to identify the prescription used to treat the most common disease each month. The disease name and accompanying drug code are both included in this drug dataset. The precise drug utilised for therapy can be identified by comparing the relevant drug code with the prevalent ailment obtained in step 3.2.

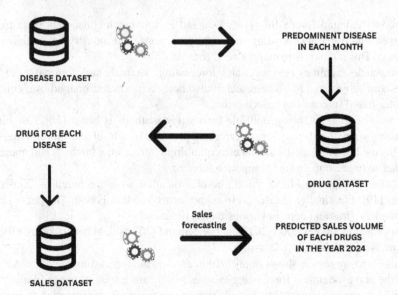

Fig. 1. Block Diagram for Seasonal Disease-based Demand Forecasting for Pharmaceutical Medications

3.4 Sales Forecasting

Sales forecasting is an essential phase in the demand forecasting process. The Random Forest method and the Decision Tree algorithm are two algorithms that are used for this task and is implemented in Python [21]. The sales dataset used is present in Kaggle [22]. The dataset is divided into training and testing samples. About 70% of the dataset is used for training the models, which are then fit to historical sales data. The remaining 30% of the data are used as the testing dataset to evaluate the models. The root mean square error (RMSE) is the evaluation metric used. By providing businesses with information on the projected amount of demand and sales for a specific prescription, these predictions help them plan ahead for manufacture, and manage inventory (Tables 1 and 2).

Table 1. Training Parameters for Decision Tree

Training Parameters	Decision Tree
max_depth	5

4 Result Analysis

The work examined the accuracy of the decision tree and Random Forest models in predicting drug need after conducting an extensive investigation of disease identification. The paper's objectives were to test the performance of these models and identify variations in accuracy among various pharmaceutical classifications. The results

Table 2. Training Parameters for Random Forest

Training Parameters	Random Forest
max_depth	5
n_estimator	100
random_state	42

offer useful information for improving pharmaceutical supply through better forecasting pharmaceutical demand.

Table 3, shows the illnesses that are most prevalent each month and the medications that treat them.From Table 4, It is clear that random forest outperforms decision tree because its mean RMSE value is 80.53 compared to decision tree's mean RMSE value of 97.10. This shows that the Random Forest algorithm, which mixes several decision trees, performs better in this situation at anticipating the need for pharmaceuticals.

The RMSE values range between various drug codes, showing that the models' accuracy may change based on the particular drug being analysed. While certain drug codes have higher RMSE values, suggesting greater prediction errors, others have lower RMSE values, showing that the models produce accurate predictions.

For the medicine code MA10AB, the RMSE values for the Decision Tree and Random Forest models are both reasonably low, indicating that these models are capable of accurately forecasting the demand for this specific medication.For the medication code NO2BE, the Decision Tree model displays a much higher RMSE value than Random Forest. As a result, it appears that the Random Forest model performs better and generates more precise forecasts for the demand for this medicine.The NO5B medication code

Table 3. Disease Identification

Month	Disease	Drug
Jan	Osteoarthritis	M01AE
Feb	Migraine	N02BA
Mar	Chronic obstructive pulmonary disease	R03
Apr	Rheumatoid arthritis	M01AB
May	Osteoarthritis	M01AE
Jun	Bronchitis	R06
Jul	Epilepsy	N02BE
Aug	Chronic obstructive pulmonary disease	R03
Sep	Osteoarthritis	M01AE
Oct	Bipolar disorder	N05C
Nov	Obsessive-compulsive disorder	N05C
Dec	Psoriatic arthritis	M01AB

Table 4. RMSE for Decision Tree and Random Forest

Drug Code	Root Mean Squared Error	
	Decision Tree	Random Forest
MA10AB	30.65	26.10
MA10AE	48.22	32.94
NO2BA	8.1148	16.52
NO2BE	469.42	380.89
NO5B	49.94	52.63
NO5C	8.60	6.49
RO3	139.35	106.25
RO6	21.29	23.42

shows similar RMSE values for the Decision Tree and Random Forest models, showing equivalent performance of these models in predicting the demand for this drug. For the NO5C medication code, the Random Forest model has a marginally lower RMSE value than the Decision Tree model, indicating that it may provide somewhat more accurate predictions. The RMSE values for the medication codes RO3 and RO6 are greater in both Decision Tree and Random Forest models. This suggests that, when compared to the other drug codes looked at, the prediction accuracy for these medications is substantially lower.

Figure 2 shows the sales volume predicted for the year 2024 for the predominant diseases of each month using Random Forest. High sales denotes the demand for the product will be high similarly, low sales denotes the demand for that product will be low.

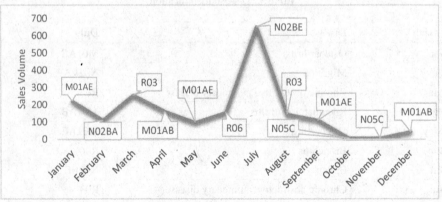

Fig. 2. Predicted sales volume for the year 2024

5   Conclusion and Future Works

The results showed that the Random Forest algorithm outperformed the Decision Tree method in anticipating pharmaceutical demand. The work also showed that different medicinal classifications varied in their accuracy. The RMSE values fluctuated according to the particular drug under analysis, indicating that the models' efficacy might vary based on the drug. The current model has limitations as it only forecasts medicine sales for a particular frequent disease, neglecting the demand for alternative medications. Future research will require gathering a larger dataset with more features from hospital databases to address this constraint. The goal is to increase the model's precision and applicability while providing thorough sales projections for a wider range of ailments and medications.

References

1. Manocha, A., Afaq, Y., Bhatia, M.: Intelligent analysis of irregular physical factors for panic disorder using quantum probability. J. Exp. Theor. Artif. Intell. (2022). https://doi.org/10.1080/0952813X.2022.2121426

2. Charbuty, B., Abdulazeez, A.: Classification based on decision tree algorithm for machine learning. J. Appl. Sci. Technol. Trends 2(01), 20–28 (2021). https://doi.org/10.38094/jastt20165

3. Segal, M.R.: UCSF recent work title machine learning benchmarks and random forest regression publication date machine learning benchmarks and random forest regression (2003)

4. Keny, S., Nair, S., Nandi, S., Khachane, D.: Sales prediction for a pharmaceutical distribution company. Int. J. Eng. Appl. Phys. (IJEAP) 1(2), 186–191 (2021). https://ijeap.org/

5. Graham, B., Bond, R., Quinn, M., Mulvenna, M.: Using data mining to predict hospital admissions from the emergency department. IEEE Access 6, 10458–10469 (2018). https://doi.org/10.1109/ACCESS.2018.2808843

6. Lucini, F.R., et al.: Text mining approach to predict hospital admissions using early medical records from the emergency department. Int. J. Med. Inform. 100, 1–8 (2017). https://doi.org/10.1016/j.ijmedinf.2017.01.001

7. Bellazzi, R., Ferrazzi, F., Sacchi, L.: Predictive data mining in clinical medicine: a focus on selected methods and applications. Wiley Interdiscip. Rev. Data Min. Knowl. Discov. 1(5), 416–430 (2011). https://doi.org/10.1002/widm.23

8. Yucesan, M., Gul, M., Celik, E.: A multi-method patient arrival forecasting outline for hospital emergency departments. Int. J. Healthc. Manage. 13(S1), 283–295 (2020). https://doi.org/10.1080/20479700.2018.1531608

9. Muriithi, I.A., Muchemi, L.: School of computing and informatics a data mining approach to private healthcare services demand forecast in Nairobi County (2014)

10. Cheng, C.Y., Chiang, K.L., Chen, M.Y.: Intermittent demand forecasting in a tertiary pediatric intensive care unit. J. Med. Syst. 40(10) (2016). https://doi.org/10.1007/s10916-016-0571-9

11. İmece, S., Beyca, Ö.F.: Demand forecasting with integration of time series and regression models in pharmaceutical industry. Int. J. Adv. Eng. Pure Sci. 34(3), 415–425 (2022). https://doi.org/10.7240/jeps.1127844

12. Sengupta, S., Dutta, R.: Identification of demand forecasting model considering key factors in the context of Healthcare products. Int. J. Appl. Innov. Eng. Manage. (2014)

13. Sudarshan, V.K., Brabrand, M., Range, T.M., Wiil, U.K.: Performance evaluation of Emergency Department patient arrivals forecasting models by including meteorological and calendar information: a comparative study. Comput. Biol. Med. **135** (2021). https://doi.org/10.1016/j.compbiomed.2021.104541

14. Amalnick, M.S., Habibifar, N., Hamid, M., Bastan, M.: An intelligent algorithm for final product demand forecasting in pharmaceutical units. Int. J. Syst. Assur. Eng. Manage. **11**(2), 481–493 (2020). https://doi.org/10.1007/s13198-019-00879-6

15. Armstrong, J.S., Green, K.C.: Demand forecasting: evidence-based methods. No. 24/05. Monash University, Department of Econometrics and Business Statistics (2005). http://dx.doi.org/10.2139/ssrn.3063308

16. Archer, B.: Demand forecasting and estimation. SSRN Electr. J. 77–85 (1987)

17. Dalrymple, D.J.: Sales forecasting practices: results from a United States survey. Int. J. Forecast. **3**(3–4), 379–391 (1987). https://doi.org/10.1016/0169-2070(87)90031-8

18. Doyle, P., Fenwick, I.: Sales forecasting—using a combination of approaches. Long Range Plan. **9**(3), 60–64 (1976). https://doi.org/10.1016/S0024-6301(76)80010-6

19. Ali, J., et al.: Random forests and decision trees. Int. J. Comput. Sci. Issues (IJCSI) **9**(5), 272 (2012)

20. Ghalehkhondabi, I., Ardjmand, E., Weckman, G.R., et al.: An overview of energy demand forecasting methods published in 2005–2015. Energy Syst. **8**, 411–447 (2017). https://doi.org/10.1007/s12667-016-0203-y

21. Ali, M.A., Matubber, M.L., Sharma, V., Balamurugan,B.: An improved and efficient technique for detecting Bengali fake news using machine learning algorithms. In: 2022 6th International Conference On Computing, Communication, Control And Automation (ICCUBEA), Pune, India, pp. 1-4. (2022). https://doi.org/10.1109/ICCUBEA54992.2022.10011096

22. https://www.kaggle.com/datasets/milanzdravkovic/pharma-sales-data?select=salesmonthly.csv

Hybrid Optimal Fine Tuning Approach in Deep Learning for Identifying Early Parkinson's Disease

S. Sivakumar[1]📷, S. Anita[2]📷, and S. Jothi[3(✉)]📷

[1] Cardamom Planters' Association College, M. K. University, Madurai, India
sivakumar_s@cpacollege.org
[2] St. Anne's College of Engineering and Technology, Anna University, Chennai, India
sranitaa@stannescet.ac.in
[3] Jayaraj Annapackiam College for Women, M. K. University, Madurai, India
srjothics@annejac.ac.in

Abstract. Introduction: The neurodegenerative Disorder called Parkinson's disease (PD) is incurable when it is indicated at an initial stage by weakening dopamine generating nerve cells. These cells are projected by the instrument of capturing DaTscan images. The qualitative analysis of DaTscan is carried out by the different deep learning algorithms. The present investigation deals with three facets of deep learning algorithms in diagnosing PD at an initial stage by analysing Volume Containing DaTscan Image Slices (VCDIS) for higher diagnostic accuracy.

Methods: The contribution includes 3 facets. The first facet classifies the people suffering from PD from healthy individuals (HI) using transfer learning of conventional deep learning networks like Alex Net, Inception, Mobile Net, ResNet, Xception, VGG16, VGG19. Second facet fine tunes the layer of conventional networks carefully. Fine tuning is done by selecting random and optimal layers of conventional networks to tune the network to learn the features of VCDIS for better performance. Third facet hybrids the predictions of previous facets to accomplish highest proficiency in predicting PD at an early stage.

Results: The results of the transfer learning facet are dictated as around 92.22% of predictive accuracy for the Inception V3 network. The diagnostic accuracy of optimal fine tuning is around 99.17% considerably. The best results of detecting EPD (Early Parkinson's Disease) is achieved as 99.95% of accuracy for the Inception V3 network by the hybrid technique.

Conclusion: The proposed hybrid technique runs remarkable performance as a supporting tool in analysing Parkinson's disease which helps the neurologist in analysing the neurodisorders.

Keywords: Parkinson's disease · Healthy individuals · transfer learning · fine tuning · hybrid technique

S. Kadry and R. Prasath (Eds.): MIKE 2023, LNAI 13924, pp. 269–282, 2023.
https://doi.org/10.1007/978-3-031-44084-7_26

1   Introduction

Parkinson's disease (PD) is a human brain related degenerative disorder for people around 60 years old, chiefly affecting the motor system. As of now, globally billons of people (all ages) suffer from PD [1–3]. The prime signs of PD are movement related: rigidity, tremor, impaired in coordination and other signs are cognitive related: sleep disturbances, depression and olfactory [4,5]. PD affects human due to the death or decreases of vital neurons called dopamine, which leads to more difficulty in voluntary movement. Once the dopaminergic neurons are decreased, there is no way to increase these neurons. This leads to worsening quality of life, increased mortality, and reduced productivity. The accurate diagnosis of the disease is a challenging task when the symptoms are not recognised at an initial stage even today. For, similar symptoms are present in all the neurodegenerative disorders. Though the cardinal symptoms fall on motor systems, PD is associated with non-motor symptoms also. Hence an early diagnosis is necessary to manage the early stage of the disease [6].

Neuroimaging technique called Single Photon Emission Computed Tomography (SPECT) approved by Food and Drug Administration (FDA) [7,8] in the U.S. serves well to analyse the images in diagnosing the EPD. SPECT neuroimages are the chief accessing tool as it clearly displays the dopamine transmitters present in the striatum [9] of human brain. It also highly identifies the progression of the disease. Normal SPECT images of the brain depict symmetric and comma shaped dopamine binding in the nucleus of the striatum as shown in Fig. 1a. An abnormal image shows asymmetric and dot shaped dopamine binding as shown in Fig. 1b.

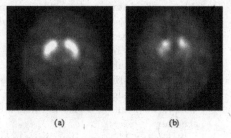

Fig. 1. DaTscan image of (a) Healthy Individual (b) Parkinson's Disease.

Different methods are utilised for PD diagnosis by (i) evaluating SPECT images which include three-dimensional analysis of SPECT images in DICOM format (consisting of 91 slices) [9], (ii) investigating single image slices by averaging the slices that have a high striatal uptake region [10], (iii) evaluating Volume Rendering Image Slices (VRIS) which consist slices number from 36 to 47 [11]. The methods have the limitations of complexity and fail to comprehend the shape of the striatum. Demand grows to make the system simple and efficient in order to understand the continuity of the changes in shape of the striatum.

In the early days, experts assessed neuroimages through visual investigations. In the early days of the deficit, the visual investigation is subjected to the skill and experience of the experts which leads to misdiagnosis of the disease [10]. Utility of the machine learning techniques in identifying and monitoring the progression of the disease resolved the problem of misdiagnosis [12,13]. Deep learning techniques [14] are introduced where it extracts the features based on the requirements by self-learning. Hence it is recognized as a powerful tool for classifying images [15,16].

The Convolutional Neural Networks (CNNs) of deep learning are applied in discriminating PD patients from healthy individuals (HI) [17]. The Parkinson's Progression Markers Initiative (PPMI) repository [18] is used to download the SPECT images for the analysis. The bounding box method is adapted to extract the striatum region alone for classifying the images by 10-fold cross validation [19]. The ensemble techniques used retrospective data, however, bring forth appreciating results, suffer certain inconveniences pointed out in the literature review section and the proposed system is free from such difficulties. It outperforms the existing systems with a large margin of diagnostic accuracy. The highlights of the present system are briefed as follows.

1. The system proposes three facets that include (i) transfer learning of conventional networks such as InceptionV3, Xception, VGG16, MobileNet, ResNet50, and VGG19, (ii) the random, optimal fine tuning of the same conventional networks (iii) mixing up of transfer learning and fine tuning facets, called hybrid technique. The proposed hybrid technique minimizes the number of parameters and tuning time as it selects a few layers for tuning to learn significant features of Volume Containing DaTscan Image Slices (VCDIS) images.
2. Optimal fine tuning takes the averaged output (accuracy) of conventional networks as the layer number for fine tuning to offer higher predictive accuracy by dropping the risk of modeling bias and overfitting.
3. The network uses VCDIS from SPECT images chosen from Parkinson's Progression Markers Initiative (PPMI) database for learning spatiotemporal relationships among them. The VCDIS comprises slices from 34 to 49 to learn the continuity of the changes in shape of the striatum.

The rest of the proposed paper is systemized as follows. Section 2, speaks on the review of the literature. Section 3, elaborates on the materials and methodology used in the study and explains the proposed deep learning technique for classifying PD from HI. Sections 4, and 5 highlight results and discussions. Finally, conclusions are drawn.

2   Review of the Literature

An interpretation technique of SPECT images has been proposed by Choi et al. using PDnet. The network shows appreciable accuracy in PD classification [20].

Kevin H. Leung et al. have adopted a three-stage, deep learning, ensemble approach for predicting PD. The ensemble approach combines spatiotemporal features from SPECT images, temporal features from clinical motor scores and additional clinical measures to diagnose PD [21].

Khosro Rezaee et al. have utilized pre-trained deep transfer learning (AlexNet, VGG-f, and CaffeNet) structures and conventional machine learning models as an automated approach to diagnose PD from sEMG signals. The hybrid deep transfer learning-based approach to PD classification could lead to hitting rates higher than 99% [22]. Yang Y et al. proposed a two-layer stacking ensemble learning framework with fusing multi-modal features. Support vector machine (SVM), random forests (RF), K-nearest neighbor (KNN) and artificial neural network (ANN) classifiers are used at the first layer and a logistic regression (LR) classifier was applied at the second layer, and achieved an accuracy of 96.88%, for identifying PD and HC [23].

Diego Castillo-Barnes, et al. developed an ensemble classification model which combines Support-Vector-Machine (SVM) with linear kernel classifiers for a different biomedical group of tests and pre-processed neuroimages features from PPMI database subjects. Using this weighted ensemble classification model, 96% of accuracy was obtained [24].

The ensemble techniques discussed in this section used retrospective data which are prone to misclassification bias and to find out temporal relationships still suffer by overfitting. This issue is addressed in the proposed system using SPECT images alone for analysing PD.

3   Computational Methodology

The three facets of the hybrid optimal fine tuning approach (HOF tuning) is designed to discriminate early stage of Parkinson's disease (EPD) from HI. The technical details of the proposed model are explained clearly.

3.1   Hybrid Optimal Fine Tuning Technique

Hybrid optimal fine tuning (HOF tuning) technique is used to diagnose Parkinson's disease at an inception stage using volume containing DaTscan image slices (VCDIS) images which are taken from the international PPMI depository. The overall workflow of hybrid optimal fine tuning technique is elaborated in Fig. 2. VCDIS of SPECT images are considered as inputs that are given to transfer learning of conventional models such as Inception V3, Xception, VGG16, Mobile Net, ResNet50, and VGG19 in the first facet. The fine tuning (random and optimal) is done for the same conventional neural networks in the second facet and finally hybrid technique mixes outputs of all the facets to improve the performance of the proposed CNN with reduced bias and improved prediction. The input images of the system are processed and tuned to offer best performance in classifying EPD from HI.

3.2   Preparation of Input Images

The input images of HOF tuning technique are taken from the multicentre PPMI database. A total of 640 pre-processed images are chosen for the analysis, including 202 HI images and 448 EPD images. Since it is early detection of PD, the baseline (BL) images are chosen which has the Hoehn and Yahr (H&Y) rating scale of stages 1 and 2 with mean±standard deviation (SD) of 1.50 ±

0.50. The raw SPECT images are reconstructed and normalized to present as a three-dimensional volume in the shape of $91 \times 109 \times 91$. The volume containing DaTscan image slices (VCDIS) are selected from the preprocessed images based on the high DAT bindings in the striatal regions. It is identified that the VCDIS starts from slice number 34 to slice no 49 as it provides the continuity in shape and volume information of the striatum in two dimensions clearly. The VCDIS is highly recommended to analyse the neural disorders using deep learning techniques.

3.3 Architectural Design of HOF Tuning

Facet 1: Transfer Learning of Conventional Model. The novel Hybrid optimal fine tuning is proposed from the ensemble method of Kevin H. Leung et al. [21]. HOF tuning is comprised of three facets: Transfer learning of conventional networks, fine tuning (random and optimal) and hybrid technique. The network uses the conventional models of Inception V3, Xception, VGG16, Mobile Net, ResNet50, and VGG19. Each conventional model has fundamental layers like convolutional layer, batch normalization, Rectified Linear Unit (ReLU) activation function, max pooling and fully connected or dense layers which play a major role in classifying EPD images from HI images in the proposed work. In transfer learning, the basic layers are kept frozen and the fully connected layers and prediction layers are tuned for better accuracy.

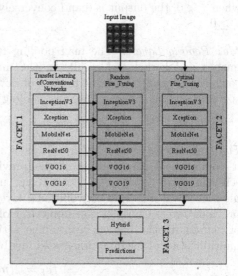

Fig. 2. Overall Workflow of the Hybrid Optimal Fine Tuning Technique.

Convolutional Layers. The Convolution layer (CL) applies a filter over the processed image by applying the convolutional operation to learn basic information about the input images. The filter slides over the input images and performs dot product to compute the feature map by neurons. The size of the filter is small compared with an image. Basically, the size of the filter is (f $\times$ f $\times$ 2), where f = 3, 5, 7, and so on. Each convolutional layer consists of precise neurons, weights, and bias to learn the important features of the input images. Consider I (a x b) as an input image and the filter is represented as $f_i(pxq)$ where $i \in M$ with the size of (p x q). M denotes filter depth. The output feature map is given as

$$D_{conv.} = \Sigma_{u=1}^{p} \Sigma_{v=1}^{q} f_i(u, v) \otimes I(a - u, b - v) \tag{1}$$

The proposed hybrid method uses different convolutional layers with different stride values to capture the basic information of an image.

The Activation Function: Rectified Linear Unit. ReLU layer is an activation function to map linear input with a nonlinear function by applying the monotonic function. It is designed to improve the training phase of the classification process. The base CNN model called Alexnet [25] applies ReLU to create the sparse feature output by replacing tanh activation. The output expression for the ReLU function is given below as a max function

$$R(x) = max(0, x) \tag{2}$$

when $x < 0$, the output is 0 and conversely, its output is a linear function when $x \geq 0$.

Max Pooling Layers. The max pooling fuses the pixel values of the SPECT images and kernel and results in the maximum value of the area where it convolves. It approaches the process of downsampling which reduces the spatial size of the input images. The max pooling is done to reduce overfitting.

$$Output = max[inputimage] \tag{3}$$

Fully Connected Layer (FCL). It is a multilayer perceptron (MLP) where it connects all the neurons of the previous layer with the neurons of the self-generative layer. FCL learns the non-linear relationship between the high-level features from convolutional layers and it flattens the input from the previous layer. Consider the input is I of size 1 and W denotes the number of neurons in the hidden layers. The activation function is estimated by the multiplication of the input images adding bias to it and it is given in the equation

$$fc_I = \varphi(m * I) \tag{4}$$

where, φ and m denote activation function and the resultant matrix respectively. The proposed system uses InceptionV3, Xception, VGG16, MobileNet, ResNet50, and VGG19.

Facet 2: Fine Tuning Techniques. Fine tuning uses the conventional networks where some network layers are tuned and others are kept frozen to minimize the training time and usage of parameters. The weights of the tuned layers are adjusted to achieve highest efficiency in detecting PD at an early stage. The proposed network consists of two fine tuning techniques such as random and optimal fine tuning. Random fine tuning chooses the layer's number (from which network needs to be tuned) randomly, whereas optimal fine tuning chooses the layer's number from the accuracy of the conventional layer (the previous layer). The layers of random and optimal fine tuning and total number of layers in each conventional networks are given in Table 1. The design procedure of transfer learning and fine tuning techniques are summarised in Fig. 3. The figure dictates that fine tuning technique freezes some of the layers and tunes from a particular layer as given in Table 1.

Table 1. Layers and Parameters of Transfer Learning of Conventional and Fine Tuning Models.

Name of Networks	No. of Layers	Parameters	Fine Tuning Layer		Parameters
			Random	Optimal	
InceptionV3	311	2K	95	80	2.0 M
Xception	132	2K	100	55	9.5 M
VGG16	19	70K	12	19	5.0 M
MobileNet	154	1K	82	74	2.0 M
ResNet50	175	2K	90	69	2.0 M
VGG19	22	2K	18	20	2. 0 M

3.4   Hybrid Optimal Fine Tuning Technique

Facet 3: Hybrid Optimal Fine Tuning Technique. The third facet of the proposed model is hybrid model where the previous facets are fused together to obtain best diagnostic accuracy in discriminating EPD from HI. It fuses its predictions and make a best prediction in diagnosing EPD from HI as it is depicted in Fig. 1.

3.5   Data Augmentation

Data augmentation is the process of increasing the size of the dataset artificially by modifying the version of the image to improve the performance of the neural network and reducing the fitting problem. The augmentation techniques used in the proposed work are scaling, cropping, horizontal flipping, padding and rotation.

3.6   Evaluation Metric

The VCDIS images are divided into training, testing and validation dataset as 60%, 20% and 20% of the split. The proposed network is trained using training and validation datasets whereas testing data is taken for predictions. The used parameters in the proposed system are batch size 32, epoch 500, learning rate 0.001 and dropout 0.5 for transfer learning. The learning rate for fine tuning is further reduced to 0.0001 to avoid losing the learned parameters. The Adam optimization algorithm is used to revise every network weight. Evaluation is done by 10-fold cross-validation where the dataset is divided into ten among them one is taken as testing and the remaining is considered as training. This proposed work considers diagnostic accuracy, sensitivity, and specificity as the evaluation metrics.

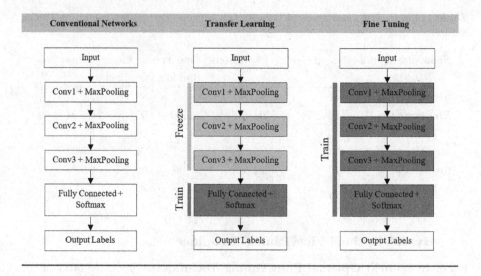

Fig. 3. The design procedure of transfer learning and fine tuning techniques.

4 Results

4.1 Analysis of Hybrid Optimal Fine Tuning Technique

Sixteen Volume Containing DaTscan Image Slices (VCDIS) are selected and taken for the analysis of early Parkinson's disease. These image slices are analysed using transfer learning of Inception V3, Xception, VGG16, Mobile Net, ResNet50, and VGG19. These conventional models are fine tuned randomly and optimally selected layers and performance is also analysed. The facets are fused to make the proposed method outperforms in identifying EPD. The results of each stage are presented below for the original data and augmented data set.

FACET 1: Transfer Learning Of Conventional Models. The analysis is conducted for a total of 640 data and assessed by 10-fold cross-validation methodology in order to attain unbiased performance. Each fold contains 64 volume containing DaTscan image slices. Among these, one-fold is used as testing data and the remaining folds are considered training data. The data set is divided into three datasets like training (train the model), validation (optimize the parameters) and testing (generalize the model) datasets. The predictive performance of the transfer learning of conventional models for the original and augmented data is given in Table 2. The table shows the diagnostic accuracy and its losses for training, validation and testing which dictates that the augmented data set of the InceptionV3 network offers better accuracy (a maximum of 92.22%) than the original dataset (offers a maximum of 90.00) of the same network and other networks like Xception, VGG16, MobileNet, ResNet50, and VGG19. The augmented data overcomes the 7 significant problems of overfitting hence improving accuracy, variability and flexibility of the data. InceptionV3,

Xception, VGG16, MobileNet, ResNet50, and VGG19 offer diagnostic accuracy of 90%, 80%, 74.62%, 69,19%,55.35% and 50.58% for original data and 92.22%, 81.94%, 76.57%, 72.32%, 57.75% and 58.27% for augmented data.

Table 2. Performance of Transfer Learning Function of the Conventional models.

Models	Original Data						Augmented Data					
	Training		Validation		Testing		Training		Validation		Testing	
	Acc.	Loss	Acc.	Loss	Acc.	Loss	Acc.	Loss	Acc.	Loss	Acc.	Loss
InceptionV3	85.64	0.3538	87.8	0.4127	90	0.274	87.41	0.2875	88.3	0.3537	92.22	0.244
Xception	60.13	0.8167	63.09	0.5989	80	0.5771	75.58	0.7147	75.06	0.4762	81.94	0.4615
VGG16	64.53	1.1286	66.05	0.6012	74.62	0.6333	79.38	1.0278	76.75	0.4954	76.57	0.4876
MobileNet	66.32	0.5838	38.21	0.7104	69.19	0.5333	77.25	0.4738	54.71	0.6103	72.32	0.4122
ResNet50	57.25	0.6873	55.43	0.6461	55.35	0.7333	75.47	0.5138	67.36	0.5722	57.75	0.6541
VGG19	82.61	0.3791	77.9	0.5544	50.58	0.7	85.12	0.3119	83.1	0.4264	58.27	0.552

FACET 2: Fine Tuning Models. The second facet of the HOF tuning technique is fine tuning methods. The fine tuning methods are done by freezing weights of the particular layers and train the remaining layers to learn the significant features of the VCDIS. Two different methods are implemented for selecting fine tuning layers like random selection and optimal selection. In random selection, fine tuning layers are selected randomly, whereas in optimal selection, the output of the transfer learning of conventional models are taken as the layer to be tuned. Table 3 and 4 show the classification output and losses of fine tuning methods (random and optimal) for original data and augmented data respectively. By comparing two tables, the optimal fine tuning method offers better performance than the random method for augmented data. In particular, the Inception V3 network gives 99.17% and 0.0410 of accuracy and losses respectively. The other network offers 96.92%, 95.22%, 98.82%, 97.59% and 95.73% of accuracy and 0.3400, 0.0601, 0.1008, 0.0617, and 0.0507 of losses. Figure 4 shows predictive outputs of transfer learning, optimal, random tuning and hybrid optimal tuning techniques.

Table 3. Performance of Random and Optimal Fine Tuning for Original Data.

Models	Random Fine Tuning						Optimal Fine Tuning					
	Training		Validation		Testing		Training		Validation		Testing	
	Acc.	Loss	Acc.	Loss	Acc.	Loss	Acc.	Loss	Acc.	Loss	Acc.	Loss
InceptionV3	91.78	0.1295	90.91	0.2113	94.35	0.1567	95.05	0.1129	96.01	0.1103	97.41	0.091
Xception	94.4	0.1436	91.55	0.2899	93.33	0.1464	91.12	0.1006	92.99	0.1	93.99	0.104
VGG16	83.56	1.1286	81.87	0.3905	90.76	0.1	89.6	1.0196	90.74	0.9107	92.52	0.1001
MobileNet	75.98	0.4489	61.82	0.5264	92.38	0.1333	89.18	0.2844	91.25	0.1211	92.3	0.1211
ResNet50	72.7	0.4792	68.14	0.5501	93.6	0.1333	89.07	0.3272	90.36	0.2015	91.98	0.1613
VGG19	84.23	0.3418	78.72	0.4308	91.33	0.2	88.27	0.1981	89.27	0.1034	91.36	0.197

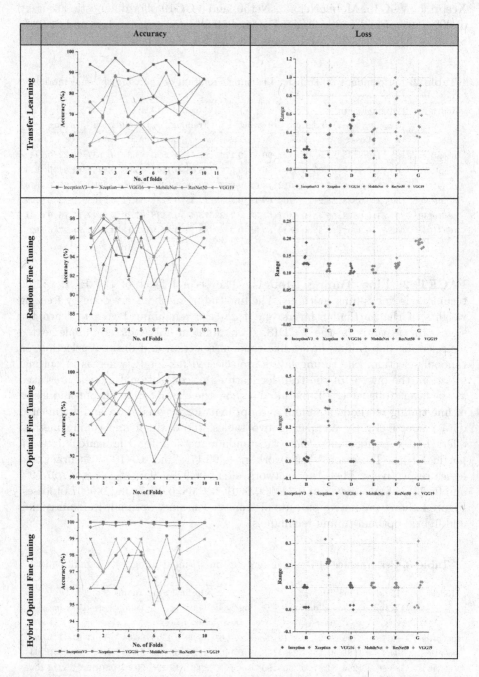

Fig. 4. Predictive outputs of transfer learning, optimal, random tuning and hybrid optimal tuning techniques.

Table 4. Performance of Random and Optimal Fine Tuning for Augmented Data.

Models	Random Fine Tuning						Optimal Fine Tuning					
	Training		Validation		Testing		Training		Validation		Testing	
	Acc.	Loss	Acc.	Loss	Acc.	Loss	Acc.	Loss	Acc.	Loss	Acc.	Loss
InceptionV3	93.54	0.1102	91.19	0.2113	96.25	0.1374	95.59	0.1009	96.58	0.1003	99.17	0.041
Xception	96.9	0.1	94.21	0.2899	96	0.1055	92.78	0.1698	95.79	0.05	96.92	0.34
VGG16	85.27	0.987	85.47	0.3905	92.16	0.0987	90.09	1.0521	93.58	0.215	95.22	0.061
MobileNet	77.57	0.2954	76.57	0.5264	95.3	0.1004	91.56	0.1542	93.74	0.1001	98.82	0.1008
ResNet50	75.3	0.5972	79.54	0.5501	96.07	0.1258	92.51	0.2332	92.48	0.1005	97.59	0.0617
VGG19	87.78	0.1147	80.6	0.4308	93.98	0.187	91.57	0.0911	95.57	0.094	95.73	0.0507

FACET 3: Hybrid Optimal Fine Tuning Technique. The third facet is called hybrid techniques where it combines all the averaged output of the facets (1 & 2) and outperforms than the individual networks. The classification accuracy and losses of the hybrid technique is given in Table 5. The table gives 99.95% of accuracy for Inception V3 network and the loss is 0.066.

Table 5. Performance of Hybrid Optimal Fine Tuning Technique.

Models	Training		Validation		Testing	
	Accuracy	Loss	Accuracy	Loss	Accuracy	Loss
InceptionV3	95.32	0.1069	96.29	0.1053	99.95	0.066
Xception	91.95	0.1352	94.39	0.075	97.46	0.222
VGG16	89.84	1.0358	92.16	0.5629	95.87	0.0801
MobileNet	90.37	0.2193	92.49	0.1106	97.56	0.1109
ResNet50	90.79	0.2802	91.42	0.151	99.78	0.1115
VGG19	89.92	0.1446	92.42	0.0987	97.54	0.1239

5   Discussions

The proposed hybrid optimal fine tuning technique yields the highest performance in classifying early Parkinson's disease from healthy individuals. The significant accuracy like 99.95% is achieved when the hybrid model is implemented with optimal fine tuning approach for augmented data in Inception V3. The selected volume containing DaTscan image slices (VCDIS) help to understand the striatum's shape as it strongly related to the progression of Parkinson's disease. Hence, it aids people who suffer from the disease. In addition, the system uses a novel approach called optimal fine tuning to improve the classification accuracy. The optimal fine tuning approach tunes the network layer from the classification output of conventional networks. The classification accuracy of the conventional networks are taken as the layer of fine tuning for the same network to learn the significant features of the VCDIS. This will improve its performance. The proposed system gives a significant performance in such a way that the parameters of the networks are selected carefully. The parameters of the networks ensure reliability in the diagnostic process. The proposed method offers 99.95% of accuracy for Inception V3 network and the loss is 0.066.

The performance of the system is compared with the existing convolutional networks and it offers the highest accuracy than the other networks in diagnosing EPD. Table 6 depicts the outcome obtained from different networks in the present work cited in the paper. Diagnostic accuracy of the proposed work is related to the present work. Though it is closely related to the present work, the proposed work uses an optimal method for learning the deep features of the image and tenfold cross-validation is used to limit unbiased output, high variance, and overfitting. This leads to stable output to minimize the high variances. The proposed approach is a feasible and easy approach to the experts in discriminating PD from HI. Hence, it offers the highest results than the systems presented in the literature.

Table 6. Comparative Analysis of the Associated Work.

Details	Methodology Used	Performance (%)
Anita et al [11]	Volume Rendering Image Slices with RBF-ELM	98.23
Prashanth et al. [13]	Averaged Image Slices (35–48) and SVM Classifier	97.29
Sheibani Ret al. [26]	Ensemble-based method for voice signal	90.60
Krzysztof Wrobel [27]	Ensemble-based method for voice signal	96.90
Najmeh Fayyazifar et al. [28]	Bagging + Genetic Algorithm	98.28
Proposed Work	Optimal Fine Tuning Approach	99.95

6   Conclusion

Parkinson's disease in its early stage is identified using a three-stage hybrid deep learning method utilizing volume containing DaTscan image slices. The stages are transfer learning of conventional networks like InceptionV3, Xception, VGG16, MobileNet, ResNet50, and VGG19, Fine tuning methods (random and optimal) and hybrid method which combines 10 the output of all the networks. The hybrid method offers the best classification accuracy of 99.95% with minimum losses for the augmented dataset which removes the overfitting problems. Thus, the proposed method aids people to diagnose PD at an early stage easily.

Acknowledgements. PPMI - a public-private partnership - is funded by the Michael J. Fox Foundation for Parkinson's Research and funding partners, including [list of all of the PPMI funding partners found at www.ppmi-info.org/fundingpartners].

References

1. Elkouzi, A.: Understanding Parkinson: statistics. https://www.parkinson.org/understanding-parkinsons/statistics. Accessed 25 Apr 2023
2. Armstrong, M.J., Okun, M.S.: Diagnosis and treatment of Parkinson disease: a review. JAMA **323**, 548–60 (2020)
3. Ball, N., Teo, W.-P., Chandra, S., Chapman, J.: Parkinson's disease and the environment. Front. Neurol. **10**, 218 (2019)
4. Durga, P., Jebakumari, V.S., Shanthi, D.: Diagnosis and classification of Parkinson's disease using data mining techniques. Int. J. Adv. Res. Trends Eng. Technol. **3**, 86–90 (2016)
5. Michael, J.: Fox foundation for Parkinson research, Parkinson's disease causes. https://www.michaeljfox.org/understanding-Parkinson's/living-with-pd.html. Accessed 20 Apr 2023
6. Jin, H., Kanthasamy, A., Ghosh, A., Anantharam, V., Kalyanaraman, B., Kanthasamy, A.G.: Mitochondria-targeted antioxidants for treatment of Parkinson's disease: pre- clinical and clinical outcomes. Biochimica et Biophysica Acta (BBA)-Molecular Basis of Disease. **1842**(8), 1282–1294 (2014)
7. Cummings, J.L., Henchcliffe, C., Schaier, S., Simuni, T., Waxman, A., Kemp, P.: The role of dopaminergic imaging in patients with symptoms of dopaminergic system neurodegeneration. Brain **134**, 3146–3166 (2011)
8. Hopkins, J.: How Parkinson's disease is diagnosed. https://www.hopkinsmedicine.org/health/treatment-tests-and-therapies/how-parkinson-disease-is-diagnosed. Accessed 25 Apr 2023
9. Prashanth, R., Roy, S.D., Mandal, P.K., Ghosh, S.: High-accuracy classification of Parkinson's disease through shape analysis and surface fitting in 123I-Ioflupane SPECT imaging. IEEE J. Biomed. Health Inform. **21**(3), 794–802 (2016)
10. Oliveira, F.P.M., Faria, D.B., Costa, D.C., Castelo-Branco, M., Tavares, J.M.R.S.: Extraction, selection and comparison of features for an effective automated computer-aided diagnosis of Parkinson's disease based on [123 I] FP-CIT SPECT images. Eur. J. Nucl. Med. Mol. Imaging **45**, 1052–1062 (2018)
11. Anita, S., Aruna, P.P.: Diagnosis of Parkinson's disease at an early stage using volume rendering SPECT image slices. Arab. J. Sci. Eng. **45**, 2799–2811 (2020)
12. Orru, G., Pettersson-Yeo, W., Marquand, A.F., Sartori, G., Mechelli, A.: Using support vector machine to identify imaging biomarkers of neurological and psychiatric disease: a critical review. Neurosci. Biobehav. Rev. **36**, 1140–1152 (2012). https://doi.org/10.1016/j.neubiorev.2012.01.004
13. Prashanth, R., Roy, S.D., Mandal, P.K., Ghosh, S.: Automatic classification and prediction models for early Parkinson's disease diagnosis from SPECT imaging. Expert Syst. Appl. **41**, 3333–3342 (2014)
14. Schmidhuber, J.: Deep learning in neural networks: an overview. Neural Netw. **61**, 85–117 (2015). https://doi.org/10.1016/j.neunet.2014.09.003
15. Bengio, Y.: Learning deep architectures for AI. Found. Trendsa® Mach. Learn. **2**, 1–127 (2009). https://doi.org/10.1561/2200000006
16. LeCun, Y., Bengio, Y., Hinton, G.: Deep learning. Nature **521**, 436 (2015). https://doi.org/10.1038/nature14539
17. Ong, S.Q., et al.: Comparison of pre-trained and convolutional neural networks for classification of jackfruit Artocarpus integer and Artocarpus heterophyllus. In: Abualigah, L. (ed.) Classification Applications with Deep Learning and Machine Learning Technologies. Studies in Computational Intelligence, vol. 1071, pp. 129–141. Springer, Cham. (2023). https://doi.org/10.1007/978-3-031-17576-3_6

18. Parkinson's Progression Markers Initiative (2017). http://www.ppmiinfo.org/
19. Martinez-Murcia, F.J., et al.: A 3D convolutional neural network approach for the diagnosis of Parkinson's disease. In: Ferrández Vicente, J.M., Álvarez-Sánchez, J.R., de la Paz López, F., Toledo Moreo, J., Adeli, H. (eds.) IWINAC 2017. LNCS, vol. 10337, pp. 324–333. Springer, Cham (2017). https://doi.org/10.1007/978-3-319-59740-9_32
20. Choi, H., Ha, S., Im, H.J., Paek, S.H., Lee, D.S.: Refining diagnosis of Parkinson's disease with deep learning-based interpretation of dopamine transporter imaging. Neuroimage Clin. **16**, 586–594 (2017)
21. Leung, K.H., Rowe, S.P., Pomper, M.G., Du, Y.: A three-stage, deep learning, ensemble approach for prognosis in patients with Parkinson's disease. EJNMMI Res. **11**, 52 (2021). https://doi.org/10.1186/s13550-021-00795-6
22. Rezaee, K., Savarkar, S., Yu, X., Zhang, J.: A hybrid deep transfer learning-based approach for Parkinson's disease classification in surface electromyography signals. Biomed. Signal Process. Control **71**, Part A, 103161 (2022). ISSN 1746–8094
23. Yang, Y., Wei, L., Hu, Y., Wu, Y., Hu, L., Nie, S.: Classification of Parkinson's disease based on multi-modal features and stacking ensemble learning. J. Neurosci. Meth. **350**, 109019 (2021)
24. Castillo-Barnes, D., Ramírez, J., Segovia, F., Martínez-Murcia, F. J., Salas-Gonzalez, D., Górriz, J. M. : Robust Ensemble Classification Methodology for I123-Ioflupane SPECT Images and Multiple Heterogeneous Biomarkers in the Diagnosis of Parkinson's Disease. Front. Neuroinform. (2018). https://doi.org/10.3389/fninf.2018.00053. 30154711
25. Krizhevsky, A., Sutskever, I., Hinton, G.E.: ImageNet classification with deep convolutional neural networks. In: Advances in Neural Information Processing Systems, pp. 1097–1105 (2012)
26. Sheibani, R., Nikookar, E., Alavi, S.E.: An ensemble method for diagnosis of Parkinson's disease based on voice measurements. J. Med. Signals Sens. **9**(4), 221–226 (2019)
27. Wrobel, K.: Diagnosing Parkinson's disease by means of ensemble classification of patients' voice samples. Procedia Comput. Sci. **192**, 3905–3914 (2021)
28. Fayyazifar, N., Samadiani, N.: Parkinson's disease detection using ensemble techniques and genetic algorithm. In: Artificial Intelligence and Signal Processing Conference (AISP), Shiraz, Iran, pp. 162–165 (2017)

Cryptocurrency Price Prediction Using Deep Learning

S. V. Tharun[1], G. Saranya[1], T. Tamilvizhi[2], and R. Surendran[3]([✉])

[1] Department of Computer Science and Engineering, Amrita School of Computing, Amrita Vishwa Vidyapeetham, Chennai, India
g_saranya@ch.amrita.edu

[2] Computer Science and Engineering, Panimalar Engineering College, Chennai, India

[3] Department of Computer Science and Engineering, Saveetha School of Engineering, SaveethaInstitute of Medical and Technical Sciences, Chennai, India
surendranr.sse@saveetha.com

Abstract. As technology advances daily, are advancing our lives into the digital sphere. The introduction of these cryptocurrencies aims to prevent the financial crisis. Due to its decentralized nature, high level of security, and restrictions on the number of coins that may be created, cryptocurrencies have attracted investors. Predicting the future price includes several limits and determinants because it involves capital. It varies according to market share. Using block chain technology and encryption, the transactions are encrypted from the beginning to end. Predicting prices to encourage consumers to invest during a specific period and earn a profit. They include a variety of elements, such as market analysis, sentiment analysis on Twitter, trading volume, and open and closing prices. Typical models that can be used to forecast bitcoin prices include regression techniques, neural networks, and support vector machines. Predictions of cryptocurrency prices based on their closing prices give investors additional insight into whether to wait until the closing period if prices are low for the entire day or to invest the following day. Using deep learning and bidirectional Long Short-Term Memory (LSTM) suggested this model to anticipate the price of digital currencies including Bitcoin, Litecoin, Ethereum, and Cardano. In this model, predictions are made using historical price statistics, and the graph is created by evaluating several performance indicators.

Keywords: Cryptocurrency · Deep Learning · Bidirectional LSTM · Exploratory Data Analysis (EDA) · Price Prediction

1 Introduction

People started to invest in cryptocurrency due to the high security, it has been decentralized, the rapid growth in the price of cryptocurrency has it been limited to a certain amount of coins. Since the cryptocurrency market has a lot of fluctuations in their price, it is hard to predict them with high accuracy [1]. Thus, there is a need for highly accurate forecasting models using modern techniques. The overall aim is to construct a deep

S. Kadry and R. Prasath (Eds.): MIKE 2023, LNAI 13924, pp. 283–300, 2023.
https://doi.org/10.1007/978-3-031-44084-7_27

learning that can predict price trends with superior results. The prediction offers great potential and provides motivation for research in the area [2]. The prediction helps users and investors, so that they can invest and use it more efficiently. To develop an application which will predict the four major cryptocurrency (Bitcoin, Litecoin, Cardano, Ethereum) prices in the future with decent accuracy. This allows the investors to invest wisely in cryptocurrency trading as the prices of cryptocurrencies have gone up to an exaggerating amount in the last ten years [3]. Investments in cryptocurrencies can be profitable. Predicting the price of a cryptocurrency is difficult because it is quite volatile and prone to market fluctuations. Using a Bi-directional Long Short-Term Memory (Bi-LSTM) neural network is one method for creating a Cryptocurrency price prediction model. A Recurrent Neural Network (RNN) variant known as a Bi-LSTM may recognize temporal connections in sequential inputs. For time-series prediction, it is advantageous since it can learn both forward and backward dependencies [4].

Bi-LSTM is considered over RNN as they have shorter memory states than LSTM. The sequential data in this instance is cryptocurrencies past price data. It is necessary to consider factors such as social media sentiments, market trends, and news events. Because cryptocurrencies depend on one another, it can be difficult to predict their prices. It offers real-time data such as financial market analyses. Our model has excellent performance and various performance metrics are done [5]. Thus, the forecasting model with the high accuracy. As cryptocurrency is one of the major investments in the world and it takes a lot of labor to predict the future prices by doing the analysis. Building a deep learning model to forecast the price of cryptocurrency is not beneficial to the companies and other investors who invest in them, and it also reduces the human labor., equation etc. does not have an indent, either Subsequent.

2 Related Works

Sudeep Tanwar and Gulshan Sharma developed the Prediction model with the classification of two phases. [6] This model involves the Deep Learning and Sentiment Analysis. In the first phase, the Model uses Twitter application programming interface (API) and valence aware dictionary and sentiment reasoner (VADER) to calculate the sentiments from a tweet. After the sentiment analysis, it utilizes the price history along with the extracted features from the first phase to predict the price of cryptocurrency. The major limitation in this model is analyzing and predicting the price of cryptocurrency of one coin and relating that coin and predicting the other cryptocurrency coin. [7] Yung-Cheol Byun improved the performance of the cryptocurrency prediction model based on Reinforcement Learning. The steps include use of the raw data information which is preprocessing, feature engineering, transformation, and feature selection. The drawback of this model is that it predicts the price only for the two coins, namely, Litecoin and Monero with the loss metrics such as mean absolute percentage error (MAPE) is 4.0048 for Litecoin and 5.1838 for Monero. These MAPE performance is high for the three days' price prediction. [8] Mahir Iqbal and the authors proposed the time-series prediction model with machine learning techniques. The model uses three different machine algorithms such as autoregressive integrated moving algorithm (ARIMA), XG Boosting algorithm, and Facebook Prophet to predict the future price of bitcoin. The best algorithm in this model is ARIMA whose R square is 189.6 since the value should range close to 1 for the perfect model and it shows the prediction only for bitcoin. [9] Burggraf et al. looked at the relationship between investor emotions on Bitcoin returns by taking into account of the household and market levels. [10] The effect of social factors on the bitcoin market was discussed by Aggarwal et al., Social Media coverage of a particular cryptocurrency and celebrity social media posts may also have an impact on cryptocurrency pricing. [11] The researchers suggested a prediction methodology for utilizing NARX to determine the closing price of Bitcoin. Certain attributes other than closing price are taken into account as external inputs in this model. As a result, the number of input attributes determines the number of input nodes. This model's primary flaw is that it displays the closing price for the following day. [12] Telegraph statistics and trends were examined for their capacity to forecast short-term patterns in cryptocurrency price fluctuations. The limitation of this model is obtained with an accuracy of 63% for Bitcoin and 56% for Ethereum with LSTM. [13] By using correlation analysis of several considerations to study the cryptocurrency market, Saad et al. suggested the model, and eventually a machine learning model was created. The method used in the model were Multivariate Regression. [14] According to the results of Nishant Jagannath, self-adaptive algorithm-based LSTM models can anticipate Ethereum prices more quickly and accurately. It could be used along with the off-chain variables (Reddit, Twitter) to predict prices. The major drawback of this work is that it can only predict the Ethereum prices. Sentiment analysis model uses Twitter API and VADER to assess the sentiments of the text in the tweets. Price prediction model uses LSTM and GRU recurrent neural network to train with the history of prices. Reinforcement learning is used to create the price prediction model (Litecoin & Monero). The researchers suggested a prediction methodology for utilizing NARX to determine the closing price of Bitcoin [15]. Certain attributes other than closing price are taken into account as external inputs in this model. LSTM algorithm is used to predict the price of Ethereum. Onchain metrics are utilized to predict the price of ethereum. LSTM and GRU algorithms are used to build to perform sentiment analysis of the tweets on crytocurrency [16]. Various feature extraction techniques such as BoW, TF-IDF, and Word2Vec are used [17].

3 Proposed Work

The proposed system involves a deep learning model with Bi-LSTM that forecasts the price of cryptocurrency. Using this technique makes it easier to create a more accurate forecasting model. The steps involved in building this deep learning forecast model involve the following: Forecast of the cryptocurrency price by their closing price gives more idea to the investors to wait for the closing period to invest if it's low than the whole day or to invest on the next day [18]. The proposed model to forecast the price of cryptocurrency coins such as Bitcoin, Litecoin, Ethereum, and Cardano using Deep Learning with Bidirectional LSTM. In this model, the prediction is made using the historical price datasets and various performance metrics are evaluated to produce the graph.

Figure 1 shows the steps involved in building the cryptocurrency price prediction model. To enhance the visualization of the graph for our model, have been displayed

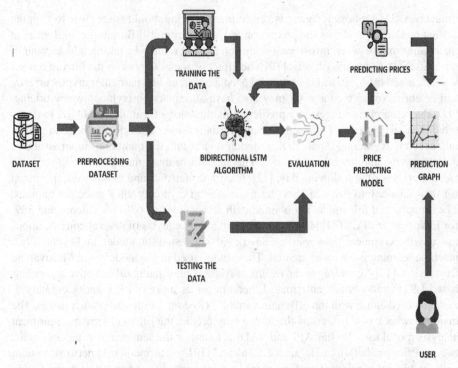

Fig. 1. Steps involved in building the cryptocurrency price prediction model.

through a website where one can view the next 15 days' prediction graph for each coin. Tools used to implement this proposed work are Anaconda with Jupiter Notebook, Python and Visual Studio Code.

Dataset Collection

The first step involved in this model is collecting the historical price data for cryptocurrency coins such as Bitcoin, Cardano, Ethereum, and Litecoin [19]. The data can be retrieved from a variety of sources including APIs, Yahoo Finance and cryptocurrency exchanges. It filters the daily price information for Bitcoin, Ethereum, Cardano, and Litecoin using the following criteria in Fig. 2. The dataset has one csv file for each of the top 50 crypto coins by Market Capitalization. Price history is available on a daily basis from Jan 1, 2021. Here 4 cryptocurrencies and predict the short term price movements of them.

Close: Closing price of each cryptocurrency.

Open: The opening price of each cryptocurrency High: Highest price of each crypto currency.

Low: Lowest price of each cryptocurrency.

Adj Close: Adjusted closing price of a cryptocurrency.

Volume: Turnover in the price of each cryptocurrency.

Date	Open	High	Low	Close	Adj Close	Volume
01-09-2021	47099.77	49111.09	46562.44	48847.03	48847.03	3.91E+10
02-09-2021	48807.85	50343.42	48652.32	49327.72	49327.72	3.95E+10
03-09-2021	49288.25	50982.27	48386.09	50025.38	50025.38	4.32E+10
04-09-2021	50009.32	50545.58	49548.78	49944.63	49944.63	3.75E+10
05-09-2021	49937.86	51868.68	49538.6	51753.41	51753.41	3.03E+10
06-09-2021	51769	52700.94	51053.68	52633.54	52633.54	3.89E+10

Fig. 2. Input datasets

Preprocessing Dataset

After collecting the data, the data should be preprocessed. Pre-processing the data includes identifying and removing outliers, noisy data, other missing values, and erroneous values.

Exploratory Data Analysis

Exploratory data analysis (EDA) is done to analyze the data and clean the data. With the use of statistical summaries and graphical representations, it is used to identify trends, patterns or to verify assumptions. The next step is to normalize the data. Normalization is usually done using min-max scaling, robust scaling, or other standard scaling techniques by in Fig. 3.

Building the Model

Create training and test sets from the data. The model is trained using the training set and its effectiveness is assessed using the testing set. As cryptocurrencies have large fluctuations in market value, applying Bi-LSTM would be more appropriate because networks can learn long-term dependencies better. Building the architecture of the model by adding the Bi-LSTM layer with the appropriate neurons in addition to the other Recurrent Neural Network (RNN) such as dropout and dense layers. Dropout layers are added to improve the efficiency and accuracy of the model. A series of historical cryptocurrency values are used as the model's input.

Bi-LSTM Algorithm

A bidirectional LSTM, often known as a bi-LSTM, is a sequence processing model that is made up of two LSTMs, one of which takes the inputs forward and the other of which receives it backward. By efficiently boosting the network's informational capacity, Bi-LSTMs enhance the context that the algorithm has access to. A bidirectional LSTM differs from a conventional LSTM in that our input flows in two directions. Make the input flow in one way, either backwards or forward, using the standard LSTM. To maintain both past and future information, bidirectional inputs can be made to flow in both ways.

```
# Input sequence
input_sequence = [x_1, x_2, ..., x_T]
# Forward LSTM iteration
for t in range(T):
    # Concatenate input and previous hidden state for forward LSTM
    input_forward = concatenate(x_t, hidden_state_forward)
    # Compute forward LSTM gate values
    forget_gate_forward = sigmoid(W_f_forward * input_forward + b_f_forward)
    input_gate_forward = sigmoid(W_i_forward * input_forward + b_i_forward)
    candidate_cell_forward = tanh(W_c_forward * input_forward + b_c_forward)
    output_gate_forward = sigmoid(W_o_forward * input_forward + b_o_forward)
    # Compute cell state and hidden state for forward LSTM
    cell_state_forward   =   forget_gate_forward   *   cell_state_forward   +   in-
put_gate_forward * candidate_cell_forward
    hidden_state_forward = output_gate_forward * tanh(cell_state_forward)
    # Append current output to the forward output list
    output_forward.append(hidden_state_forward)
# Backward LSTM iteration
for t in range(T-1, -1, -1):
    # Concatenate input and previous hidden state for backward LSTM
    input_backward = concatenate(x_t, hidden_state_backward)
    # Compute backward LSTM gate values
    forget_gate_backward   =   sigmoid(W_f_backward   *   input_backward   +
b_f_backward)
    input_gate_backward   =   sigmoid(W_i_backward   *   input_backward   +
b_i_backward)
    candidate_cell_backward   =   tanh(W_c_backward   *   input_backward   +
b_c_backward)
    output_gate_backward   =   sigmoid(W_o_backward   *   input_backward   +
b_o_backward)
    # Compute cell state and hidden state for backward LSTM
    cell_state_backward  =  forget_gate_backward  *  cell_state_backward  +  in-
put_gate_backward * candidate_cell_backward
    hidden_state_backward = output_gate_backward * tanh(cell_state_backward)
    # Prepend current output to the backward output list
    output_backward.insert(0, hidden_state_backward)
# Concatenate forward and backward outputs
output_sequence = concatenate(output_forward, output_backward)
# Continue with further layers or downstream tasks
```

Note that this pseudocode assumes that you already have defined and initialized the parameters (weights and biases) for the forward and backward LSTMs. The input sequence is represented as input_sequence, where x_t denotes the input at time step t. The outputs of the forward and backward LSTMs are stored in output_forward and output_backward, respectively. Finally, the outputs from both directions are concatenated to form the output_sequence.

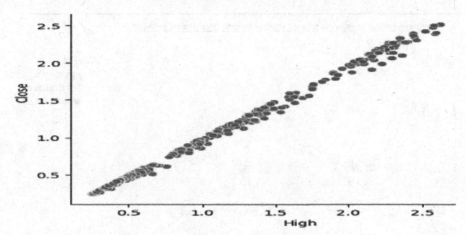

Fig. 3. Visualization for close and high price.

Training the Forecast Model

The model is trained with the train data and the accuracy is evaluated using the test data. Using the training data to train the model. The model gains the ability to recognize temporal connections in the data during training [20]. By using the testing data, the model's performance is evaluated. The train data and test data should be transformed back to the original form [21]. Here trained the model separately for cryptocurrency coin such as Bitcoin, Cardano, Ethereum, and Litecoin. Figure 4 shows the Real Closing Price vs Predicted Closing Price model for Bitcoin. Figure 5 shows the Real Closing Price Vs Predicted Closing Price model for Cardano. Figure 6 shows the Real Closing Price Vs Predicted Closing Price model for Ethereum. Figure 7 shows the Real Closing Price Vs Predicted Closing Price model for Litecoin.

Fig. 4. Real Closing Price Vs Predicted Closing Price model for Bitcoin.

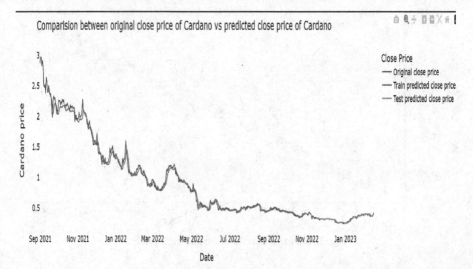

Fig. 5. Real Closing Price Vs Predicted Closing Price model for Cardano.

Fig. 6. Real Closing Price Vs Predicted Closing price model for Ethereum.

Fig. 7. Real Closing Price Vs Predicted Closing price model for Litecoin.

Prediction Module

The parameters for the future price prediction are initialized. Then the price prediction for the next 15 days is done by the model for each cryptocurrency coin. The model's output forecasted price is represented in the form of graph. Figure 8 shows the Predicted Close Price of Bitcoin for the next 15 days. Figure 9 shows the Predicted Close Price of Cardano for the next 15 days. Figure 10 shows the Predicted Close Price of Ethereum for the next 15 days. Figure 11 shows the Predicted Close Price of Litecoin for the next 15 days.

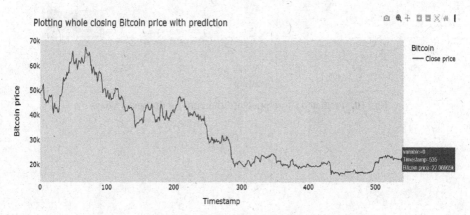

Fig. 8. Predicted Close Price of Bitcoin for the next 15 days.

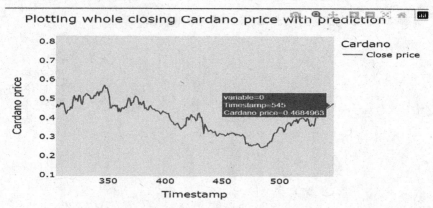

Fig. 9. Predicted Close price of Cardano for the next 15 days

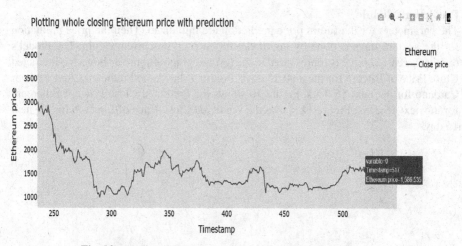

Fig. 10. Predicted Close price of Ethereum for the next 15 days

Fig. 11. Predicted Close price of Litecoin for the next 15 days

Performance Evaluation

The efficiency of various deep learning models can be assessed using performance metrics such as accuracy, R-squared (R2), mean absolute percentage error (MAPE), variance regression score, and mean squared error (MSE). Accuracy can be done using statistical metrics such as mean squared error, root mean squared error that quantifies the degree to which a prediction or estimate matches the actual or expected value. It is often used to evaluate the performance of a model. Mean Absolute Percentage Error is a commonly used metric for measuring the accuracy of a forecast or prediction. MAPE measures the average percentage difference between the forecasted and actual values of a time series.

$$\text{MAPE} = \left(\frac{1}{n}\right) * \left(\frac{|A_t - f_t|}{|A_t|}\right) * 100\% \tag{1}$$

"A_t" represents the actual values or observations.

"f_t" represents the forecasted or predicted values.

"|A_t - f_t|" denotes the absolute difference between the actual and forecasted values.

"|A_t|" represents the absolute value of the actual values.

R-squared (R2 score) is also known as the coefficient of determination, which is a statistical measure that represents the proportion of variance in the dependent variable (y) that is explained by the independent variables (X) in a linear regression model. It is commonly used to evaluate the performance of a regression model.

$$R_2 = 1 - \frac{\text{RSS}}{\text{TSS}} \tag{2}$$

"RSS" represents the residual sum of squares, which is the sum of the squared differences between the actual values and the predicted values.

"TSS" represents the total sum of squares, which is the sum of the squared differences between the actual values and the mean of the actual values.

Variance Regression score is also known as explained variance score, which is a statistical measure that quantifies the proportion of the variance in the target variable that can be explained by the independent variables in a regression model (Table 1).

Table 1. Comparison table for each cryptocurrency with their performance metrics.

CRYPTO COINS	ACCURACY	MAPE	R2 Score	Variance Regression Score
BITCOIN	95.00	3.38	0.87	0.90
CARDANO	96.30	2.73	0.93	0.93
ETHEREUM	97.07	2.09	0.95	0.95
LITECOIN	95.01	3.78	0.87	0.87

Analysis of Predicted vs Actual Price of Each Cryptocurrency

Figure 12 shows the Actual price vs Predicted price of Cardano. Figure 13 shows the Actual price vs Predicted price of Ethereum. Figure 14 shows the Actual price vs Predicted price of Litecoin.

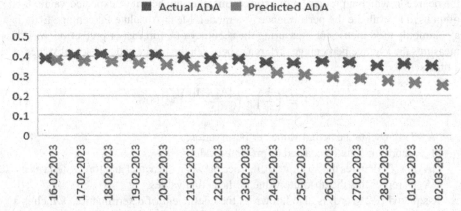

Fig. 12. Actual price vs Predicted price of Cardano.

Deploying to Website

To show our predicted model, a simple user interface website is designed which is named as "crypto". It includes a home page, prediction page, and prediction graph page for each coin, a live price page, and Frequently Asked Questions (FAQ).

Runtime Screens

This is the landing page for Cryptocurrency Price Prediction which has an embedded graph for 15 days' prices of four cryptocurrencies such as Bitcoin, Cardano, Ethereum, and Litecoin. Figure 15 shows home page of the predicted model.

Comparative analysis of Actual & Predicted Price of Ethereum

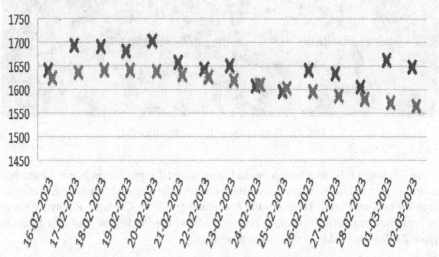

Fig. 13. Actual price vs Predicted price of Ethereum.

Comparative analysis of Actual vs Predicted price of Litecoin

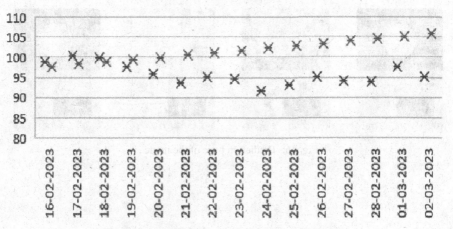

Fig. 14. Actual price vs Predicted price of Litecoin.

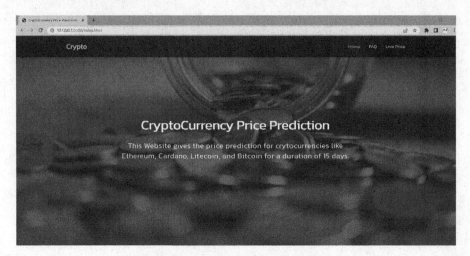

Fig. 15. Home page of the predicted model.

By Clicking on Predict button on each coin shows the prediction graph for the next 15 days. Figure 16 shows the prediction page of the model. Figure 17 shows the Prediction graph for bitcoin. Figure 18 shows the Prediction graph for Cardano. Figure 19 shows the Prediction graph for Ethereum. Figure 20 shows the Prediction graph for Litecoin. Figure 21 display the Live Price of cryptocurrency.

Fig. 16. Prediction page of the model.

15-DAYS
price prediction for
BITCOIN.

Fig. 17. Prediction graph for bitcoin.

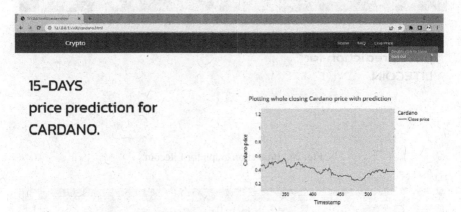

Fig. 18. Prediction graph for Cardano.

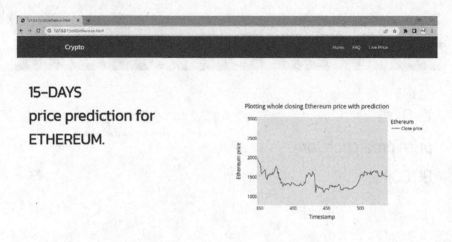

Fig. 19. Prediction graph for Ethereum.

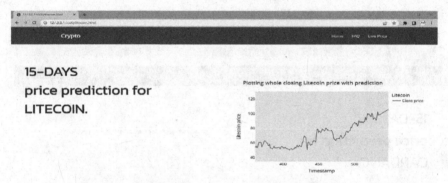

Fig. 20. Prediction graph for Litecoin.

Fig. 21. Live Price of cryptocurrency

4 Conclusion

Thus, this forecasting model predicts the prices of cryptocurrencies such as Bitcoin, Cardano, Ethereum, and Litecoin using Bi-LSTM algorithm with high accuracy. The accuracy of the price prediction ranges from 95 to 97% and the other various performance metrics are evaluated by this forecast model. These predictions are represented as a graph which is embedded in a website. Thus, the prediction and live prices can be viewed through the website by the users. Updating our datasets to ensure our forecast model to be relevant. The proposed work can use APIs for data collection or else it's better to fetch the live price of the cryptocurrency data from their exchanges. In future, the proposed work can be extend in the form of real-time data for each country based on country regulations using blockchain technology.

References

1. Jagannath, N., et al.: A self-adaptive deep learning-based algorithm for predictive analysis of bitcoin price. IEEE Access **9**, 34054–34066 (2021)
2. Patel, N.P., et al.: Fusion in cryptocurrency price prediction: a decade survey on recent advancements, architecture, and potential future directions. IEEE Access **10**, 34511–34538 (2022)
3. Selvin, S., Vinayakumar, R., Gopalakrishnan, E.A., Menon, V.K., Soman, K.P.: September. Stock price prediction using LSTM, RNN and CNN-sliding window model. In: 2017 International Conference on Advances in Computing, Communications and Informatics (ICACCI), pp. 1643–1647. IEEE (2017)
4. Surendran, R., Alotaibi, Y., Subahi, A.F.: Wind speed prediction using chicken swarm optimization with deep learning model. Comput. Syst. Sci. Eng. **46**(3), 3371–3386 (2023)
5. Sujatha, R., Mareeswari, V., Chatterjee, J.M., Abd Allah, A.M., Hassanien, A.E.: A Bayesian regularized neural network for analyzing bitcoin trends. IEEE Xplore **9** (2021). https://doi.org/10.1109/ACCESS.2021.3063243
6. Agarwal, A., Keerthana, S., Reddy, R., Moqueem, A.: Prediction of Bitcoin, Litecoin and Ethereum trends using state-of-art algorithms. IEEE Access (2021). https://doi.org/10.1109/MysuruCon52639.2021.9641735
7. Khedr, A.M., Arif, I., Raj, V.P., El-Bannany, M., Alhashmi, S.M., Sreedharan, M.: 'Cryptocurrency price prediction using traditional statistical and machine-learning techniques: a survey. Intell. Syst. Accounting, Finance Manage. **28**(1), 3–34 (2021)
8. Benjamin, J.J., Surendran, R., Sampath, T.: A professional strategy for Bitcoin and Ethereum using Machine Learning for Investors. In: 2022 4th International Conference on Inventive Research in Computing Applications (ICIRCA), pp. 798–804. IEEE (2022)
9. Soni, K., Singh, S.: Bitcoin price prediction-an analysis of various regression methods. IEEE (2022). https://doi.org/10.1109/ISCAIE54458.2022.979453
10. Herskind, L., Katsikouli, P., Dragoni, N.: Privacy and cryptocurrencies—a systematic literature review. IEEE Access (2020). https://doi.org/10.1109/ACCESS.2020.298095
11. Fernandes, M., Khanna, S., Monteiro, L., Thomas, A., Tripathi, G.: Bitcoin price prediction. IEEE Xplore (2010). https://doi.org/10.1109/ICAC353642.2021.9697202
12. M. M. Patel, S. Tanwar, R. Gupta, Neeraj Kumar, "A Deep Learning-based Cryptocurrency Price Prediction Scheme for Financial Institutions" Elsevier Vol. 55, Dec (2020), 102583
13. Mudassir, M., Bennbaia, S., Unal, D., et al.: Time-series forecasting of Bitcoin prices using high-dimensional features: a machine learning approach. Neural Comput. Appl. (2020). https://doi.org/10.1007/s00521-020-05129-6

14. Aslam, N., Rustam, F., Lee, E., Washington, P.B., Ashraf, I.: Sentiment analysis and emotion detection on cryptocurrency related tweets using ensemble LSTM-GRU model. IEEE Access **10**, 39313–39324 (2022)

15. Indera, N.I., Yassin, I.M., Zabidi, A., Rizman, Z.I.: Non-linear autoregressive with exogenous input (NARX) Bitcoin price prediction model using PSO-optimized parameters and moving average technical indicators. J. Fundam. Appl. Sci. **9**(3), 791–808 (2017)

16. Parekh, R., et al.: DL-Guess: deep learning and sentiment analysis-based cryptocurrency price prediction. IEEE Access (2022). https://doi.org/10.1109/ACCESS.2022.3163305

17. Gowri, S., Surendran, R., Jabez, J., Srinivasulud, S.: IoT forensics: what kind of personal data can be found on discarded, recycled, or re-sold IoT devices. J. Discrete Math. Sci. Crypt. **25**(4), 999–1008 (2022)

18. Vinayakumar, R., Soman, K.P., Poornachandran, P.: Long short-term memory based operation log anomaly detection. In: 2017 International Conference on Advances in Computing, Communications and Informatics (ICACCI), pp. 236–242. IEEE (2017)

19. Lahmiri, S., Bekiros, S.: Deep learning forecasting in cryptocurrency high-frequency trading. Cognit. Comput. **13**, 485–487 (2021)

20. Tanwar, S., Patel, N.P., Patel, S.N., Patel, J.R., Sharma, G., Davidson, I.E.: "Deep learning-based cryptocurrency price prediction scheme with inter-dependent relations. IEEE Access **9** (2021). https://doi.org/10.1109/ACCESS.2021.3117848

21. Phaladisailoed, T., Numnonda, T.: Machine learning models comparison for Bitcoin price prediction. IEEE Access (2018). https://doi.org/10.1109/ICITEED.208.8524911

Ensemble Learning Based Social Engineering Fraud Detection Module for Cryptocurrency Transactions

Vishvesh Pathak[1], B. Uma Maheswari[1(✉)] [iD], and S. Geetha[2] [iD]

[1] Department of Computer Science and Engineering, Amrita School of Computing, Bengaluru, Amrita Vishwa Vidyapeetham, Bengaluru, India
vishveshpathak1999@gmail.com, b_uma@blr.amrita.edu
[2] School of Computer Science and Engineering, Vellore Institute of Technology, Chennai, India
geetha.s@vit.ac.in

Abstract. The practical applications of blockchains can far supersede the widely known trading and cryptocurrency realm. If any service provider is looking for a consistent, immutable, and multitenancy-supported ledger, then blockchain is the promising solution. Nowadays, social engineering attacks are prevalent. And the attackers deceive cryptocurrency traders. This work investigates various ensemble learning, neural network, and machine learning algorithms for fraud detection and identifies the best decision-making algorithm. It is observed that Adaptive Boosting (AdaBoost) algorithm outperforms with an accuracy of 98.92%. Further, the fraud detection module is integrated with an application developed for cryptocurrency transactions. Before a new transaction is committed to blockchain, The fraud detection module intervenes and alerts the user. We have also designed a test bed of deployable Peer-to-Peer (P2P) network to simulate cryptocurrency transaction.

Keywords: Blockchain · Ensemble Learning · XGBoost · AdaBoost · Logistic Regression · proof of work consensus · Ethereum · Cryptography · Fraud Detection · Social Engineering Attacks

1 Introduction

The inclusion of blockchain for any application is a wiser solution where immutability and consistency are concerned [1]. However, there is an assumption that the transactions being committed to blockchain are not fraudulent. In other words, blockchain does not automatically determine whether the end user is paying the intended party or an impostor party; therefore, blockchain is prone to social engineering attacks. Often blockchain subscribers pay fraud parties by being misled using a fake User Interface(UI). There is a time gap between the payer broadcasting a new transaction to the network and the same transaction being published as a block in the blockchain. Hence, there is a need for a system to detect fraudulent transactions before transactions are committed to the blockchain.

© The Author(s), under exclusive license to Springer Nature Switzerland AG 2023
S. Kadry and R. Prasath (Eds.): MIKE 2023, LNAI 13924, pp. 301–311, 2023.
https://doi.org/10.1007/978-3-031-44084-7_28

In this paper, we propose a solution against the above scenario, which employs the Ethereum fraud transactions dataset and machine learning, ensemble learning, and neural network algorithms are trained with the dataset to classify transactions as fraudulent or legitimate. The legitimate transaction is then broadcasted into the test environment. Even though blockchain environments are available in the cloud as a service [2], they have additional security mechanisms to provide a real-world implementation. These security mechanisms have no direct relation to blockchain technology itself. Hence, there is a need to create a blockchain environment to test all the test cases [3]. In the simulation environment, there is a server node to establish communication. The server node is not used for the storage of data or computation. It is not a part of the blockchain network and acts analogous to an Internet Service Provider (ISP). Our proposed system embeds a fraud detection module in the application, which is triggered when the end user initiates a payment. Furthermore, we designed a UI after referring to [4] where the end user can perform operations like authentication, view transactions, and commit transactions to demonstrate the working of our fraud detection module. If the user pays to a party with a relatively higher chance of fraud, it alerts the user.

The proposed study in this article introduces a way to merge the domains of blockchain and machine learning. The best-performing model is used in the fraud detection module of the application. The main contributions of this article are as follows:

- Compared the performance and evaluation metrics of popular decision-making models [5] and selected the best model.
- Trained the selected model by hyperparameter tuning.
- Reduced the number of features required for decision-making by eliminating less correlated features to the target variable [6]. The correlation is measured using Chi-squared.
- Built a test bed for a blockchain network integrated with the best-performing model and created an application to simulate cryptocurrency transactions.

2 Literature Survey

This section details the literature survey conducted on the application of machine learning algorithms in performing predictions, evaluating machine learning models, and understanding the architecture and workings of blockchain.

Authors Soni et al. in work [1] have described various concepts of the blockchain along with applications, challenges, threats, and possible attacks. Wang et al. in [2] have designed a platform consisting of an interface for trading and a blockchain with proof of interaction as the consensus mechanism. Authors Banno et al. in work [3] have simulated a blockchain network with a SimBlock simulator, which is flexible in the behaviour of network and blockchain blocks. The significant parameters which can be varied for blocks are block size, the difficulty of generation, and number of nodes. Authors Tharatipyakul et al. in [4] reviewed papers to evaluate the contribution of blockchain integration into various industries such as agriculture, foods, and supply chain management. Bhowmik et al. in [5] have presented a performance comparison of various machine learning algorithms on the detection of fraudulent transactions in blockchain transactions. Sivaranjani et al. in [6] have performed feature selection, dimensionality reduction, and supervised learning classification in order to predict diabetes from the PIMA dataset. Shakbulatov et al. in [7] propose a record-keeping system for keeping track of the carbon footprint of food supply chain industries. Results show faster block creation than average blockchain and linear increment in storage space required to store every subsequent block.

Lin et al. [8] have suggested a blockchain solution to record X.509 certificates to achieve transparency and immutability to prevent phishing websites from getting legitimate certificates. This work also proposed solutions against compromised certification authority and compromised publishing key pairs. Manikumar et al. in [9] proposed a method to detect a node that transmits malicious packets in P2P network, and blacklist the IP address of the node and store in blockchain. Sharma et al. in [10] have discussed various security threats on blockchain, including 51% double spending attacks, replay attacks, eclipse attacks, forking problems, malware, etc. and briefly discussed the possible solutions to some of these problems. Zhou et al. in [11] presented a hierarchical scheme for providing confidentiality to the nodes and offers tampering protection to the blockchain environment. The participation node's hierarchy is divided into managerial, worker, and user nodes. Wang et al. in [12] provide a detailed description of smart-contract applications in the area of decentralization, enforceability, and verifiability of blockchain. It elaborates on the applications in industries like management, finance, Internet of Things (IOT), energy, and other industries. The authors Rekha et al. in their work [13] have presented a solution for evidence preservation in blockchain for IOT-related cyber-crime. Innocent et al. in [14] have proposed combining concepts of blockchain and secured computation for an improved solution of the millionaire's problem, i.e., determination of common information without disclosing any details. Shree et al. in work [15] have proposed moving the supply chain management functions to the blockchain in order to facilitate the delivery of goods and payments at the leisure of the involved parties and minimizing the waiting times and theft. Authors Lakshminarayan et al. in [16] have proposed solution using blockchain against the malpractices in blood donation activities, and data related to quality of blood, important days, and usage are stored in blockchain for the transparency and immutability.

The study on state-of-the-art blockchain systems shows that existing blockchain systems do not discriminate between genuine and fraudulent transactions. If the validations of the blockchain are satisfied, even the fraudulent transactions will get committed. In this work, we embed the best-performing decision-making model in the blockchain structure to prevent fraudulent transactions from being committed in the blockchain.

3 Proposed Methodology

This section describes the architecture design, process flow, and implementation of the blockchain network, including the fraud detection module and its integration into the blockchain.

3.1 Process Flow Overview

The process flow of the network architecture of our proposed system is illustrated in Fig. 1. It includes a server analogous to ISP, and the server is not involved in cryptocurrency transactions. Moreover, it will not perform any computation or store data related to blockchain. The client runs in a dual-threaded structure, as shown in Fig. 1, with one thread for sending messages to the server and another for receiving communication from other nodes in the network.

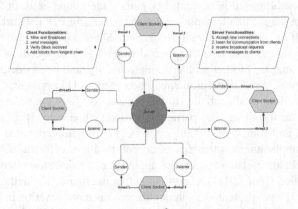

Fig. 1. Illustration of the blockchain network.

The integration of the fraud detection module in the blockchain architecture is shown in Fig. 2. The first part describes the process of building the memory pool, and the second part illustrates the working of the blockchain network. When the subscribers try to make payments, the data sent in the payment is collected in the backend of the trading application. This data is then given to a pre-trained model which predicts the transaction's legitimacy. If the model predicts the transaction as fraud, then the app reloads the payment window with an alert. In this way, we ensure that no transactions are broadcasted in the network which are identified as fraud by the fraud detection module.

In the blockchain network, each client simultaneously tries to collect valid transactions from the memory pool and then performing a process called mining. The information stored in a block is perpetually hashed to find a nonce during mining. This nonce generates a string called proof of work which should have four leading zeros. The client who finds this proof of work broadcasts the block it has mined to the entire network. The receivers accept new blocks until one of the senders has sent six mined blocks. The probability that the same miner keeps getting correct blocks is infinitesimally low. Each client writes the first block from the longest chain in their local copy of the blockchain ledger. This is the point when the transaction is committed in the blockchain, and then the immutability is achieved. A new transaction called Coinbase, exempt from validations, is added to the block to increase the pool of the cryptocurrency in existence and reward the miner for successfully creating a block. Each received block is stored in its sender's buffer on the receiver's side.

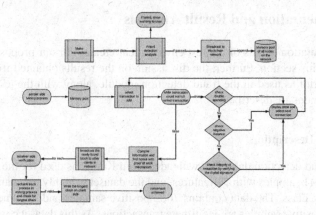

Fig. 2. Process flow diagram of the proposed approach.

3.2 Design of the Fraud Detection Module

The fraud detection module is illustrated in Fig. 3. It follows the standard Machine Learning procedure of imputation of missing data, followed by normalization to prevent bias due to differences in the magnitude of features, and comparison of algorithms concerning performance metrics such as accuracy, precision, and recall.

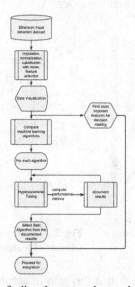

Fig. 3. Flow chart for finding the most relevant decision-making model.

Hyperparameter tuning ensures that the best-performing parameters are selected in any given model. The best-selected algorithm is then trained with more extensive training data and validated using K-fold cross-validation. Evaluation metrics are computed for each machine learning algorithm.

4   Implementation and Result Analysis

The implementation details of the test environment created for our proposed work are described in this section. Further, the discussion on the results obtained from machine learning algorithms used in the fraud detection module, along with the test cases used in our experiment are described.

4.1   Dataset Description

Ethereum fraud detection dataset downloaded from Kaggle is used for the experiment. It contains 9841 samples with 51 features, and the dataset is highly imbalanced in favor of the positive class. The data contains 7662 positive samples and 2179 negative samples, where positive samples are legitimate transactions. As this dataset contains a large number of features, our proposed method followed feature reduction methodologies.

4.2   Feature Reduction

A correlation matrix is plotted using the Python seaborn library, as shown in Fig. 4. The features are sorted based on the chi-squared technique in the order of most dependent to least dependent on the target variable. The accuracy is plotted against the number of features selected. The Elbow Technique is used to determine the minimum number of features for obtaining relatively high accuracy.

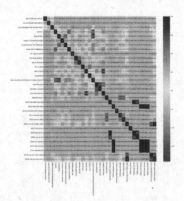

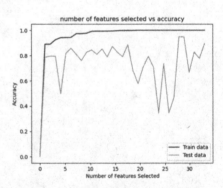

Fig. 4. Heatmap showing correlation between the features in dataset.

Fig. 5. Training and testing accuracy against number of features selected

It is very evident from Fig. 5 that the gain in accuracy becomes insignificant by increasing the number of features beyond 12 from the dataset. Hence, we have shortlisted features ['Avg min between sent tnx', 'max val sent', 'total transactions', 'Avg min between received tnx', 'total val received', 'Time Diff between first and last', 'Number of Created Contracts', 'total ether balance', 'Unique Received From Addresses', 'Unique Sent To Addresses', 'Received Tnx', 'Sent tnx'] respectively. The model is trained with this reduced number of features and stored in a pickle file. When the end user initiates the payment using the application, the backend loads the trained model from the pickle file. The payment details are given as input to the fraud detection module, and the prediction is made using the model stored in this pickle file.

4.3   Implementation of the Blockchain Network

The blockchain network is constructed among the subscribers using socket programming in our proposed system. The server.accept() is called recurrently until all the clients are connected. Each time a new client connects, a new thread is started in the server for listening to the client. The server may receive several other messages, such as broadcast requests, reception of client messages, delivery of broadcast messages to every client, assembling the block when a block broadcast request is generated by a client for a mined block, etc.; these messages will prompt the server to perform some predefined tasks. The client is programmed to connect to the server and has a couple of threads, one for listening to the user and the other for listening to the server. When the server receives a block broadcast request, it broadcasts the message to other clients in the network. Once the clients receive the block broadcast message, they separate the values from the block, rerun the checks performed during mining on the sender-side, and write the longest chain to the client-side ledger if no inconsistency is found.

4.4   Results and Performance Analysis of Fraud Detection Module

We have investigated five models, viz., Logistic Regression, Extreme Gradient Boosting (XGBoost), Adaptive Boosting (AdaBoost), Recurrent Neural Network (RNN), and Long and Short Term Memory (LSTM) Artificial Neural Network, with respect to the performance metrics of accuracy, precision, and recall. The results obtained are tabulated in Table 1. The results show that the ensemble model AdaBoost outperforms other models with an accuracy of 98.992%, precision of 99.134%, and recall of 99.486%. It is also observed that logistic regression with gridsearchCV has the lowest performance with accuracy 85.572%. Hence, it has been decided to embed AdaBoost algorithm in our proposed work. Root Mean Square Error (RMSE) and Mean Absolute Error (MAE) have also been calculated for the test data.

The time taken for predicting the results of new input from a deployed model is constant and very low because the model training finds the equation of the assumed curve, and prediction only puts the values in the equation formed and calculates the result. The time taken for training the dataset is model dependent; hence we assumed it as Tmodel. Various processes in the prediction require $O(n^2)$, $O(n^2)$, $O(n^2 \log n)$, $O(n^m)$, and $O(n)$ respectively. Therefore, the overall complexity for the module will be $O(\max(n, n^2, n^2 \log n, n^m, Tmodel))$, which essentially becomes $O(\max(n^m, Tmodel))$. RNN and LSTM take 45ms each, and the number of iterations is the same as the number of epochs. The Blockchain functions have a maximum of $O(n^2)$ complexity. Execution time for mining proof of work is the longest because they are perpetually hashed using SHA256 to find proof of work. A total of 18 transactions are performed using the fraud detection algorithm, twelve are classified as legitimate and six as fraud. The correctly added transactions are used to form three blocks in the blockchain (Table 2).

Table 1. Evaluation metrics comparison for various models in fraud detection.

Model	Best parameters	Accuracy	Precision	Recall	RMSE	MAE
Logistic Regression	C: 100 Penalty: 12 Solver: newton-cg	0.85572	0.85035	0.99168	0.37714	0.14224
XGBoost	Min_child_weight: 3 Max_depth: 8 Learning_rate: 0.2 Gamma: 0.1 Colsample_bytree:0.3	0.98399	0.98472	.99485	0.12894	0.01662
AdaBoost	N_estimators: 500 Learning_rate: 0.1 Algorithm: SAMME	0.98922	0.99134	0.99486	0.14362	0.02062
RNN	Dropout: 0.2 Layers: 1000*500*300*2 Optimizer: Adam Activation function: ReLU	0.96624	0.96498	0.99327	0.19954	0.03981
LSTM	Dropout: 0.3 Layers: 100*50*30*2 Activation: Sigmoid Optimizer: Adam Threshold: 0.96	0.91202	0.90143	0.99175	0.29014	0.08418

Table 2. Performance comparison of various models in fraud detection.

Function	Complexity(in terms of big O)	Actual Time taken (milliseconds)
Preprocessing	n^2	1.59
Feature Selection + PCA	$n^2.\log n$	101.45
Logistic Regression	Max(Tmodel, n^m)	1.71
AdaBoost	Max(Tmodel, n^m)	5.04
XGBoost	Max(Tmodel, n^m)	5.0
Random Search CV(for each model)	n^m	5.02. 6.06, 5.32
Network + Blockchain functions	$n^2.m$(m increasing iteratively until loop is stopped)	N/A due to real time
Time for prediction of a new entry	Constant time	Negligible

4.5 Test Cases for the Proposed System

This section explains how the blockchain code responds to various input scenarios or attempts to bypass the validations.

Test case 1: Adding a transaction in the memory pool with someone else's name.

Attackers cannot generate digital signatures because they do not know the sender's private key. Our blockchain system rejects this kind of transaction, and a new block containing such transactions is not added to the transaction list buffer.

Test case 2: Repeating the same transaction, i.e., double spending.

An attacker may try to copy an already completed transaction. As the copied digital signature matches an existing digital signature of the original transaction, this transaction is rejected.

Test case 3: Proof of work is not found during mining.

This test case occurs when no suitable nonce is found before the end of execution. When this test case arises in our proposed system, we restart the mining process after replacing the last transaction in the block with the following valid transaction in the pool.

Test case 4: Previous block's proof of work does not match the next block's previous address.

Two variables are maintained, one for holding the current block's address and the other for holding the previous block's proof of work. We iterate through the entire Blockchain assigning these values to the respective variables, and the block is rejected if there is a mismatch.

Test case 5: Negative balance after a transaction added.

Valid transactions should be selected from the memory pool such that no subscriber will have a negative balance when all those transactions are committed to the blockchain. A hash map is used to track the balance of all the users.

Test case 6: User trying to add a fraud transaction into the memory pool.

When the user initiates the payment, the Fraud Detection module running in the background detects the fraud and alerts the user with a warning message.

Test case 7: Six valid blocks are added to a receiver's buffer.

When six valid transactions are pending to be written on the receiver's side, it writes the first block to its local copy of the blockchain. The buffer for storing received blocks is cleared, and a message is broadcasted to every node that a new block is added. The mining process is restarted with new data.

4.6 Application User Interface

The payment screen shown in Fig. 6 has a template that accepts data from the end user. The first three inputs are for the sender ID, receiver ID, and the amount the sender wants to send, and the next 12 inputs are for the shortlisted features. On pressing submit, the data is given to the model running in the backend of the application.

Based on the result given by the fraud detection module, the payment screen is reloaded with the verdict, as can be seen in Fig. 7. If the transaction is legitimate, it is broadcasted to the memory pool of each node in the blockchain network. If failure, transaction is not broadcasted, and the verdict is shown as fraud.

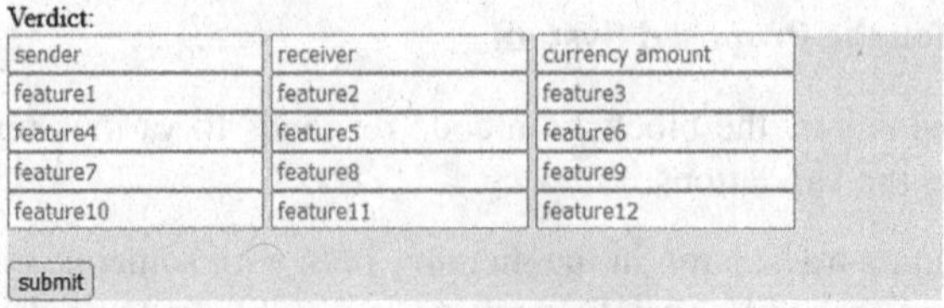

Verdict:

sender	receiver	currency amount	
feature1	feature2	feature3	
feature4	feature5	feature6	
feature7	feature8	feature9	
feature10	feature11	feature12	

submit

Fig. 6. Template for the payment screen of the user.

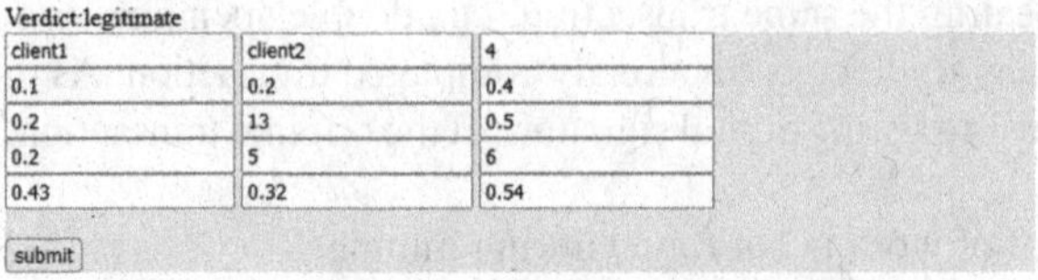

Verdict:legitimate

client1	client2	4	
0.1	0.2	0.4	
0.2	13	0.5	
0.2	5	6	
0.43	0.32	0.54	

submit

Fig. 7. Successful validation of entered transaction.

5   Conclusion and Future Work

This research work integrated fraud detection functionality in cryptocurrency payment application. A testing environment consisting of a blockchain network is built. An ensemble learning algorithm called AdaBoost is used in the fraud detection module. The computation time of the fraud detection module is significantly minimized by selecting the most impactful features. This work demonstrates how the machine learning and blockchain domains can be combined to achieve improved security in cryptocurrency payments. This integration provides an extra line of defense against social engineering attacks. This work can be extended further for a peer-to-peer network with enhanced visualization and analytics.

References

1. Soni, S., Bhushan, B.: A comprehensive survey on blockchain: working, security analysis, privacy threats and potential applications. In: 2019 2nd International Conference on Intelligent Computing, Instrumentation and Control Technologies (ICICICT), Kannur, India, pp. 922–926 (2019). https://doi.org/10.1109/ICICICT46008.2019.8993210
2. Wang, Z., Yang, L., Wang, Q., Liu, D., Xu, Z., Liu, S.: ArtChain: blockchain-enabled platform for art marketplace. In: 2019 IEEE International Conference on Blockchain (Blockchain), vol. 2019, pp. 447–454 (2019). https://doi.org/10.1109/Blockchain.2019.00068. [3] https://scet.berkeley.edu/wp-content/uploads/BlockchainPaper.pdf
3. Banno, R., Shudo, K.: Simulating a blockchain network with SimBlock. In: 2019 IEEE International Conference on Blockchain and Cryptocurrency (ICBC), pp. 3-4 (2019). https://doi.org/10.1109/BLOC.2019.8751431
4. Tharatipyakul, A., Pongnumkul, S.: User interface of blockchain-based agri-food traceability applications: a review. IEEE Access **9**, 82909–82929 (2021). https://doi.org/10.1109/ACCESS.2021.3085982
5. Bhowmik, M., Sai Siri Chandana, T., Rudra, B.: Comparative study of machine learning algorithms for fraud detection in blockchain. In: 2021 5th International Conference on Computing

Methodologies and Communication (ICCMC), pp. 539–541 (2021). https://doi.org/10.1109/ICCMC51019.2021.9418470

6. Sivaranjani, S., Ananya, S., Aravinth, J., Karthika, R.: Diabetes prediction using machine learning algorithms with feature selection and dimensionality reduction. In: 2021 7th International Conference on Advanced Computing and Communication Systems (ICACCS), pp. 141–146 (2021). https://doi.org/10.1109/ICACCS51430.2021.9441935

7. Shakhbulatov, D., Arora, A., Dong, Z., Rojas-Cessa,R.: Blockchain implementation for analysis of Carbon footprint across food supply chain. In: 2019 IEEE International Conference on Blockchain (Blockchain), pp. 546-551 (2019). https://doi.org/10.1109/Blockchain.2019.00079

8. Wang, Z., Lin, J., Cai, Q., Wang, Q., Zha, D., Jing, J.: Blockchain-based certificate transparency and revocation transparency. IEEE Trans. Depend. Secure Comput. **19**(1), 681–697 (2022). https://doi.org/10.1109/TDSC.2020.2983022

9. Manikumar, D.V.V.S., Maheswari,B.U.: Blockchain based DDoS mitigation using machine learning techniques. In: 2020 Second International Conference on Inventive Research in Computing Applications (ICIRCA), pp. 794-800 (2020). https://doi.org/10.1109/ICIRCA48905.2020.9183092

10. Sharma, S., Shah, K.: Exploring security threats on blockchain technology along with possible remedies. In: 2022 IEEE 7th International Conference for Convergence in Technology (I2CT), pp. 1–4 (2022). https://doi.org/10.1109/I2CT54291.2022.9825123

11. Zhou, R., Lin,Z.: A privacy protection scheme for permissioned blockchain based on trusted execution environment. In: 2022 International Conference on Blockchain Technology and Information Security (ICBCTIS), pp. 1-4 (2022). https://doi.org/10.1109/ICBCTIS55569.2022.00012

12. Wang, S., Ouyang, L., Yuan, Y., Ni, X., Han, X., Wang, F.-Y.: Blockchain-Enabled Smart Contracts: Architecture, Applications, and Future Trends. IEEE Transactions on Systems, Man, and Cybernetics: Systems **49**(11), 2266–2277 (2019). https://doi.org/10.1109/TSMC.2019.2895123

13. Rekha, G., Maheswari, B.U.: Raspberry Pi forensic investigation and evidence preservation using blockchain. In: 2021 International Conference on Forensics, Analytics, Big Data, Security (FABS), Bengaluru, India, pp. 1–5 (2021). https://doi.org/10.1109/FABS52071.2021.9702622

14. Innocent, A.A.T., Prakash,G.: Blockchain applications with privacy using efficient multiparty computation protocols. In: 2019 PhD Colloquium on Ethically Driven Innovation and Technology for Society (PhD EDITS), pp. 1-3 (2019). https://doi.org/10.1109/PhDEDITS47523.2019.8986954

15. Shree, J., Kanimozhi, N.R., Dhanush, G.A., Haridas, A., Sravani, A., Kumar, P.: To design smart and secure purchasing system integrated with ERP using Block chain technology. In: 2020 IEEE 5th International Conference on Computing Communication and Automation (ICCCA), Greater Noida, India, pp. 146–150 (2020). https://doi.org/10.1109/ICCCA49541.2020.9250767

16. Lakshminarayanan, S., Kumar, P.N., Dhanya, N.M.: Implementation of blockchain-based blood donation framework. In: Chandrabose, A., Furbach, U., Ghosh, A., Kumar, M.A. (eds.) Computational Intelligence in Data Science. ICCIDS 2020. IFIP Advances in Information and Communication Technology, vol. 578, pp. 276–290. Springer, Cham (2020). https://doi.org/10.1007/978-3-030-63467-4_22

Analysis and Prediction of Cryptocurrency Using Deep Learning Algorithms

Megha Pathak and Alka Chaudhary[✉]

Amity Institute of Information Technology, Amity University Noida, Noida, Uttar Pradesh, India
meghapathak.876@gmail.com, achaudhary4@amity.edu

Abstract. There are numerous titles for crypto currencies. Most likely, you must have heard about the most well-known crypto currencies, including Bitcoin, Tether, Ethereum. Crypto currencies are increasingly popular alternative for online payments. A digital currency, often known as a crypto currency, is a different type of payment system created utilising encryption methods. By utilising encryption algorithms, crypto currencies may serve as a virtual accounting system and an exchange medium. Typically, no government or any other central body issues or controls crypto currencies. They are managed via peer-to-peer networks oficomputers running open-source, free software. Generally, anyone who wants to join them in welcome to do so.

Keywords: Crypto currency · Bitcoin · Tether · Ethereum

1 Introduction

Virtual or digital currency known as crypto currency for security uses cryptography and is not controlled by a central bank. It is built on decentralised technology, which means that no single authority, like financial institution or a government, has control over it. Crypto currencies are of different types:

Bitcoin: Bitcoin was created by Satoshi Nakamoto. It was founded in 2009. It was the first crypto currency created.

Ethereum: Ether (ETH), often known as Ethereum, is the name ofithe digital money used by the blockchain platform Ethereum, which was launched in 2015. It is the second-most used crypto currency after Bitcoin.

Tether: Tether was launched by an organization called Tether Limited Crypto currencies, unlike conventional money, are not backed by tangible goods like gold or silver and are not issued by any government or financial organisation. Instead, the market and their availability determine their value, and they can be bought and sold on a various exchanges.

The crypto currency market is highly volatile, and there are many factors that can impact prices beyond what can be captured in data. Therefore, it is important to use deep learning a part of a large investment strategy that takes into account market trends, and other factors.

S. Kadry and R. Prasath (Eds.): MIKE 2023, LNAI 13924, pp. 312–318, 2023.
https://doi.org/10.1007/978-3-031-44084-7_29

A dynamic project created for studying and forecasting the price of crypto currency. This project is highly endorsed for our generation because someday we might all contemplate investing in crypto currencies. This project aids in predicting the prices for crypto currency as well as their worth. This project is created using deep learning technique; LSTM (Long Short Term Memory). The project is created using Python programming language incorporation. It was launched in 2014. It is a stablecoin with asset backingBNB: Binance coin is a crypto currency that runs on the Ethereum blockchain. It is issued by the Binance exchange Proceedings in Information and Communication Technology (PICT) (Table 1).

Table 1. Summary of abbreviations used

Abbreviation	Meaning
LSTM	Long Short Term Memory

2   Literature Review

In their article of Bitcoin Forecasting, Minakhi Rout, Suresh Chandra Satapathy, Temesgen Awoke and Lipika Mohanty discussed about a digital kind of money known as crypto currencies that holds all transactions electronically. It is a soft currency that doesn't actually exist as hard notes. However, because of their extreme price volatility, using these crypto currencies has an influence on trade and international relations. In their study they mainly paid attention to a well-known crypto currency called bitcoin. To deal with the price volatility of bitcoin and to achieve high accuracy, their goal is to implement effective deep learning - based prediction models, particularly long short - term memory (LSTM)and gated recurrent unit (GRU). Their research also compares these two time series deep learning approaches and demonstrated which one ismore effective in price prediction. Stelios Bekiros and Salim Lahmiri goal was to predict the price ofithe three most popular crypto currencies— digital Cash, Ripple and Bitcoin using deep learning algorithms. In their study according to the results of testing the existence ofiinonlinearity they revealed that the time series of all crypto currencies display long memories, self-similarity, and fractal dynamics. They also found that the LSTM model has a higher computational cost than brute force in nonlinear pattern recognition, and deep learning has ultimately proven to be quite effective at predicting the naturally chaotic dynamics of crypto currency markets. S.K. Rath and Deepak Kumar conducted a study that aims to use deep learning algorithms to forecast the price patterns of Ethereum, taking into account those of its time series in particular. Their paper basically examines the use of deep learning methods to forecast Ethereum price trends, specifically the long short-term memory (LSTM) and multi-layer perceptron (MLP) techniques. They used these methods in accordance with historical data that was calculated minutely hourly and daily, and. The CoinDesk repository is where the dataset is found. Utilising statistical measures like mean absolute error (MAE), mean square error (MSE), and root mean square error (RMSE), the performance of the produced models is carefully evaluated. In their paper, Hyeonseung Kim, Suhwan Ji and Jongmin Kim investigate and contrast numerous cutting-edge deep learning techniques for predicting the price of bitcoin, including long short - term memory (LSTM) models, deep residual networks, deep neural networks (DNN), convolutional neural networks, and their combinations. The results from experiments indicated that while DNN - based models performedthe best for price ups and downs prediction (classification), LSTM-based models still slightly outperformed the other prediction models for Bitcoin price prediction (regression). Additionally, a straightforward examination of profitability revealed that for algorithmic trading, classificationmodels outperformed regression models. Models of prediction based on deep learning all performed similarly in terms of overall effectiveness.

3 Proposed Methodology

To create this model a five step procedure is used:

3.1 Data Preparation

Take the dataset for the crypto currency you want to analyse, clean it up, put it through some preliminary processing, and then divide it into training, validation, and testing sets.

3.2 Feature Selection

Identify the characteristics that are important for predicting crypto currency prices. Market capitalization,trading volume, and historical prices are a few common elements.

3.3 Model Selection

Select a deep learning system that can manage the intricate nature ofibitcoin pricing information. A few examples of well-liked algorithms are long short - term memory (LSTM) networks, Recurrent neural networks (RNNs), and convolutional neural networks (CNNs).

3.3.1 Deep Learning

A division of machine learning is called deep learning. It has the ability to recognise intricate links and patterns in data. In deep learning, nothing needs to be explicitly programmed. Due to improvements in processing power and the accessibility of massive datasets, it has grown in popularity recently.

But unlike Machine learning, deep learning is built on an artificial neural network. These neural networks are built to learn from massive quantities of data and are modelled after the structure and operation ofiorganic neurons in the human brain.

The main feature ofideep learning is the utilisation ofideep neural networks, which include numerous layers of connected nodes. By identifying hierarchical patterns and features in the data, these networks can develop complex representations of the data.

Without explicit feature engineering, deep learning algorithms may automatically learn from data and get better.

Deep neural network training often calls for a lot ofidata and processing power. However, the development ofispecialised technology, such as Graphics Processing Units (GPUs), and the accessibility of cloud computing have made it simpler to train deep neural networks.

3.3.2 Long Short Term Memory

A deep learning technique called Long Short Term Memory is made to handle sequential data,includingtime series, speech, and text. LSTM networks are well suited for different tasks such as recognising speech, translating different languages, and time series forecasting.

Unlike Recurrent Neural Network (RNN), LSTM addresses the issue ofilong - term dependencies. Unlike RNN, LSTM is able to provide an efficient performance as the gap length increases since it can bydefault maintain information for a lengthy period ofitime. LSTM has a chain structure that is made up of numerous memory blocks known as cell and four neural networks.

A memory cell, or storage unit, is a component ofiLSTM that has the capacity to store data for a long period of time. Three gates regulate the memory cell: the forget gate, the output gate & the input gate. These gates determine what should go into, be taken out of, and be output from the memory cell. The input gate selects the input values that should be utilised to modify the memory. In output gate, the block's input and memory are used to determine the output. Lastly, the forget gate finds the details thatshould be removed from the block. As a result, LSTM networks can learn long-term dependencies by selectively retaining or discarding information as it passes through the network.

3.4 Model Evaluation

Analyse the outcomes after testing the deep learning algorithm's performance on the training dataset.

3.5 Interpretation

Understanding the outcomes of the deep learning algorithm can help you make wise decisions about your investments in crypto currencies.

4 Implementation

To make the coding simple and understandable I have divided the code into two parts; analysing the dataand predicting the data This will help us to understand the code and easily fix if there is any error in the code.

4.1 Analysing the Data

Our first part of implementation would be of analysing the data. To being with this part our first step would be to import libraries. In this step we will first list down all the required libraries and then we willstart importing them. So, the required libraries are:

1.NumPy: Numerical Python is referred to as NumPy. It is an array manipulation library for Python. Additionally, it offers functions for working with matrices, linear algebra, and manyother areas.
2.Pandas: Working with data sets is made possible by the Python package called Pandas.It also provides tools for modifying, examining, and cleaning up data.
3.Plotly: Plotly is a library that is used for visualizing the data.
4.Keras: A deep learning API called Keras is employed to simplify the implementation ofineural network
5.Sklearn: Data analysis is performed using Sklearn, an open - source library.

After importing libraries, the next step would be to import the data set. Using the read method we willimport the data and later print it. After importing the data we will select only those columns which are required in analysing the Bitcoin pricing. After selection ofirequired column we will analyse the data.Then our last step ofithis part would be to visualize the analysed data and print it. After this we will move on to the second part of the project that is to predict the data.

4.2 Predicting the Data

The second part of implementation would to predict or forecast the data. To begin with the second part ofithe implementation the first step would be the pre - processing ofithe data. After pre-processing the data, the second step would be data split. In this step we will split the data so that we can use easily the data according to our requirement for predicting it.

After this, the third step would be to convert the shape of the data so that it can fit into the model using the LSTM algorithm. The fourth step to predict the data would be to start coding for the model. After completing the code for the model now, we will find the optimal epoch for the model. Next we will predict the results for that epoch.

Now, with the help of this prediction we will create data frame. After creating the data frame we will find the mean absolute error and save it's result and then we will convert the predicted price into valid price. After this conversion we will find the optimal size ofithe time from for the model.

Now, we will again repeat the same steps as earlier; finding the mean absolute error and then converting the predicted price into valid price. Next we will print the result ofiprice prediction of the data. After this, we will visualize the result of both the analysis and prediction of the data.

5 Result Analysis

After getting the result for the first part ofithe implementation we see that there is a perfect graph which has analysed our data and is showing the increase and the decrease in the price of bitcoin (Fig. 1).

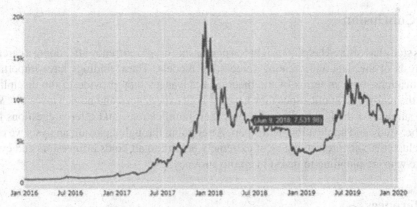

Fig. 1. Analysed graph of Bitcoin pricing

Here, the price ofiBitcoin is on the Y axis while the time period is on the X axis.

Next, the result of second part of the implementation compares the predicted value and the analysed value side by side (Fig. 2).

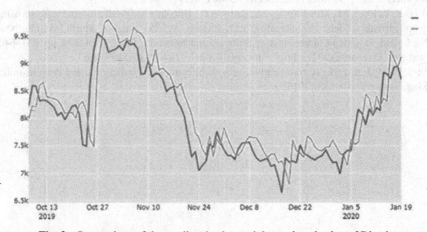

Fig. 2. Comparison of the predicted value and the analysed value of Bitcoin

In this graph, the predicted value of the price and the analysed value of the price are being compared. The blue colour line represents the real price value of bitcoin while red colour represents the predicted value of bitcoin price. Here, price is on the Y axis while the time period is on the X axis.

In the graph, we can see that the predicted value and quite similar to the actual price of the bitcoin. We can also see that the predicted price is also increasing where the actual value is increasing and decreasing at the same time as the actual value.

6 Conclusion

This study has offered helpful insights on purchasing crypto currency after doing a careful analysis of the data and creating forecasting models. These findings have important implications, both in terms of the theoretical advances they provide to the discipline and the real-world applications they have for both experienced and novice investors. We will summarise the key findings, go over their ramifications, and offer suggestions for further study and application in this part. After going through the result analysis we can conclude that this project can be of extremely helpful to all kinds ofiinvestors and even those who are planning to invest in crypto currency.

References

1. Kumar, D., Rath, S.K.: Predicting the trends of price for ethereum using deep learning techniques. In: Dash, S., Lakshmi, C., Das, S., Panigrahi, B. (eds.) Artificial Intelligence and Evolutionary Computations in Engineering Systems. AISC, vol. 1056, pp. 103–114. Springer, Singapore (2020). https://doi.org/10.1007/978-981-15-0199-9_9
2. Lahmiri, S., Bekiros, S.: Cryptocurrency forecasting with deep learning chaotic neural networks. Chaos, Solitons Fractals **118**, 35–40 (2019)
3. Awoke, T., Rout, M., Mohanty, L., Satapathy, S.C.: Bitcoin price prediction and analysis using deep learning models. In: Satapathy, S.C., Bhateja, V., Ramakrishna Murty, M., Gia Nhu, N., Jayasri, K. (eds.) Communication Software and Networks. LNNS, vol. 134, pp. 631–640. Springer, Singapore (2021). https://doi.org/10.1007/978-981-15-5397-4_63
4. Ji, S., Kim, J., Im, H.: A comparative study of bitcoin price prediction using deep learning. Mathematics **7**(10), 898 (2019)

Detecting the Attacks Using Blockchain-Based Decentralized Security Architecture in IoT Environment

M. Rudra Kumar, R. V. S. S. S. Tarun Teja[✉], A. Venkata Rakesh Reddy,
S. Vamshi Krishna, and P. Venkata Koushik

Department of Computer Science and Engineering, G.Pullaiah College of Engineering and
Technology, Kurnool, India
tarunteja1212@gmail.com

Abstract. There has been a surge in the demand for embedded devices as people prepare for the arrival of the Internet of Things (IoT), which plans for the autonomous interaction of sensors and actuators to provide various forms of smart services. Due to their limited processing power, data storage, and connectivity, Internet of Things gadgets are ripe targets for cybercriminals. Engineering scalable security solutions that are best suited to the IoT ecosystem is essential for ensuring the safe growth of IoT. In recent years, blockchain skill has arose as a central tenet of IoT-based app creation. The blockchain may be used to address concerns about data privacy, security, and reliance on a third party in IoT programmes. Individuals and communities alike stand to gain from blockchain's incorporation into the Internet of Things. Nevertheless, a DDoS assault on a mining pool in 2017 revealed serious vulnerabilities in the blockchain-enabled Internet of Things (IoT). The data that this programme creates is also massive. As Deep Learning (DL) allows for fully autonomous large data analysis and decision-making skills, it is employed as a decision-making and analytical tool. This research presents a unique distributed (IDS) to identify DDoS assaults on a financial institution's statement as a solution to the problems mentioned above. Long Short-Term Memory (LSTM) is trained to identify attacks against banks and other financial institutions using the Banking Dataset, and its effectiveness is then measured. The findings produced utilising banking-IoT data flow prove the technique's efficacy.

Keywords: Internet of Things · Distributed Denial of Service · Internet of Things Blockchain technology

1 Introduction

The proliferation of communications networks in recent years has enabled users to stay connected from almost everywhere, leading to an increase in the need for data transfer. Data traffic is increasing at an unprecedented rate due to the apps and the maturation of various network technologies. In 2019, we anticipate worldwide Internet traffic increase of about 200 Exabytes per month, rising to 396 Ex-abytes per month by 2022 [1]. The

© The Author(s), under exclusive license to Springer Nature Switzerland AG 2023
S. Kadry and R. Prasath (Eds.): MIKE 2023, LNAI 13924, pp. 319–329, 2023.
https://doi.org/10.1007/978-3-031-44084-7_30

growth of the IoT is inspiring the development of cutting-edge new services that place greater demands on factors like reaction speed and energy efficiency. The network must be prepared to provide the appropriate network capabilities and to deal with various security concerns [2, 3] in order to support IoT-enabled critical infrastructures like those in the industrial IoT, the motorised IoT, and e-health.

Moreover, for IoT to be successful, we need to create cutting-edge mechanisms that can guarantee adequate security levels, identify cyber-attacks when they occur in the managed IoT network, and counteract their effects [4]. The fact that IoT devices may deal with sensitive information presents a significant challenge, and the fact that many profitable IoT low-end devices do not usually support robust to imitate the malicious network of devices for various attacks like DoS and DDoS [5]. Future IoT-based system growth is constrained by the underlying working model [6]. Thus, a decentralised or distributed storage approach is required to solve these problems. Blockchain technology is one of the new decentralised systems that is beginning to emerge [7].

Initiated by a peer node in the system, the blockchain is a dispersed and immutable ledger of all transaction data. Distributed ledger is a term used to describe the idea of decentralised data storage [8]. The consensus of the network is required for each ledger-processed transaction to be considered legitimate. Bitcoin is the most well-known use of blockchain knowledge in the real world at the present moment [9]. As an example, a decentralised blockchain storage architecture may synchronise IoT devices and offer real-time data to each IoT node, among other advantages. The reliance on other parties and the risk of a bottleneck are both removed thanks to this underlying integrated app-roach [10]. And without the requirement for a centralised client-server approach [11], IoT integration with blockchain may allow for decentralised.

While blockchain is immutable and verifiable, it may be attacked in a number of ways [12]. Massive advancements have been made in changing standalone IoT applications via the merging of IoT and blockchain [13]. Unfortunately, this trend has been accompanied with a rise in the frequency of assaults. When a blockchain network is subjected to a distributed denial of service attack, the effects for genuine users may be devastating [14]. Thus, a reliable and strong security system is essential for DDoS attack detection. This study presents an artificial intelligence (AI) (IDS) for use by mining pools in blockchain-enabled Internet of Things (IoT) networks.

The following is a list of the major scholarly advances made by this study: Based on their impressive results, this research proposes LSTM for DDoS attack categorization using the Banking Dataset. We looked at how adjusting the training settings affected the precision, recall, F1-score, and Kappa index.

The rest of the paper is designed as follows: Sect. 2 presents the related works and brief explanation of proposed model is depicted in Sect. 3. The validation analysis with its conclusion is provided in Sect. 4 and 5.

2 Related Works

A safe and smart fuzzy blockchain architecture is designed and implemented by Yazdine-jad et al. [15]. For network intrusion detection, this framework employs a unique fuzzy DL model, as well as improved attack detection using an inference system (ANFIS), fuzzy matching (FM), and a (FCS). Fuzzy Choquet integral is used in the proposed fuzzy DL to provide a robust nonlinear aggregation function for detection. Take use of metaheuristic methods to fine-tune ANFIS's error function for attack detection. Use FM verification for blockchain transaction validation to improve efficiency and combat fraud. When it comes to detecting and identifying security threats, this fuzzy blockchain framework that also takes into account uncertainty issues in IoT networks and provides greater leeway in terms of making decisions and accepting transactions at the blockchain layer. The findings of the evaluation confirm the efficacy of the blockchain layer in throughput and dormancy metrics and the layer in performance metrics for threat detection on the blockchain and IoT network sides. Moreover, FCS indicates that a reliable model may be created to effectively identify threats in blockchain-based IoT systems.

In this study, Masood et al. present a Tolerant Control (BB-DD-FTC) architecture for smart factories. Blockchain's immutable ledger and decentralised design guarantee the accuracy of data records. Also, DD-FTC is realised and mitigation action is taken on in the event of cyber-attacks thanks to the blockchain smart contract capability incorporated with a (DD-IDS) and reconfiguration conditions. For attack detection and component identification, DD-IDS makes use of principal educated by neural networks. To evaluate the efficacy of the proposed framework, the Tennessee Eastman (TE) industrial benchmark process is used as a case study. Simulation findings show that the approach is successful in reducing performance of the system when two types of integrity attacks are performed to the sensors of the TE process. Due of the detrimental effects that feedback delays might have on performance, a thorough delay analysis is carried out by means of network calculus. Finally, the completed security analysis highlights the benefits and drawbacks of the suggested approach with regards to security. The findings provide hope for the widespread implementation of the idea of centralised blockchain management.

For the (IoT), Mitra et al. [17] propose a AI/ML-enabled big data analytics method. Extensive experimental findings have been presented both for the ML model's performance in the face of data poisoning assaults and in the absence of such attacks. Demonstrate the impact of data poison attacks on an ML model in two scenarios: the first involves cloud storage (and hence is not affected by blockchains) and the second involves blockchains. In the absence of data poisoning assaults, the experimental findings show considerable improvements in accuracy, recall, precision, and F1 score. In addition, the feasibility of the proposed security architecture has been shown by a comprehensive blockchain simulation.

A hybrid centralized/blockchain-based verification construction for IoT schemes is proposed by Khashan and Khafajah [18]. Centralized authentication for connected IoT devices is made possible via the use of edge servers. Next, to guarantee the decentralised authentication and verification of IoT devices from various and heterogeneous IoT systems, a blockchain network of centralised edge servers is formed. To provide effective authentication while minimising the load on the Internet of Things' resources, we use lightweight cryptographic approaches. The architecture is shown using an instance of the Ethereum blockchain operating locally. The results show that the suggested technique outperforms centralised and blockchain-based authentication systems for IoT in terms of computing cost, implementation time, and power consumption. The security study verifies that the architecture can prevent attacks and conform to the standards set by the IoT.

To improve the safety of IoV networks, Ayed et al. [19] suggested a blockchain-based trust management paradigm. The performance results for this framework were satisfactory. Yet it relied on a centralised model of trust. As a result, it performs less and less well as scalability increases. Also, the suggested model's energy use is not the main emphasis. Put up a supplement to fix the noted flaws. That's why we rely on a trust system that's distributed throughout the network. Create our framework with the help of a clustering algorithm. Define the clustering procedure based on several criteria the energy factor to take into consideration the enforcing security. Relevant data regarding and runtime characteristics are shown in the assessment of the proposal. The simulation findings show that using blockchain to create reliable clusters is effective in bolstering the IoV's dependability.

An AI-based system model with two goals was presented by Shah et al. [20] and [21, 22]. In the first stage, a binary classification issue is used to identify the hostile user attempting to breach the IoT infrastructure. In addition, blockchain technology is used to provide immutable storage for safekeeping of non-malicious IoT data. A bad user, however, might take advantage of the smart contract's blockchain implementation to degrade the IoT ecosystem's functionality. In order to distinguish between harmful and safe smart contracts, this article use deep learning techniques. The suggested system paradigm provides a secure channel for transmitting IoT data from end to end. Finally, the proposed system model is tested using a variety of evaluation metrics.

3   Proposed System

In-depth explanations of the data, methods, and KPIs are provided here. Raw datasets were acquired from a publicly available source. We have preprocessed the data by eliminating duplicates and cleaning up the data. After removing all of the null values from a dataset, it was balanced using several methods. After feature extraction, we've divided the data into a training set The DL models are trained on the training set and tested on the testing set.

3.1   Data Description

Intruders in banking systems are tracked by the Banking Dataset. Included in this harmful software is a Denial of Service. Table 1 displays the characteristics of the dataset along with brief explanations [23, 24, 25]. A significant DDoS assault is possible if the value in the PC's service redistribution is more than 0.5. There is less of a likelihood of an attack happening when the value from a PC is less than 0.5. In the data set, DDoS assaults occur 50,000 times.

Table 1. The Bank dataset of features account.

Feature/Attribute	Variable Category	Description
ID	Input Variable	ATM ID
State	Input Variable	National of Railway (Connectivity)
Spkts	Input Variable	Basis Packets (Directed to destination)
Dpkts	Input Variable	Terminus Packets (Received at destination)
Sbytes	Input Variable	Source Bytes (Sent from Source)
Dbytes	Input Variable	Terminus Bytes (Received from Source)
Attack_Cat	Output/Target Variable with Nine Classes	Types of Attacks Here we've implemented DDoS assaults; if the label reads "0," no such attack will occur, and if it reads "1," an assault of this kind will be launched

The proposed method is evaluated using these data sets. For deep learning, pre-processed datasets are useful. When it comes to selecting useful characteristics from both datasets, the homogeneity measure method. Estimating and enhancing the performance of deep learning models is possible using five-fold cross validation. In order to properly categorise assaults, we used a trio of machine learning models. We have divided the data set into a training set of 70% and a test set of 30%. Using training has been shown to provide the greatest outcomes in empirical investigations.

3.2 Data Pre Processing

The dataset is pre-processed to make it more suitable for the DL classifier.

(a) Elimination of Socket Info

Any possibility for bias in the identification process may be eliminated by erasing the source and destination IP addresses. By analysing packet characteristics rather than only relying on data from a single socket, it is possible to exclude potential hosts that share the same data.

(b) Remove White Spaces

White space is permitted in multi-class labels. Since the other tuples in this class have different names, it really contains two classes.

(c) Label Indoctrination

Label encoding is the process of translating labels into a machine-readable numerical format. Hence, machine learning algorithms are better able to determine how to put those labels to use. It is an essential part of preparing the structured dataset for supervised learning.

(d) Data Normalization

In order to improve the accuracy of our predictions, we have normalised the dataset using a scalar function that is widely used in statistics. Data normalisation is followed by feature ranking.

(e) Feature Ranking

Since there are so many different kinds of characters and strings, we've labelled the columns containing the feature counts as col 0 through col 111. For attribute feature ranking, we used k-means clustering, which takes into account the relative importance of each feature and ranks them accordingly.

3.3 Our Proposed Model

This section describes how a distributed IDS may be integrated with a mining pool to prevent a bank attack on AI-enabled fog computing. Data preparation and AI-enabled DL approaches for deploying blockchain-based IoT networks are both outlined in depth.

3.3.1 IDS Integration at Mining Pool in IoT Situation

To monitor blockchain-based IoT devices for suspicious transactions, a mining pool and intrusion detection system are used together. Our goal is to safeguard mining pools against DDoS assaults once they have been effectively integrated into IoT networks. The IDS built within the mining pool serves as the last line of protection. The IDS is used to inspect incoming traffic in the suggested setup. If everything goes well, miners will mine the transaction packets and add them to the blockchain network's distributed ledger. On the other hand, if the new traffic or transactions involve abnormal performance, the manager will be notified and given the opportunity to take corrective measures. Furthermore, the suggested detection system's administrator may transmit the transaction to the mining pool, where a miner would then mine it and add it to the blockchain network if a false alarm is generated. Hence, the IDS provides administrators with a second opportunity to respond and counter the threat.

3.3.2 Working of Distributed IDS in Blockchain-Enabled IoT Network by Using DL for Mining Pool

Security and data storage functions that were previously located at the network's periphery have been dispersed in favour of the suggested detection system.

(i) I Traffic Processing Engine: At this stage, fog nodes are used to process network traffic and to install an intrusion finding scheme on the edge of the fog network (IoT devices). At this phase, we prepare the training data set for further analysis. The dataset is normalised using the StandardScaler method. Then, LSTM, a deep learning approach based on artificial intelligence, is used to the distributed architecture. Both tactics for maximising shared parameters and minimising differences between them have a home in the master node's centralised command centre. By receiving updated parameter values from neighbours, this method has the potential to improve local attack detection capability via local training and parameter optimisation. Every cooperative (worker) node receives an update from the master node with updated values for a set of parameters, which the master node then corrects. For real-time responses to data with a wide range of factors, it's crucial during storage offloading and when calculating costs using IoT models.

(ii) The Intrusion Detection Engine builds a prediction model to gauge the success of the detection mechanism. In a blockchain-enabled IoT network, data is produced in vast quantities by the IoT sensors. When the incoming traffic has been preprocessed, it is next examined using the prediction model.

(iii) The Transaction Handling Engine sorts transactions into good and bad kinds based on their behaviour. The miner will execute and add it to the cloud-based blockchain network if it is valid. Nonetheless, the administrator is notified of potentially malicious transactions and given the opportunity to take defensive measures. The worldwide status of IoT devices is effectively maintained by submitting log information of these communications.

3.3.3 AI Methods Deployed for Classifying Attack Examples

A significant difficulty for blockchain-based Internet of Things systems is dealing with large amounts of data. The ever-increasing amount of information being sent via sensors. Length, memory, and processing power are typical areas where sensors fall short. Data storage and privacy concerns have emerged as one of the most pressing issues in this fast developing area. An IoT-based IDS is essential for keeping tabs on network activity and thwarting attacks before they ever begin. The suggested IDS uses DL and other forms of AI to deal with the aforementioned issues.

By default, the LSTM classifier retains data for a considerable amount of time. When using neural networks to improve the accuracy of an attack categorization, a massive amount of photos is required. The LSTM classifier has already shown to be the most effective neural network option. It is the LSTM units themselves, which hold the temporal quasi-periodic characteristics used to extract long-term and short-term relationships, that make up the LSTM classifier as a whole. There are 98 LSTM units built into the construction of the classifier.

Each LSTM unit is made up of the mathematically-expressed o n (Eqs. (1)–(3)). (4):

$$i_n = \sigma (W_{ih}h_{n-1} + W_{ia}a_t + b_i) \tag{1}$$

$$f_n = \sigma \left(W_{fh}h_{n-1} + W_{fa}a_t + b_f\right) \tag{2}$$

$$c_n = f_n \times c_{n-1} + i_n \times tanh(W_{ch}h_{n-1} + W_{ca}a_n + b_c) \tag{3}$$

$$o_n = \sigma (W_{oh}h_{n-1} + W_{oa}a_n + b_0) \tag{4}$$

where at each n-th time step denotes a distinct frequency range where the quasi-periodic characteristic is present. h_ is the result from the previous LSTM unit (n−1).

The hyperbolic tangent (tanh(.)) and the sigmoid activation function ((.)) are symbols for the work coefficients W and b, respectively. The LSTM module's output may be represented mathematically via Eq. (5):

$$h_n = o_n \times \tanh(c_n) \tag{5}$$

h n, the output of the LSM unit at the n time step, gathers the extracted features from the previous time steps, beginning with c n. The long- and short-term memories of the temporal quasi-periodic characteristics are trained into the cell state cn|n = 1,2,...,N through a dependence connection.

Before going on, it's significant to discuss a major problem with neural networks: overfitting. Overfitting occurs when a network successfully fits the training data but fails to do so with the testing data, indicating that it has not yet learnt to generalise to new scenarios. Overfitting prevents the neural network from performing at its best, and this may be the case with LSTM. One approach used to train neural networks and avoid gradient expansion and disappearance is an early halting operation a weight limitation. Using dropout, certain LSTM units will have their outputs zeroed out at random on each iteration. Just a subset of the LSTM units' output values are utilised in the error computation; the rest are ignored until it's time to do error back propagation. In order to prevent overfitting during LSTM network training, it is common practise to impose constraints on the network's parameters. The following hyper-parameter values are considered appropriate when using LSTM networks: Using a maximum of 100 epochs, a minimum of 27, a gradient threshold of 1, a learning rate of 0.001, and a layer-by-layer hidden unit count of: 200 units in the first layer, 225 in the second, 200 in the third, and 225 in the fourth.

4   Results and Discussion

4.1   Performance Measure

If you want reliable data on the efficacy of your strategy, you need a performance metric that evaluates consequences and outcomes on a even basis and has a high kappa index (Figs. 1 and 2, Table 2). The kappa index and the general formula for identifying the composite are given by the following equations: (6, 7), (8, 9).

$$Sensitivity = \frac{TP}{TP + FN} \times 100 \tag{6}$$

$$Specificity = \frac{TN}{TN + FP} \times 100 \tag{7}$$

$$Accuracy = \frac{TP + TN}{TP + TN + FP + FN} \times 100 \tag{8}$$

$$Kappa\ index = \frac{Accuracy - Accuracy_T}{1 - Accuracy_T} \tag{9}$$

Table 2. Comparative Investigation of projected classifier with existing procedures

Methodologies	Sensitivity (%)	Specificity (%)	Accuracy (%)	Kappa index (%)
AE	72.33	76.55	72.03	86
RNN	86.95	83	87.33	79.86
CNN	91.77	88.4	92	85.45
LSTM	**97.34**	**97.49**	**96.89**	**88**

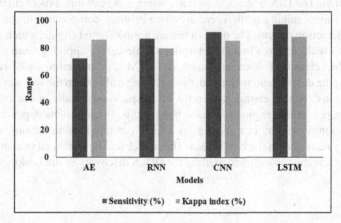

Fig. 1. Analysis of Various DL Models

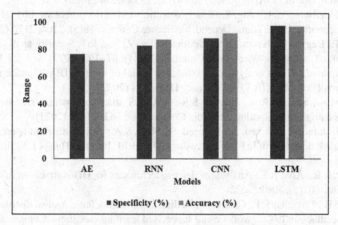

Fig. 2. Comparative Analysis of Proposed Model

5 Conclusion

Because of the high value of the information they store, financial institutions are especially susceptible to cyberattacks. Selling stolen banking credentials and other sensitive financial data may net hackers a tidy sum. The growth of banks' digital footprints has also increased the hackers. Our goal in analysing the Banking Dataset is to find evidence of DDOS attacks on banks and other businesses. This study shows how blockchain may create a decentralised network to overcome the shortcomings and security holes of standalone IoT systems. In this article, we presented the concept of an AI-integrated distributed IDS. The DDoS attack detection system was built into a blockchain-enabled Internet of Things mining pool. There are three primary components to the proposed distributed detection system. The first is a feature normalisation engine, which uses StandardScaler to scale features to a predetermined scale before applying them to network data. In a blockchain-IoT distributed context, LSTM is employed as an AI-based approach. When the data was preprocessed, the incoming traffic from the IoT was evaluated by an intrusion detection engine to identify any suspicious or malicious activity. Third, a transaction processing engine separates good and bad transactions depending on the detection findings. Miners in a mining pool carry out regular transactions, which are subsequently recorded on the blockchain. The model will be enhanced in a later project by including a feature assortment method for DDoS discovery in the banking dataset.

References

1. Gaur, V., Kumar, R.: Analysis of machine learning classifiers for early detection of DDoS attacks on IoT devices. Arab. J. Sci. Eng. **47**(2), 1353–1374 (2022)
2. Doshi, K., Yilmaz, Y., Uludag, S.: Timely detection and mitigation of stealthy DDoS attacks via IoT networks. IEEE Trans. Dependable Secure Comput. **18**(5), 2164–2176 (2021)
3. Kumar, P., Bagga, H., Netam, B.S., Uduthalapally, V.: Sad-IoT: Security analysis of DDoS attacks in IoT networks. Wirel. Pers. Commun. **122**(1), 87–108 (2022)
4. Ali, M.H., et al.: Threat analysis and distributed denial of service (DDoS) attack recognition in the internet of things (IoT). Electronics **11**(3), 494 (2022)
5. Lawal, M.A., Shaikh, R.A., Hassan, S.R.: A DDoS attack mitigation framework for IoT networks using fog computing. Procedia Comput. Sci. **182**, 13–20 (2021)
6. Bhayo, J., Jafaq, R., Ahmed, A., Hameed, S., Shah, S.A.: A time-efficient approach toward DDoS attack detection in IoT network using SDN. IEEE Internet Things J. **9**(5), 3612–3630 (2021)
7. Papalkar, R.R., Alvi, A.S.: Analysis of defense techniques for DDos attacks in IoT–A review. ECS Trans. **107**(1), 3061 (2022)
8. Gupta, B.B., Chaudhary, P., Chang, X., Nedjah, N.: Smart defense against distributed Denial of service attack in IoT networks using supervised learning classifiers. Comput. Electr. Eng. **98**, 107726 (2022)
9. Kumar, R., Kumar, P., Tripathi, R., Gupta, G.P., Garg, S., Hassan, M.M.: A distributed intrusion detection system to detect DDoS attacks in blockchain-enabled IoT network. J. Parallel Distrib. Comput. **164**, 55–68 (2022)
10. Huraj, L., Horak, T., Strelec, P., Tanuska, P.: Mitigation against DDoS attacks on an IoT-based production line using machine learning. Appl. Sci. **11**(4), 1847 (2021)

11. Kumar, P., Kumar, R., Gupta, G.P., Tripathi, R.: A distributed framework for detecting DDoS attacks in smart contract-based Blockchain-IoT Systems by leveraging Fog computing. Trans. Emerg. Telecommun. Technol. **32**(6), e4112 (2021)

12. Machaka, P., Ajayi, O., Maluleke, H., Kahenga, F., Bagula, A., Kyamakya, K.: Modelling DDoS attacks in IoT networks using machine learning (2021). arXiv preprint arXiv:2112.05477

13. Gopi, R., et al.: Enhanced method of ANN based model for detection of DDoS attacks on multimedia internet of things. Multimedia Tools Appl. **81**, 26739–26757 (2021)

14. Sharma, D.K., et al.: Anomaly detection framework to prevent DDoS attack in fog empowered IoT networks. Ad Hoc Netw. **121**, 102603 (2021)

15. Yazdinejad, A., Dehghantanha, A., Parizi, R.M., Srivastava, G., Karimipour, H.: Secure intelligent fuzzy Blockchain framework: effective threat detection in IoT networks. Comput. Ind. **144**, 103801 (2023)

16. Masood, A.B., Hasan, A., Vassiliou, V., Lestas, M.: A Blockchain-based data-driven fault-tolerant control system for smart factories in industry 4.0. Comput. Commun. **204**, 158–171 (2023)

17. Mitra, A., Bera, B., Das, A.K., Jamal, S.S., You, I.: Impact on Blockchain-based AI/ML-enabled big data analytics for cognitive internet of things environment. Comput. Commun. **197**, 173–185 (2023)

18. Khashan, O.A., Khafajah, N.M.: Efficient hybrid centralized and Blockchain-based authentication architecture for heterogeneous IoT systems. J. King Saud Univ.-Comput. Inf. Sci. **35**(2), 726–739 (2023)

19. Ramana, K., et al.: A vision transformer approach for traffic congestion prediction in urban areas. IEEE Trans. Intell. Transp. Syst. **24**(4), 3922–3934 (2023). https://doi.org/10.1109/TITS.2022.3233801

20. Chalapathi, M.M., Kumar, M.R., Sharma, N., Shitharth, S.: Ensemble learning by high-dimensional acoustic features for emotion recognition from speech audio signal. Secur. Commun. Netw. **2022**, Article ID 8777026, 10 pages (2022). https://doi.org/10.1155/2022/8777026

21. Ayed, S., Hbaieb, A., Chaari, L.: Blockchain and trust-based clustering scheme for the IoV. Ad Hoc Network. **142**, 103093 (2023)

22. Shah, H., et al.: Deep learning-based malicious smart contract and intrusion detection system for IoT environment. Mathematics **11**(2), 418 (2023)

23. Ramana, K., et al.: Leaf disease classification in smart agriculture using deep neural network architecture and IoT. J. Circuits Syst. Comput. **31**(15), 2240004 (2022). https://doi.org/10.1142/S0218126622400047

24. Kumar, V.A.K., et al.: Dynamic wavelength scheduling by multiobjectives in OBS networks. J. Math. **2022**, 10 Article ID 3806018 (2022). https://doi.org/10.1155/2022/380601

Developing a System Based on Block Chain Technology for e-Voting Mechanism

N. Parashuram, K. Bhanu Nikitha[✉], U. Jaya Sree, S. Lakshmi Prasanna, and K. Lavanya

Department of CSE, G. Pullaiah College of Engineering and Technology, Kurnool, India
nikithabhanu123@gmail.com

Abstract. Next-generation electronic voting systems have been made possible by the maturation of blockchain-based systems. Blockchain technology, the basis for electronic voting, might be used to improve online voting security. Nevertheless, due to its resource intensive nature, one of Blockchain's key principles, Work, cannot be applied in the E-voting founded blockchain. As a consequence, the information included in a transaction block may be changed and its hash easily recalculated. In totalling, the Proof-of-Work is necessary to verify the block's legitimacy. Hence, the digital signature may provide a means to address these concerns. Thus, the data is encrypted using an enhanced version of the Blowfish technique, which improves both security and efficiency. We also assess the security of the hybrid blockchain we propose and compare it to the security of the traditional blockchain via the discussion and analysis of attack execution. We learned new things about the safety, speed, of blockchain-based electronic voting schemes through our tests.

Keywords: E-voting · Blockchain · Improved Blowfish · Proof-of-Work · Transaction block

1 Introduction

E-voting, or electronic voting, is a voting method that relies on electronic technologies for both casting and tallying ballots. In the last several years, it has been the focus of intense study in the field of cryptography. By addressing security and privacy concerns in accordance with regulations [1–3], we may guarantee voter anonymity, reduce the cost of conducting polls, and guarantee polling integrity and end-to-end verification. As we reach the fourth age, it is imperative that all fields, endeavors, and facets be tested via revolutionary digitization [4]. Voting via paper ballots and other old-fashioned techniques are still in use, nevertheless. As most election fraud occurs at the polls itself, it can no longer be denied that it occurs under the current voting system (polling stations) [5]. The absence of privacy and security also makes this approach commercially untenable. Traditional offline services, such as voting, are increasingly shifting online and using Blockchain technology for economic and security reasons [6].

© The Author(s), under exclusive license to Springer Nature Switzerland AG 2023
S. Kadry and R. Prasath (Eds.): MIKE 2023, LNAI 13924, pp. 330–340, 2023.
https://doi.org/10.1007/978-3-031-44084-7_31

In today's world, security concerns about e-voting systems are always the primary factor in deciding whether or not to deploy such a system. With such weighty choices at risk, there should be no doubts regarding the system's capacity to protect data and repel attackers. Blockchain technology is being discussed as a possible solution to security problems. The blockchain is the underlying system of the bitcoin cryptocurrency [7, 8]. Records are stored in the form of transactions in this decentralised database format, and a block is a group of these transactions. We reach this conclusion due to the fact that Blockchain may be used to authenticate, authorise, and audit data created by devices and has a self-validating cryptographic framework between transactions. In addition, the transparency principle will be realised thanks to the decentralised structure of the blockchain [9]. Moreover, there is no central point of failure or trusted third party in a blockchain system. Furthermore, blockchain is immutable, which means that each proposed addition the previous version of the ledger, resulting in the establishment of an immutable chain that safeguards the integrity of the previous entries [10]. Consequently, blockchain technology may be used to create an electronic voting scheme that is both secure and reliable.

For large-scale elections, [5] proposes a (dBAME) electronic voting model that allows for the involvement of two opposing parties in order to guarantee the security and verifiability of the voting process. Small-scale e-voting using Hyperledger private Blockchain technology has been presented [6] as a way to ensure trustworthy electronic voting. An evoting system built on the Blockchain was proposed to be used with a website in [7]. With the ability to vote from afar and increase turnout, e-voting may bring democracy to new heights. There have been concerns raised about voter impersonation, fraud, and ballot duplication in electronic voting. Voter apathy has been exacerbated by the public's loss of faith in and understanding of the election process. The design of a more secure and usable electronic voting system is a hot issue in business and IT security right now [11, 12]. This research presents options for enhancing the privacy and security of electronic voting by using Blockchain technology in the design of novel voting systems. The publication is meant to acquaint curious scientists with the topic. In addition, the current security and privacy challenges in blockchain-based e-voting systems, as well as the reader's grasp of Blockchain, are discussed. This research strategy will introduce refined Blockchain technology for protecting the confidentiality of electronic voting systems.

The rest of the paper is designed as: The related work is given in Sect. 2. The brief explanation of proposed model is presented in Sect. 3. The validation analysis with its results is provided in Sect. 4. Finally, it concludes the research in Sect. 5.

2   Related Works

Data Hassan et al. [13] propose a blockchain-powered voting platform, built on Ethereum and the Truffle outline. The software's operations were codified in an contract released on the Ethereum network. The user's vote was read through a web interface and then sent to the Ethereum network using the web3.js API. To connect to the Ethereum network, ganache was used. The proposed system was implemented using Metalmark as a website wallet and the remix to deploy the smart contract on the main network; results show that the cost of each transaction is not stable, increasing with the increase in network load, and that throughput ultimately settles at 14 transactions per second.

Electronic voting methods, such as those presented by Echchaoui et al. [14], have the potential to reduce costs, increase turnout rates, and enhance traditional voting procedures. Problems with security, dependability, secrecy, and other factors prevent the widespread use of existing electronic voting methods. The voting process can be sped up, streamlined, and made more secure by using modern and reliable technology like blockchain and NFC. In this work, we present a novel electronic voting system that combines near-field communication (NFC) technology with blockchain technology to overcome existing problems with electronic voting. Voter confidentiality is protected while yet being assured by this system's public and open nature.

Mohsen et al. [15] presented a blockchain-based voting system in which votes are recorded at an Internet of Things (IoT) node. Due to the fact that the blockchain's internal data is not encrypted, Speck cypher lightweight block cypher and bakers chaotic key generator are used to enhance the safety of the information stored in each block (Vote). The lightweight approach was selected so that it may be used to encrypt data on IoT nodes that have minimal hardware resources, such as Raspberry pi. Data integrity and privacy are both significantly enhanced when the blockchain is used as a trusted public authority, thanks to its combination of a robust hashing algorithm and the lightweight speck cypher. A fingerprint logging mechanism is also available for further security. Effective and reliable fingerprint authentication system. Voting may be done in-person or remotely using the planned electronic voting system. At last, the suggested Blockchain voting system slashed the time needed to create blocks and encrypt data contained inside each block.

By using blockchain technology, Anitha et al. [17] have developed a decentralised transparent voting and examination scheme that may be applied in nations where conventional physical voting with gameable securities is utilised, increasing the likelihood of manipulated elections.

There will be no inequalities as a result of different proxies, long wait times will be reduced, prices will be reduced, scalability will be increased, and the system will be independent of physical location. In sum, a reliable voting system that can only help the democratic process. Voters may use the planned App to cast their ballots from the convenience of their own homes, cutting down on wasted time and fraudulent votes.

Users may access the voting system by entering their Aadhar Card number or one-time password, according the Singh et al. [18] method. A Block-Chain system, in which the ledger is duplicated across multiple, identical databases hosted by a different process and all other nodes are updated simultaneously if changes are made to one node and a transaction occurs, was proposed for Bit-coin, a virtual currency system in which a central authority decides on the production of money, the transfer of ownership, and the validation of transactions. If all transactions on a blockchain-based system were instantaneously resolved, while also being secure, verifiable, and transparent, then the technology would be revolutionary. Voting is the sector that is struggling from a lack of security, centralized-authority, management-issues, and many more despite the fact that transactions are stored in a distributed and secure form thanks to block-chain technology, which is the basis for Bitcoin and other digital currencies.

3 Proposed System

3.1 Problem Formulation

During contentious presidential campaigns, however, the results of any electoral process might seem to spark scepticism. Due to the immutable nature using it as a bulletin board would remove any debate about the genuineness and validity of the conclusion. Although blockchain technology has numerous advantages, it still must overcome several serious obstacles before it can be widely used in electronic voting systems. Electronic voting methods have serious issues with election integrity. There is widespread agreement that proof-of-work (PoW) algorithms are very resource- and time-intensive. As a large number of nodes must exist in the e-voting system's network, the system's throughput, latency, and scalability must be able to keep up with its rapid expansion.

In this work, we provide a revised version of the blockchain concept that fixes these problems. This approach may be used to improve the security, performance, and scalability of large-scale electronic voting systems while also reducing the potential for manipulation and fraud at the polls.

3.2 Blockchain-Based E-Voting System

Each voter only needs access to a computer or smartphone. Each voter has their own wallet where their credentials are stored. Each voter also gets a "digital coin", which stands in for one vote. To better understand the safety, scalability, and presentation concerns inherent in blockchain-based systems, we chose the electronic voting scheme as a case study.

Electronic voting systems might benefit from blockchain technology because of its decentralised design, mechanism. In addition, we tried to improve the voting procedure's efficiency, swiftness, and economy.

Interacting Entities. The first portion of this section addresses the architecture of the proposed scheme design and the functions of the various parts of the electronic voting system:

Keep track of MS: Certificates for nodes are issued by the (MS), and the network's lower layer stores and broadcasts that information to the upper layer. Access to the system is granted after authentication of each node, and user credentials are also included.

Distributed ledger system: The proposed blockchain network for electronic voting consists of many chains operating in tandem. The system's overall performance and scalability are enhanced by the structure's support for parallel execution. As each private-chain node has its own local blockchain containing the sensitive information, the lower-chains are used to store data about the nodes and the voter identification register. When some voters agree on the transactions via a method known as proof-of-stake consensus, the upper-chain (public blockchain like Ethereum) stores different blockchain states across all voters and processes transactions simultaneously. The higher chain (public blockchain) contains only verified and permanent transactions. Similar to what was described in a prior paper, routing management between lower-chains and upper-chain remains unchanged.

Audience (voters): Users not only participate in the electoral process as voters, but also as members of the election committee, who may utilise their identification ID for secure wallet access. In order to cast their ballot, voters are given a digital token. Consequently, the blockchain's top layer is where the smart contracts are first implemented (Ethereum blockchain).

Smart contract, or blockchain contract: In the proposed decentralised system, smart contracts are autonomously executed bits of code. Smart contracts' built-in functionalities build contract agreements that make it possible to monitor transactions in the blockchain's outermost layer. In our scenario, the blockchain network's nodes self-sufficiently execute the smart agreement to establish a consensus, which paves the way for the development of a multipurpose cryptosystem for e-voting systems.

Data Encryption. We employed an enhanced version of the Blowfish algorithm for the encryption's central processing unit in an effort to offer a cryptographic method that is both secure and fast. This technique employs several subkeys with lengths ranging from 32 bits to 448 bits. Before data can be encrypted or decrypted, the subkeys must be calculated in advance. This method is far more efficient and less memory intensive than its competitors. Four 256 SBOXs, for a total of 1024 32-bit entries, are used in the blowfish algorithm. To locate the corresponding entry in the first S-BOX, the first byte of the first 32-bit entry will be utilized; similarly, the second byte of the first 32-bit item will be used; and so on (Modified blowfish procedure). The converse is true for the decryption process, which also takes the cyphertext as input but uses the subkeys in a forward direction to decipher the message.

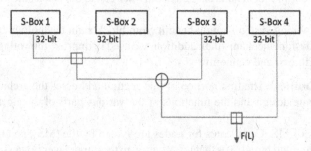

Fig. 1. F-Function Component in Blowfish.

As the Ffunction performs the major computation in all rounds of the Blowfish method, including Adder and Rotation, it is the most time-consuming portion of the encryption process. In this research, we alter the module of the F-function to shorten the Blowfish's running time. Standard Blowfish's F-function module's overall process is shown in Fig. 1, whereas the suggested model's enhanced F-function module's overall process is depicted in Fig. 2.

We enhance the algorithm by simplifying its runtime. The F-function of a typical Blowfish is given by Eq. (1).

$$F(X_L) = ((S_{(1,a)} + S_{(2,b)} \bmod 2^{32}) \text{ XOR } S_{(3,c)}) + S_{(4,d)} \bmod 2^{32} \tag{1}$$

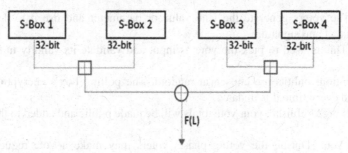

Fig. 2. The Improved F-Function Module.

Changing the F-function to (2) does not compromise the Blowfish algorithm's security.

$$F(X_L) = ((S_{(1,a)} + S_{(2,b)} \bmod 2^{32}) + S_{(3,c)}) + S_{(4,d)} \bmod 2^{32} \tag{2}$$

With this modification, we may simultaneously do the additions (S (1,a) + S (2,b) mod 232) and (S (3,c) + S (4,d) mod 232). As the time required to do two operations will be reduced to a single operation, the execution time will decrease thanks to this parallel process. As Blowfish uses 16 rounds, this modification should speed up the process of encryption and decryption by a factor of 16. And because Blowfish's security depends on its keys, this update won't compromise the algorithm's safety either.

The 64-bit entry is first divided into two 32-bit halves, one for the left and one for the right. The first 32-bit block is then subjected to an XOR operation (L). Third, the F-function will receive the computed 32 bits and XOR them with the other 32 bit block (R). Then, during the subsequent Blowfish rounds, the L and R will be switched. The only difference in the decryption process is that the P1, P2, and P18 will be utilized backwards.

If other methods that must exist in blockchain technology. A blockchain, in its most basic sense, is inconceivable without some kind of cryptographic hashing or digital signature. Every single blockchain transaction must be either digitally signed using the sender's private key or hashed using hashing methods to ensure its integrity. To ensure the legitimacy of transactions, blockchains rely heavily on digital signatures. Every node in the network must be convinced that the person submitting a transaction (such as a vote in an open poll) has the right to do so. Thereafter, the blockchain nodes (miners) will verify the voter's public key, the transaction's parameters, and the signature's validity. After a transaction's legitimacy has been verified, the corresponding block may be completed by a validator. Digital signatures are a simple way to verify the authenticity of a communication.

Process of the e-Voting System

The voting system is divided into several stages that occur in the following sequence:

By entering the security settings or parameters and producing the private (or public) pair of keys, the processes may be encrypted (or decrypted).

Sign Up: Enter the identifiers as IDs to produce the private (or public) key.

Vote: The voters generate the vote value or parameter and use it to deduce the encryption text and signature.

Valid: This is used to pick the vote as input and validate its validity in the ballot server.

An addition: ballots are chosen at random. The polling box's encryption text is randomized every time it is updated.

By clicking "Publish", your vote totals will be made public and added to the polling box.

Verify Vote: During the voting phase, voters may make a vote request in the blockchain contract during the poll's public visibility and double-check the results by providing public parameters, voter status, and privacy information. The results that were returned are either correct or incorrect.

When every vote has been cast and verified, the total is calculated by feeding in the associated private key and the polling box's parameter. The system will return False if the calculation was incorrect.

By inputting public criteria during publicising, a vote is certified as a legitimate and proper vote towards the final tally.

Method for Verifying the Credibility of a Node. The additional information for nodes consists of an ID and a Private Key, with the former serving as a unique identifier for each node and the latter serving as asymmetric encryption and a flag for node credibility verification. The Management Server (MS), which is in charge of the nodes, is the one that obtains and distributes the Private Key. The MS considers the identifier, private key, and data contained inside a block to be supplementary information. The ID is the exclusive identification for each MS. The MS is also a node in the system, it should be highlighted (except for its computational ability and storage capacity, it is the same as the other nodes). As a result, the MS should share the same attribute ID with those nodes; in other words, the MS and node IDs should be defined in the same way.

Each block in a blockchain is supposed to include some kind of blockhead and some sort of block data. The blockhead's function is to keep track of data from prior blocks, such as block identifiers and hash values. In addition to the node or MS ID, the timestamp, transaction count, Merkle root, and contract are also part of the block record's structure. The proper object's Private Key must be used to generate the corresponding Public Key in a single block record.

4 System Design

A number of pieces of equipment were needed to carry out this study, which included designing and evaluating an authoritarian open ballot electronic voting system. Node.js was selected as the backend technology, MongoDB was selected as the database, and several node package modules including sha256, tweetnacl, crypto-js, and jsonwebtoken were selected as the tokenization mechanisms Table 1.

Several aspects have been illustrated starting from vote casting, validation, and mining process.

Voting: Only users who have been verified as eligible to vote will be allowed to vote. The information must also be correct when entered.

Table 1. Node Package Modules used for system's implementation.

Node Module	Description
crypto-js	JavaScript library of crypto standards that performs SHA256 encryption function
tweentnacl	Signature encryption with public keys using the Ed25519 standard
crypto	ECDSA signature encryption using a public key
jsonwebtoken	to encrypt user information during the vote
mongoose	Modeling objects in MongoDB using a tool optimised for asynchronous processing
Express	Web outline for node

Some kind of validation procedure should be carried out before a vote is cast. First, check that you've provided all the necessary inputs. The second check ensures the existence of the previous building component. If that's not the case, then Genesis As the first block on the blockchain is a genesis block, the block mining procedure must be completed. As this origin block is never included in the tally of votes cast, its contents are essentially meaningless [20]. This block is essential because it serves as a building block, and the value it provides is passed on to the next one. The validity of the hash from the prior block is checked in the third validation. By rehashing the existing data in the block, this procedure verifies the hash. Hashing requires the user ID, user data (option), a timestamp, and a prior hash, all of which the user must provide. A new hash will be generated from this data when it is recalculated. The new hash will be compared to the one stored in the block. If correct, the following steps may be taken. If it doesn't happen, the procedure will end. The signature of the prior block is checked for validity in the fourth validation. This verification occurs after the hash has been verified [23–25]. For this verification, you'll need the hash, the signature, and the prior block's public key of the relevant user (the person who signed the block). If the transaction is legitimate, mining will proceed to the next stage. In that case, we'll have to call it quits.

Encryption: When the block has been validated, it is ready to be added to the blockchain. Certain crucial data will be hashed using a hash function before being registered. Hashing requires the index userId, data (vote), and hash timestamp from the preceding block. After the data has been hashed, the user may sign it with their private key. Index, hash, signature, timestamp, and userId [20–22] now make up the block's new structure. The last step is to save the information in a database.

5 Results and Discussion

However, the Poisson distribution was employed in this experiment to look into and contrast the attack probabilities in the hybrid blockchain and the conventional blockchain. The percentage of successful attacks in each chain is shown in Table 2.

To calculate the possibility that honest voters would create fewer blocks in the future in proportion to the attacker's assets q is a considerably bigger scenario. In this case, there

Table 2. Proposed hybrid blockchain and classical blockchain

Attacker assets (q, q1)	Classical blockchain	q2 = 0.1	q2 = 0.2	q2 = 0.3	q2 = 0.4
0.1	2.42803 E−4	2.42803 E−4	0.00232	0.01256	0.05020
0.2	0.01425	0.00322	0.01425	0.04636	0.12464
0.3	0.13211	0.02074	0.05721	0.13211	0.26724
0.4	0.50398	0.09039	0.17697	0.31310	0.50398
0.5	1.00000	0.29233	0.43879	0.62224	0.82447

is an extra requirement that the elector wait z blocks before confirming the vote. Yet, the attack probabilities of the hybrid blockchain and the conventional blockchain were studied and compared in this experiment using the Poisson distribution. Probabilities of launching a successful assault on each chain are shown in Table 2. (proposed hybrid blockchain and classical blockchain). The findings of this study verify our hypothesised increased difficulty in attacking the hybrid blockchain. As a function of the attacker's chance of success, it indicates the attacker's resources (such as their buffer size, processing power, coin holdings, and credibility score). This demonstrates the superiority of the proposed hybrid blockchain over the traditional blockchain.

6 Conclusion

Blockchain technology has the potential to solve the issues and limitations of the electronic voting technique. Recent studies have shown that blockchain-based electronic voting systems are being developed as the next generation of state-of-the-art electronic voting systems. Coins like Bitcoin employ a blockchain consensus technique called Proof-of-Work (PoW), however this method is inefficient, slow, and wastes a lot of power. Uniqueness, anonymity, integrity, invulnerability, verifiability, transparency, portability, accessibility, auditability, certification, detection, recovery, and simplicity of use are all met by the blockchain employed in this research. A blockchain-based electronic voting system does not need the Proof-of-work consensus technique. Memory and processing power are essential. Hence, the Proof-of-work technique cannot be implemented, despite its importance to blockchain systems. Because of hardware restrictions and energy cost, Proof-of-Work cannot be employed in open electronic voting systems. Using a signature technique, the transaction may be verified rapidly and securely. These results show that the proposed blockchain with collaborative features is secure and scalable. A randomizer token, a tamper-resistant source of randomness that functions as a black box, will need to be used in the future to generate the ballot in a manner that is both receipt-free and resistant to coercion.

References

1. Taş, R., Tanrıöver, Ö.Ö.: A systematic review of challenges and opportunities of blockchain for E-voting. Symmetry **12**(8), 1328 (2020)

2. Jafar, U., Aziz, M.J.A., Shukur, Z.: Blockchain for electronic voting system—review and open research challenges. Sensors **21**(17), 5874 (2021)
3. Khan, K.M., Arshad, J., Khan, M.M.: Investigating performance constraints for blockchain based secure e-voting system. Futur. Gener. Comput. Syst. **105**, 13–26 (2020)
4. Pawlak, M., Poniszewska-Marańda, A.: Trends in blockchain-based electronic voting systems. Inf. Process. Manage. **58**(4), 102595 (2021)
5. Kamil, M., Bist, A.S., Rahardja, U., Santoso, N.P.L., Iqbal, M.: COVID-19: implementation e-voting blockchain concept. Int. J. Artif. Intell. Res. **5**(1), 25–34 (2021)
6. Widayanti, R., Aini, Q., Haryani, H., Lutfiani, N., Apriliasari, D.: Decentralized electronic vote based on blockchain P2P. In: 2021 9th International Conference on Cyber and IT Service Management (CITSM), pp. 1–7. IEEE (2021)
7. Roopak, T.M., Sumathi, R.: Electronic voting based on virtual id of aadhar using blockchain technology. In: 2020 2nd International Conference on Innovative Mechanisms for Industry Applications (ICIMIA), pp. 71–75. IEEE (2020)
8. Alvi, S.T., Uddin, M.N., Islam, L.: Digital voting: a blockchain-based e-voting system using biohash and smart contract. In: 2020 third international conference on smart systems and inventive technology (ICSSIT), pp. 228–233. IEEE (2020)
9. Daramola, O., Thebus, D.: Architecture-centric evaluation of blockchain-based smart contract e-voting for national elections. In: Informatics, vol. 7, no. 2, p. 16. MDPI (2020)
10. Khan, K.M., Arshad, J., Khan, M.M.: Empirical analysis of transaction malleability within blockchain-based e-voting. Comput. Secur. **100**, 102081 (2021)
11. Abuidris, Y., Kumar, R., Yang, T., Onginjo, J.: Secure large-scale E-voting system based on blockchain contract using a hybrid consensus model combined with sharding. ETRI J. **43**(2), 357–370 (2021)
12. Li, H., Li, Y., Yu, Y., Wang, B., Chen, K.: A blockchain-based traceable self-tallying E-voting protocol in AI era. IEEE Trans. Netw. Sci. Eng. **8**(2), 1019–1032 (2020)
13. Hassan, H.S., Hassan, R., Gbashi, E.K.: E-voting system based on ethereum blockchain technology using ganache and remix environments. Eng. Technol. J. **41**(4), 1–16 (2023)
14. Echchaoui, H., Roumaissa, B., Boudour, R.: A proposal of blockchain and NFC-based electronic voting system. In: Hatti, M. (ed.) IC-AIRES 2022. LNNS, vol. 591, pp. 66–75. Springer, Cham (2023)
15. Mohsen, A.S., Shujaa, M.I., Wahhab, A.B.A.: Proposed authentication platform for E-voting IoT system. Int. J. Intell. Syst. Appl. Eng. **11**(2), 367–376 (2023)
16. Su, P.C., Su, T.C.: Secure blockchain-based electronic voting mechanism (2023)
17. Anitha, V., Caro, O.J.M., Sudharsan, R., Yoganandan, S., Vimal, M.: Transparent voting system using blockchain. Meas. Sens. **25**, 100620 (2023)
18. Singh, J., Rastogi, U., Goel, Y., Gupta, B.: Blockchain-based decentralized voting system security Perspective: Safe and secure for digital voting system. arXiv preprint arXiv:2303.06306 (2023)
19. Chalapathi, M.M., et al.: Ensemble learning by high-dimensional acoustic features for emotion recognition from speech audio signal. Secur. Commun. Netw. **2022** (2022). https://doi.org/10.1155/2022/8777026
20. Reddy, K.U.K., Shabbiha, S., Kumar, M.R.: Design of high security smart health care monitoring system using IoT. Int. J. **8** (2020)
21. Kan, L., et al.: A multiple blockchains architecture on interblockchain communication. In: Proceedings of the IEEE International Conference on Software Quality, Reliability Security Companion, Lisbon, Portugal, pp. 139–145 (2018)
22. Kishen Ajay Kumar, V., et al.: Dynamic wavelength scheduling by multiobjectives in OBS networks. J. Math. **2022**, 10 (2022). https://doi.org/10.1155/2022/3806018. Article ID 3806018

23. Dwaram, J.R., Madapuri, R.K.: Crop yield forecasting by long short-term memory network with Adam optimizer and Huber loss function in Andhra Pradesh, India. Concurr. Comput.: Pract. Exp. **34**(27), e7310 (2022)

24. Kim, S.K., Kim, U.M., Huh, J.H.: A study on the improvement of blockchain application to overcome the vulnerability of IoT multiplatform security. Energies **12**(3), 402 (2019). https://doi.org/10.3390/en12030402

25. Seok, B., Park, J., Park, J.H.: A lightweight hash-based blockchain architecture for industrial IoT. Appl. Sci. **9**(18), 3740 (2019). https://doi.org/10.3390/app9183740

A Permissioned Blockchain Approach for Real-Time Embedded Control Systems

Pronaya Bhattacharya[1], Sudip Chatterjee[1], Rajan Datt[2(✉)], Ashwin Verma[2], and Pushan Kumar Dutta[3]

[1] Department of Computer Science and Engineering, Amity School of Engineering and Technology, Amity University, Kolkata, Kolkata 700135, West Bengal, India
{pbhattacharya,schatterjee1}@kol.amity.edu
[2] Department of Computer Science and Engineering, Institute of Technology, Nirma University, Ahmedabad 382481, Gujarat, India
{rajandatt27,ashwin.verma}@nirmauni.ac.in
[3] Department of Electronics and Communication Engineering, Amity School of Engineering and Technology, Amity University, Kolkata, Kolkata 700135, West Bengal, India
pkdutta@kol.amity.edu

Abstract. In real-time embedded control (RTEC) systems, sensors collect data which is processed and sent to different control nodes. RTEC deployments have numerous applications in diverse verticals like industrial control, healthcare, and vehicular networks. In such cases, a trusted and verifiable control is required, particularly when the data is kept in a distributed manner, and is exchanged over open wireless channels. Thus, blockchain (BC) is a viable option to store the sensor data between RTEC systems, which maintains a trusted ledger of associated operations. Existing works have not focused on the integration of BC in RTEC systems. Motivated by the gap, the paper presents a systematic approach to integrating BC in RTEC ecosystems. We present a reference architecture and discuss the device registration, the hyperledger fabric set up, and the task offloading strategy between edge gateways and cloud nodes, and present the performance analysis of the architecture. The discussion of open issues and challenges also highlights the practical implications of the approach, emphasizing its importance for future deployments of real-time embedded control systems.

Keywords: Blockchain · Real Time Embedded Control · Internet-of-Things · Hyperledger Fabric · Permissioned Setup

1 Introduction

In real-time embedded control systems (RTEC), sensor devices are installed with hardware and software control that responds to any signal or event [1].

RTEC needs to respond to raised query events in a time-delay bound and is required to be precise in measurements to assure the safety and reliability of the systems. For example, in oil and gas pipelines, sensor devices measure the pressure to find leakage points in the gas pipeline. Once found, an event is generated. Moreover, precision becomes equally important to assure specific fault isolation and enhance reliability. RTEC are mainly classified into two categories, namely, the soft RTEC, and hard RTEC [2]. Hard RTEC guarantees that critical tasks are completed in a strict time-bound, mainly used in industrial processes, control systems, aviation, and other allied areas. On the contrary, a soft RTEC does not form stringent deadlines, and some task deadlines might miss, without causing damage to the control operation. Examples of soft RTEC include weather stations, mobile communication networks, and multimedia transmission. In soft RTEC, the performance of the system degrades if it is not able to complete the task as per the deadline.

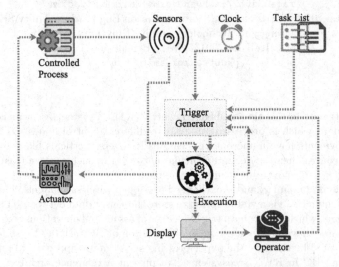

Fig. 1. A systematic overview of a RTEC system.

As indicated in Fig. 1, the sensors installed in an RTEC are connected to a control process. Based on the task list and deadlines, a master clock triggers the task execution. The RTEC performance is measured on display monitors. Based on defined logic, and control information, the actuator is instantiated to perform the process task. As an example, consider a chemical supervisory control and data acquisition (SCADA) pipeline monitoring system. The system has installed programmable logic controller, which communicates with micro-programmed remote units. In leak detection, basic testing mechanisms like hydraulic, infrared, and laser testing are applicable. Acoustic emission sensors are used in SCADA systems to measure and display pressure, flow, and temperature of valve opening and closing. The sensors are connected to hard RTEC relay systems that trigger an alarm if a leak is detected. The critical information is displayed, and actuators install clamps at the required places. The sensors exchange critical readings over open channels, at distributed sites, and thus trust in the shared information is crucial. In such cases, blockchain (BC) becomes a potential solution to support the data exchange, where the data is recorded as transactions [3]. As RTEC forms a closed-loop system, the machine-to-machine, and machine-to-human communication needs to be authorized. Thus, a permissioned BC is a favorable approach, where a shared ledger of the control system data is managed.

Thus, a permissioned BC approach in RTEC offers several benefits. Firstly, the fault tolerance of the entire system is improved as it introduces traceability points in the entire pipeline system. Secondly, permissioned BC enhances security by inducing a tamper-proof and auditable entry of control logs and transactions, which makes the system transparent for all stakeholders. Thus, owing to the potential benefits, in this article, we present a formulation framework for the BC-based RTEC ecosystem and discuss the data-sharing approach by authorized entities. The framework aims to improve fault tolerance and enhance security among sensor nodes. To address the scalability issue of data storage in BC, we consider a distributed offline storage like interplanetary file systems (IPFS), where the actual data is stored and content hash and IPFS key are stored on main BC [4].

1.1 Existing Works

Recent research is focused on the integration of BC in IoT systems [5,6]. In some schemes, to address the high computational requirements of low-powered IoT sensors, cloud services are provided that perform resource offloading and analytics [7]. Esposito et al. [8] presented the utilization of BC in the healthcare IoT domain, where patient data is secured and shared to authorized doctors, hospitals, and staff. Viriyasitavat et al. [9] presented an IoT-based smart home automation, where device (gadget) data is managed on a public BC. Authors in [10] proposed a sustainable microgrid that stores transactions in BC. A fuzzy-based optimization approach is formed to find all renewable energy distribution nodes, and energy transactions are stored on the main chain. Yu et al. [11] proposed a real-time networked system where a cyber-physical authentication scheme is developed over the BC network. Bodgan et al. [12] presented a BC-based water plant management, where water supply is distributed among different consumers, and bills are supported via SCs to the water suppliers. A significant framework is presented by Han et al. [13], where security in smart city applications (intelligent transportation) is presented. To assure auditability, rule-based access mechanisms are used, and BC stores the data on shared ledgers. Authors in [14] discussed a chaotic stream cipher on the Xilinx Virtex-6 FGPA processor. Guo et al. [15] designed a multi-robot control system. In particular, amphibious multi-robot control on BC in a WAN setup.

1.2 Novelty

BC immutable and transparent nature allows distributed and trusted data sharing, which is a crucial requirement in RTEC ecosystems. In the proposed framework, we present a low-powered computational offloading edge paradigm, where the edge gateways fetch required services from cloud nodes to meet the peak demands at high loads. This paradigm ensures that effective and scalable processing is maintained at RTEC components, where the speed of data transfer is maintained at a steady rate without drops. The framework also ensures that any CRUD query operations reference data from BC. This step ensures that each node gets a stored copy from BC, linked to IPFS, which increases trust in the entire system.

1.3   Article Contributions and Layout

Following are the contributions of the article.

- The BC-assisted RTEC management is presented, where the control system and the associated device identifiers are stored and accessed via the web.
- A permissioned hyperledger fabric control line information (CLI) channel is set up for devices to communicate with each other, and the data is shared with the BC handler via the edge gateway and is accessed uniformly over the web via the representational state transfer (REST) APIs.
- Between the edge cloud, a computational offloading mechanism is presented, which makes the scheme lightweight and scalable in operation.

The rest of the article is presented as follows. Section 2 presents the proposed approach, where RTEC device registration, control processes, and data sharing via hyperledger fabric are managed via BC. Section 3 presents the performance evaluation of the framework. Section 4 presents the open issues and challenges of the deployment of the approach in practical setups. Finally, Sect. 5 concludes the article.

2   The Proposed Approach

In this section, we present the proposed approach of BC-based RTEC system. Firstly, the IoT sensor node registration process is presented, followed by data storage on IPFS nodes, and accessed via BC. The data sharing is governed over the web via the REST APIs. We then present the resource offloading scheme between edge and cloud and present the data sharing process. Figure 2 presents the proposed approach for the BC-based RTEC control system. The components and details are presented in the following subsections.

2.1   The IoT Device (Sensor Nodes) Registration Process

In the proposed approach, we consider that n IoT devices, represented as $D = \{D_1, D_2, \ldots, D_n\}$ are present in the BC-RTEC system. For each $D_i \in D$, a unique device identifier ID_{D_i} is generated and is stored in local IPFS. A content key $CK(IPFS)$ is generated, which is hashed and stored as hash-key pairs. The

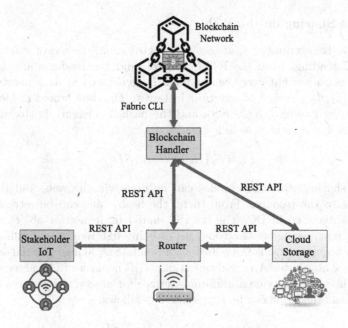

Fig. 2. The proposed BC-based RTEC system.

hash values are stored in the BC in encrypted form using symmetric encryption key S_k. The process is represented as follows.

$$E_k = S_k(H_{CK}, ID_{D_i}, CK(IPFS)) \tag{1}$$

where E_k represents encryption with key S_k. To ensure the integrity and authenticity of the device identifiers, an elliptic curve digital signature algorithm (ECDSA) is applied, and public-private key pairs (pk_i, sk_i) are generated. The public key pk_i is stored in BC along with E_k, while the private key sk_i is securely stored in the device itself. To verify the authenticity of ID_{D_i} stored on BC, we use pk_i to verify the ECDSA signature. Along with the device data, the registration process also considers device attributes A_{D_i} which contains its mode of operation OM_i, data fields DF_i, user list U_i (which access the device), permissions P_i, and owner details O_i. Mathematically, it is represented as follows.

$$A_{D_i} = \{OM_i, DF_i, U_i, P_i, O_i\} \tag{2}$$

where OM_i can be set to functional or error mode, DF_i is the list of data fields associated with D_i (like operation cycle, codes, and instruction sets), U_i returns the list of users who can access the device (similar to the access control list (ACL) mechanism), P_i is read, write, and execute permission, and O_i is the owner ID.

2.2   Data Sharing on the Web

Once the device attributes A_i are set post-registration, the sensor starts recording critical readings from the RTEC control, and the readings are stored on IPFS. We consider that every sensor node D_i generated m data points, represented as $\{d_1, d_2, \ldots, d_m\}$ at any time instant t. The data stored in local IPFS can be accessed using $CK(IPFS)$, and the hashed reference is stored in BC. The process is represented as follows.

$$E_k = S_k(H_{CK}, H_{SD_i}, SD_i) \tag{3}$$

The data sharing with other nodes can be done via the web, and the IPFS URI points to the resource. From there, the hash value can be retrieved, and the data is searched in BC. The data is shared in Javascript object notation (JSON) format, owing to its advantages of being lightweight and uniform [16]. The JSON object contains the sensor data points SD_i along with the device ID ID_{D_i}, device attributes A_{D_i}, and timestamp T_i. The data can be accessed using REST APIs, which provide a uniform interface for accessing resources over the web. Mathematically, it can be represented as follows.

$$JSON_i = ID_{D_i}, A_{D_i}, T_i, SD_i \tag{4}$$

REST API provides a uniform interface to access and manipulate data resources using standard HTTP methods such as GET, POST, PUT, and DELETE. Mathematically, the process of sharing sensor data SD_i using REST API can be represented as follows.

$$D_i \rightarrow IPFS \rightarrow URL_{CK} \rightarrow JSON \rightarrow RESTAPI \tag{5}$$

where SD_i represents the sensor data generated by D_i, URL_{CK} represents the IPFS URL generated using the content key $CK(IPFS)$, and $JSON$ represents the JSON format used for data sharing.

2.3   The Hyperledger Fabric Setup

The IoT sensor node data stored in IPFS is accessed via the IPFS URL, and the data is shared in JSON format. We consider this as D_{shared}. The REST API then sends D_{shared} to execute chaincode operations over the hyperledger fabric, denoted by H_{fabric}. The hyperledger fabric network is used to store and manage IoT sensor data. Let N be the set of nodes in the fabric network, where $N = P, O, C$ and P represent the set of peers, O represents the set of orderers, and C represents the set of clients. The chaincode for the BC-RTEC system is written in the Go programming language, denoted by CC_{Go}. Let TX be a transaction that includes D_{shared} as its input data, which is validated and executed by CC_{Go} according to the rules and logic defined for managing the data. The execution of TX is recorded in the ledger maintained by the peers in P, denoted by L_P.

The process of storing the IoT sensor data in the BC handler can be mathematically represented as follows:

$$D_{shared} =\rightarrow H_{fabric} \tag{6}$$

$$H_{fabric} = N, CC_{Go}, TX, L_P \tag{7}$$

2.4   The Offloading Process Between Edge and Cloud Nodes

The data from chain codes D_{shared} is sent to all nodes (IoT stakeholders) via an edge gateway (router) in this scenario. As the edge gateway might require significant computational resources to carry out analytics-related tasks, the scheme might not be scalable with an increase in the number of nodes. Thus, to reduce the computational bottlenecks, we consider that a task offloading process takes place, where compute-intensive tasks are sent to the cloud nodes for analysis. To model the requirement, we consider that let G denotes the edge gateways in the BC-RTEC system, and $C = \{C_1, C_2, \ldots, C_k\}$ is the set of available cloud nodes, which offer service sets SS. Any task can be offloaded to C_i by the gateway node, represented as $T : E \rightarrow C$. A task is subdivided into smaller tasks t, and is sent to cloud nodes for processing.

Overall, the data at edge nodes g is D_g, and $F(D_g)$ represent the processing function applied to the data D_g. The function F can represent any analytical operation, such as statistical analysis or ML algorithms. The output of the function $F(D_g)$ can be denoted as O_g. The task offloading process decides the portion of the task to be processed locally at g, and the portion to be executed globally at cloud.

Let L_g denote the local processing load at an edge gateway g, and O_c denote the output of the function $F(D_c)$ processed at a cloud node c. The load at the edge gateway can be reduced by offloading a portion of the data to a cloud node, such that the local processing load is given by.

$$L_g = D_g - D_{c,g} \tag{8}$$

where $D_{c,g}$ represents the portion of data offloaded to a cloud node c by the edge gateway g. The output of the function processed at the cloud node is combined with the local output of the edge gateway to obtain the final output O_g as follows.

$$O_g = F(L_g) + O_c \tag{9}$$

2.5   Algorithms and Discussion

In Algorithm 1, the input is the JSON data obtained from IPFS and the chaincode C. We consider an empty variable $txID$ which is initialized to store the transaction ID, and its value is returned at later stages. The data is extracted from the JSON file and stored in the variable $data$. The $result$ variable is set to the output of the execution of the chaincode function $C.store(data)$. If the execution of the function is successful (i.e. the $result$ variable is set to $success$), the transaction ID is assigned to the $txID$ variable. The transaction ID is returned as output from the algorithm.

Algorithm 1. Hyperledger Fabric Code for Storing IoT Data

Input: JSON data from IPFS, chaincode C
Output: Transaction ID

```
1: data ← JSON data from IPFS
2: txID ← empty
3: result ← execute C.store(data)
4: if result == success then
5:     txID ← transaction ID
6: end if
7: return txID
```

Algorithm 2. IoT Data Update and Edge Offloading

Input: IoT device hash h, field f, new value v, edge gateway G, cloud node N
Output: Transaction ID

```
1: if h ∉ registered devices then
2:      return Authentication Error
3: end if
4: D ← device entry with hash h
5: if f ∉ fields(D) then
6:      return Field Not Found Error
7: end if
8: D.f ← v
9: txID ← empty
10: if G has sufficient resources then
11:     result ← execute C.update(D)
12:     if result == success then
13:         txID ← transaction ID
14:     end if
15: else
16:     txID ← execute N.offload(D.f, v)
17: end if
18: return txID
```

Algorithm 2 considers the IoT data update and the edge-based task offloading process. Lines 1–2 take the input which considers the device hash h, the field to be updated f, the new value v, the edge gateway G, and the cloud node N. In line 3, a condition checked that if the device hash h is not in the list of registered devices, then an authentication error is returned. Otherwise, lines 4–5 suggest that the entry for the device with hash h is retrieved. In lines 6–7, if the field to be updated f is not in the list of fields for the device, then a field not found error is returned. In line 8, we consider that the value for the specified field in the device entry is updated to v. Lines 9–10 initialize the transaction ID to an empty value. In lines 11–12, if the edge gateway g has sufficient resources, the updated device data is stored in the BC using the chaincode function $C.update(D)$. In lines 13–14, if the transaction is successful, the transaction ID is set to the ID of the new transaction. Lines 15–16 present the condition when the transaction is offloaded to the cloud noes when edge do not have sufficient resources to execute the task. In line 17, the offloaded transaction ID is returned as the output of the algorithm.

3   Performance Analysis

In this section, we discuss the performance analysis of the proposed approach in terms of latency of storing IoT transactions on BC, query throughput from edge offloading, and response time. The details are presented as follows.

In registration, any i^{th} device D_i details are entered, and identifier ID_D is generated which is stored in a local IPFS. The data is then encrypted via S_k, hashed and content key $CK(IPFS)$ refers to the record in the IPFS. We compare the transactional latency of the captured data from embedded sensors during the registration phase on IPFS off-chain (proposed), against a conventional BC storage (on-chain) [17]. Figure 3a presents the results. For a transaction size of 100 KB, in IPFS, the latency is ≈3.87 ms, compared to 14.92 ms in BC. On average, we obtain an improvement of 45.67% in transaction latency via off-chain storage. As IPFS stores the registration details of sensors, and only meta-information is stored in BC, the latency to fetch the record from BC decreases significantly, which makes it suitable to accommodate large number of registration requests in the RTEC ecosystem.

Next, we analyse the read-write and query throughput measured in TPS. Figure 3b shows the results. As indicated, the query throughput varies proportionally for both query and read/write operations. A bottleneck is observed at 26 and 23 TPS for read-write and query information. When transaction rate is low the submitted transactions need to wait to be included in the block. Thus, many transactions timeout, and are then included in the next round of block proposal. With high TPS, the delay of addig transactions reduces, but the rate (of new and old transactions) also increases, which leads to the bottleneck condition.

Figure 3c shows the performance variation of the file read with integration of blockchain with IPFS [18], and IPFS alone [19]. As we integrate BC, The result shows different file size is uploaded and respective time is measured which indicate while using BC and IPFS together the read operation is slightly slower as time influenced by network speed and the computing resources used.

4   Open Issues and Future Directions

To make permissioned BC technology a practical choice for RTEC systems, there are still unresolved problems and unexplored directions that need to be addressed. The details are presented as follows.

1. *Resources requirements*: BC is a distributed ledger that requires significant computational and storage resources to maintain, process, and validate transactions. This can be a significant barrier to using BC technology in RTEC systems, where the devices may have limited processing power and storage capacity. One approach is to use lightweight consensus mechanisms that require fewer computational resources, such as proof-of-stake, delegated proof-of-stake, or practical Byzantine fault tolerance. These consensus mechanisms rely on a small number of nodes to validate transactions, reducing the computational load on individual devices [20]. Another approach is to use off-chain scaling solutions, such as state channels or sidechains, to reduce the number of on-chain transactions and minimize the load on the BC [21].

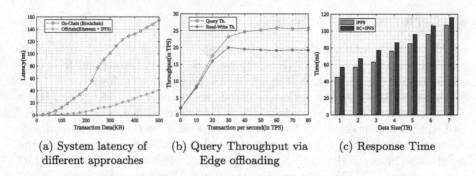

(a) System latency of
different approaches

(b) Query Throughput via
Edge offloading

(c) Response Time

Fig. 3. Performance Evaluation of the BC-based RTEC framework.

2. *Security Considerations*: One of the primary security concerns with BC technology is the risk of 51% attacks, where a group of nodes control more than 50% of the network's computing power and can manipulate transactions on the BC. The possibility for malicious actors to take over specific system devices and tamper with transaction integrity is another issue. Access control systems and identity management protocols can be put in place to limit access to the BC network to approved users and gadgets in order to solve this problem [22].
3. *Network channel considerations*: When using permissioned BC technology for RTEC, latency and network congestion is one of the main problems with network channels. The amount of data that needs to be transmitted and processed likewise grows when more devices are added to the network, which may cause delays and inefficiencies. Researchers are investigating novel network protocols and optimization methods, such as multipath routing, dynamic load balancing, and content caching, to address this problem [23].

5   Conclusion and Future Work

In this paper, we proposed a reference approach of RTEC ecosystem, where the IoT sensor information, signal readings, and measurements are stored in permissioned BC ledgers, to assure trust and verifiability among the stakeholders. Via permissioned BC, chain codes are executed on dockers on the hyperledger fabric channel setup. As potential findings of the work, we proposed algorithms on device registration, query on target data fields of the device, and the correct updates in the system.

As part of the future work, the authors would explore permissions and access control setups on permissioned BC setups and would propose an end-to-end RTEC scheme that would be generic and applicable to diverse verticals for trusted ownership, control, and exchange of sensor data among open channels.

References

1. Frikha, T., Chaabane, F., Aouinti, N., Cheikhrouhou, O., Ben Amor, N., Kerrouche, A.: Implementation of blockchain consensus algorithm on embedded architecture. Secur. Commun. Netw. **2021**, 1–11 (2021)
2. Dorri, A., Kanhere, S.S., Jurdak, R.: Towards an optimized blockchain for IoT. In: Proceedings of the Second International Conference on Internet-of-Things Design and Implementation, IoTDI 2017, pp. 173–178. Association for Computing Machinery, New York (2017). https://doi.org/10.1145/3054977.3055003
3. Saraswat, D., et al.: Blockchain-based federated learning in UAVs beyond 5G networks: a solution taxonomy and future directions. IEEE Access **10**, 33154–33182 (2022). https://doi.org/10.1109/ACCESS.2022.3161132
4. Verma, A., Bhattacharya, P., Saraswat, D., Tanwar, S.: NyaYa: blockchain-based electronic law record management scheme for judicial investigations. J. Inf. Secur. Appl. **63**, 103025 (2021). https://doi.org/10.1016/j.jisa.2021.103025, https://www.sciencedirect.com/science/article/pii/S2214212621001873
5. Bhatttacharya, P., Patel, K., Zuhair, M., Trivedi, C.: A lightweight authentication via unclonable functions for industrial internet-of-things. In: 2022 2nd International Conference on Innovative Practices in Technology and Management (ICIPTM), vol. 2, pp. 657–662 (2022). https://doi.org/10.1109/ICIPTM54933.2022.9754198
6. Trivedi, C., Rao, U.P., Parmar, K., Bhattacharya, P., Tanwar, S., Sharma, R.: A transformative shift toward blockchain-based IoT environments: consensus, smart contracts, and future directions. Secur. Priv. e308 (2023). https://doi.org/10.1002/spy2.308, https://onlinelibrary.wiley.com/doi/abs/10.1002/spy2.308
7. Verma, A., Bhattacharya, P., Bodkhe, U., Zuhair, M., Dewangan, R.K.: Blockchain-based federated cloud environment: issues and challenges. Blockchain Inf. Secur. Priv. 155–176 (2021)
8. Esposito, C., De Santis, A., Tortora, G., Chang, H., Choo, K.K.R.: Blockchain: a panacea for healthcare cloud-based data security and privacy? IEEE Cloud Comput. **5**(1), 31–37 (2018)
9. Viriyasitavat, W., Anuphaptrirong, T., Hoonsopon, D.: When blockchain meets internet of things: characteristics, challenges, and business opportunities. J. Ind. Inf. Integr. **15**, 21–28 (2019)
10. Tsao, Y.C., Thanh, V.V., Wu, Q.: Sustainable microgrid design considering blockchain technology for real-time price-based demand response programs. Int. J. Electr. Power Energy Syst. **125**, 106418 (2021). https://doi.org/10.1016/j.ijepes.2020.106418, https://www.sciencedirect.com/science/article/pii/S014206152030911X
11. Yu, Y., Liu, G.P., Xiao, H., Hu, W.: Design of networked secure and real-time control based on blockchain techniques. IEEE Trans. Industr. Electron. **69**(4), 4096–4106 (2022). https://doi.org/10.1109/TIE.2021.3071705
12. Pahontu, B., Arsene, D., Predescu, A., Mocanu, M.: Application and challenges of blockchain technology for real-time operation in a water distribution system. In: 2020 24th International Conference on System Theory, Control and Computing (ICSTCC), pp. 739–744 (2020). https://doi.org/10.1109/ICSTCC50638.2020.9259732
13. Han, D., Zhu, Y., Li, D., Liang, W., Souri, A., Li, K.C.: A blockchain-based auditable access control system for private data in service-centric IoT environments. IEEE Trans. Industr. Inf. **18**(5), 3530–3540 (2022). https://doi.org/10.1109/TII.2021.3114621

14. Pande, A., Zambreno, J.: A chaotic encryption scheme for real-time embedded systems: design and implementation. Telecommun. Syst. **52**, 551–561 (2013)

15. Guo, S., Cao, S., Guo, J.: Study on decentralization of spherical amphibious multi-robot control system based on smart contract and blockchain. J. Bionic Eng. **18**(6), 1317–1330 (2021)

16. Bhattacharya, P., Patel, F., Tanwar, S., Kumar, N., Sharma, R.: MB-MaaS: mobile blockchain-based mining-as-a-service for IIoT environments. J. Parallel Distrib. Comput. **168**, 1–16 (2022). https://doi.org/10.1016/j.jpdc.2022.05.008, https://www.sciencedirect.com/science/article/pii/S0743731522001228

17. Chopade, M., Khan, S., Shaikh, U., Pawar, R.: Digital forensics: maintaining chain of custody using blockchain. In: 2019 Third International conference on I-SMAC (IoT in Social, Mobile, Analytics and Cloud) (I-SMAC), Palladam, India, pp. 744–747 (2019). https://doi.org/10.1109/I-SMAC47947.2019.9032693

18. Saraswat, D., Patel, F., Bhattacharya, P., Verma, A., Tanwar, S., Sharma, R.: UpHaaR: blockchain-based charity donation scheme to handle financial irregularities. J. Inf. Secur. Appl. **68**, 103245 (2022). https://doi.org/10.1016/j.jisa.2022.103245, https://www.sciencedirect.com/science/article/pii/S2214212622001144

19. Shen, J., Li, Y., Zhou, Y., Wang, X.: Understanding I/O performance of IPFS storage: a client's perspective. In: Proceedings of the International Symposium on Quality of Service, IWQoS 2019. Association for Computing Machinery, New York (2019). https://doi.org/10.1145/3326285.3329052

20. Lin, W., Yin, X., Wang, S., Khosravi, M.R.: A blockchain-enabled decentralized settlement model for IoT data exchange services. Wirel. Netw. 1–15 (2020)

21. Darbandi, M., Al-Khafaji, H.M.R., Hosseini Nasab, S.H., AlHamad, A.Q.M., Ergashevich, B.Z., Jafari Navimipour, N.: Blockchain systems in embedded internet of things: systematic literature review, challenges analysis, and future direction suggestions. Electronics **11**(23) (2022). https://doi.org/10.3390/electronics11234020, https://www.mdpi.com/2079-9292/11/23/4020

22. Volety, T., Saini, S., McGhin, T., Liu, C.Z., Choo, K.K.R.: Cracking bitcoin wallets: i want what you have in the wallets. Future Gener. Comput. Syst. **91**, 136–143 (2019). https://doi.org/10.1016/j.future.2018.08.029, https://www.sciencedirect.com/science/article/pii/S0167739X18302929

23. Kumar, R., Kumar, P., Tripathi, R., Gupta, G.P., Islam, A.N., Shorfuzzaman, M.: Permissioned blockchain and deep learning for secure and efficient data sharing in industrial healthcare systems. IEEE Trans. Industr. Inf. **18**(11), 8065–8073 (2022)

An Empirical Study of Machine Learning for Business Enterprises Management of Cloud Computing Services

D. Jayanarayana Reddy[1], D. Vamshi Krishna[2(✉)], S. Sharmas Vali[1,2], E. Tharun[1,2], and M. Vamsi Kumar[1,2]

[1] Department of CSE, G. Pullaiah College of Engineering and Technology, Kurnool, India
[2] G. Pullaiah College of Engineering and Technology, Kurnool, India
vamshikrishnadevireddy30121@gmail.com

Abstract. Without a question, two of the maximum significant technologies to reach conventional IT in recent years are computing and big statistics analytics. The surprising convergence of the two technologies is yielding potent outcomes and advantages for enterprises. The delivery of IT services by so-called cloud firms and the relationship between enterprises and IT resources are already being affected by cloud computing. Recent developments in information and communication technology have made possible a new approach to data analysis known as "Big Data". Nevertheless, the large quantity of computer resources needed for big data analysis means that many small and medium-sized businesses cannot afford to embrace big data technologies at this time. Affordances in business analytics, cloud computing data security are all part of the concept. Technique (KPCA-LDA-XGB) is used to conduct the empirical research. Using a structural equation model built using Partial Least Squares, this theory is experimentally evaluated with data from 316 businesses. Business analytics and the decision-making affordances of safety are positively moderated by data-driven ethos and IT business process integration. The findings of this research provide practical guidelines for organisations looking to advance their computing data safety organisation with the usage of analytics.

Keywords: Cloud computing · Extreme gradient boosting algorithm · Linear discriminant analysis · Big Data · Business analytics affordances

1 Introduction

There is increasing demand on businesses to establish and scale up their business intelligence initiatives rapidly and affordably in today's dynamic business environment. Cloud computing, a relatively new concept, is altering how both enterprises and end users engage with IT infrastructure and services [1]. It's a radical departure from the status quo since it allows businesses to pay only for the services they really utilise. The amount of data available in the globe is expanding rapidly. The phrase "big data" refers to ever-increasing amounts of data, whether organised, semi-structured, or unstructured, that

© The Author(s), under exclusive license to Springer Nature Switzerland AG 2023
S. Kadry and R. Prasath (Eds.): MIKE 2023, LNAI 13924, pp. 353–364, 2023.
https://doi.org/10.1007/978-3-031-44084-7_33

may be mined for insights. When there is too much information to be stored in a conventional database, we refer to it as "big data" [2, 3]. There is just too much information for a single computer to handle. The rapidly developing subject of big data analytics focuses on analysing massive datasets in search of in computing power, along with algorithms and methods tailored to huge data, have made big data technology a reality [4].

The commercial plans of individual companies are the primary motivators for the adoption of such technology. Each company has its own business analytics strategy (BAS), which is its plan for analysing data and using the insights gained from doing so in order to further its own strategic goals. Nevertheless, several firm-level antecedents of SMC technology adoption [5] impact the connection between business analytics strategy and adoption of SMC technologies. One possible driver for businesses to adopt SMC is an interest in better serving external clients and streamlining internal operations. Nevertheless, slow adoption of these technologies is caused by issues with integration, such as a lack of technical talent and concerns about the safety of the new technologies [6, 7]. A company's level of adoption of new technologies will be determined by a number of factors, counting the company's business plan and the company's attention to either technological enablers or organisational hurdles. Based on existing literature and empirical data from a worldwide corporate survey, we propose a model to determine the relative importance of an organization's current endogenous strategic position and external drivers of technology adoption in determining the extent to which it has adopted SMC technologies [8].

The logical judgement is preferable to the intuitive decision when more data is accessible, and business analytics play a larger role in decision making than intuition does [9]. As a result, business analytics provide a chance to better MCCDS via the use of data-driven insights. Although the MCCDS contributes significantly to management effectiveness, the company is also considering applying business analysis to it in order to logically improve it [10, 11]. Unfortunately, there is a dearth of theoretic study on how to improve the MCCDS by using the affordances provided by business analytics (BAAs). Therefore, this study develops from the vantage point of cloud computing enterprise users, in order to probe the mechanism by which BAAs affect the MCCDS. This research will examine the impact of data-driven cultures (DDCs) and IT-integrated business processes (IBPs) simultaneously (IBI). The data collected in this research will be used by the company to hone its own business analytics for use in enhancing their MCCDS.

As machine learning techniques advance, more and more models are being used to analyse spectral data. Ensemble learning is a recent development in machine learning that combines the advantages of several learning techniques to boost each one's generalizability and predictive efficacy. The unique KPCA-LDA-XGB model construction was suggested using machine learning. It consists of analysis, and the boosting technique. Our KPCA-LDA-XGB model uses an enhanced version of the technique for supervised learning, together with KPCA and LDA to extract features from the data and lower its dimensionality. In Sect. 2, we discuss the relevant literature. In Sects. 3 and 4, we provide the findings and a short explanation. In Sect. 5 we provide our final findings.

2 Related Works

In [14], Park et al. provide a new method of quantifying cloud computing at the business level by using IT facilities delivered over the internet. This study, based on an analysis of 57 different businesses throughout the U.S. economy from 1997 to 2017, suggests that cloud- IT services lead to greater energy competence for their end users. Only with the commercialization of cloud computing in 2006 and its subsequent strengthening in 2010 is this impact seen to be meaningful. Although SaaS has a positive correlation with increased electric and nonelectric energy efficiency in all sectors, IaaS has a positive correlation with increased electric energy efficiency only in sectors with a high intensity of IT hardware. To shed light on the mechanisms at play, we conducted a survey analysis at the company level and found that SaaS helps businesses save money by making it easier for them to produce in a way that uses less energy, while IaaS primarily serves to reduce the energy consumption of in-house information technology resources. We predict that in 2017 alone, the United States economy as a whole may save between USD 2.8 and 12.6 billion in user-side energy costs thanks to cloud computing, which would be equal to a lessening in power usage of between 31.8 and 143.8 billion kilowatt-hours. This approximation is almost on par with the overall power usage in American data centres and much higher than the entire energy expenditure in the cloud service vendor industry.

Using Kruskal Wallis to test the first hypothesis, Pazhayattil et al. [15] found statistically significant differences between the two groups, supporting Rogers' diffusion of innovation theory's relevance to the pharmaceutical sector. Kendall's was used to test the second hypothesis and found that the convergence of machine learning and AI is soon. For the third hypothesis, we employed correlation test to provide light on the high return-on-investment areas of maintenance. Using Spearman's test, the fourth hypothesis established that delays in the pharmaceutical industry can be caused by the five factors of behavioural change using the output.

Tu et al. [16] attempt a micro viewpoint analysis of how digital revolution affects M&A. Construct a measure of corporate transformation using machine learning techniques, companies between 2010 and 2019 to determine that business digital alteration can significantly indorse mergers and acquisitions.

The idea of data network effects is used by Costa-Climent et al. [17], and [18, 19, 20] it provides a potential explanation of how to generate value via learning. Yet, it does not detail how to extract value from machine learning. Yet, theory does not rely on machine learning to describe how businesses employ technology to generate and capture value. The statistical correlations between these two ideas are investigated using a sample of 122 new businesses.

A model of the main elements affecting the chain founded on cloud computing knowledge is developed by Amini et al. [21–23]. Relative security worries, cost savings, technological readiness, top management backing were theorised to have a major influence on the rate of technology adoption. The study's goal is to pinpoint these elements and assess their impact on SMEs' decisions to adopt CC. As a result, the study provides an explanation of a research model based on the technology, organisation, and environment (TOE) framework and the theory of the diffusion of innovations (DOI). A total of 22 SMEs, all of which are customers of the same cloud service provider, responded to a survey questionnaire used to gather the data. Seventy-seven IT professionals from among those SMEs were chosen to fill out the surveys. For this data analysis, we utilised Smart PLS. The data analysis confirms all of the assumptions and provides broad support for the model. In sum, this study's findings indicate that relative advantage, compatibility, security worries, cost savings, technological readiness, support from top management, competitive pressure, and support all play a important role in determining whether or not SMEs adopt supply chain management based on CC.

Using SmartPLS 4.0 software, we performed a equation modelling (PLS-SEM) analysis of the DT, SCI, and OSSCP framework created by Oubrahim et al. [24]. The results showed that DT significantly improves SCI and OSSCP. In addition, SCI has been shown to have direct and beneficial effects on OSSCP, mediating in part the link between DT and OSSCP. Specifically, the research showed that DT adoption leads to a more ethical supply chain from the standpoint of sustainability and operational competence. In the industrial sector, this research is the first to examine the impact of numerical alteration and supply chain addition on long-term supply chain presentation.

3   Proposed System

3.1   Depolying Bid Data Analytics in the Cloud

When parts of the big data analytics process are made available through a public or private cloud, this service model is known as analytics. It employs a number of analytical tools and approaches to assist companies mine enormous data sets for useful insights and make those findings readily accessible through a web browser. Subscription and utility (pay-as-you-go) pricing models predominate for these types of cloud-based data analytics software and services. Cloud analytics as a service is the name for this kind of service delivery (CLAaaS). With this configuration, analytics are easily available through a cloud service. Businesses will be able to automate procedures at any time, from any location, with the help of this cloud-based data analytics tool. Hosted data warehouses, software as a service business database.

Cloud-based big data provides analysts with not just additional data to work with, but also the computing ability to deal with very huge datasets and several characteristics per record. The potential for increased predictability is clear here. New behavioural data, such as daily website visits or location, may be explored by analysts thanks to the integration of big data and cloud computing.

3.2   Major Benefits for Business Organisations

The term "on-demand self-service" refers to the ease with which businesses may add more space or more services with no intervention from an employee. Companies may set up their big data systems as soon as feasible.

Information and data gleaned via the web: Data is stored centrally and can be accessed from anywhere and at any time using the internet and various devices (computers, mobile phones, tablets, etc.).

Multi-tenant models enable resource pooling by effectively grouping and using provider resources. The term "resources" may refer to a wide variety of things, like as

Quick elasticity: Hardware and software resources can be scaled up or down quickly with little impact. The materials are available for purchase by customers at any time and in any quantity.

Little overhead since resource use can be tracked and priced accordingly. The high level of openness in this system encourages trust from both the service provider and the end user. Hadoop and cloud-based analytics are two examples of big data technologies that provide considerable cost savings when it comes to storing massive volumes of data and may also reveal more effective methods of doing business.

3.3 Questionnaire Design and Measurement Tools

Based on prior research, a questionnaire was created for this investigation. In addition to providing background about the company, we also employed a 7-point Likert scale, with scores ranging from 1. The research collects the observables from the various sources listed in Table 1 [20].

Table 1. Variable sources.

Codes	Variables
IBI	IT business process integration
DAC	Data security in the cloud and its implications for decision Making
DRC	The Logic Behind Cloud Computing's Data Safety Decsions
DDC	Data-driven ethos
BAAs	Business analytics affordances

3.4 Data Collection

Data was gathered via online questionnaires conducted for the research. After eliminating 327 questionnaires that did not match the threshold, 316 valid surveys were collected. The percentage of success was 96.6%. Beijing, Mongolia are only some of the provinces, autonomous regions, and municipalities represented in the example businesses. Companies in the sample hail from a wide variety of industries, with manufacturing accounting for 99, information technology for 34, textiles and apparel for 9, chemicals for 16, pharmaceuticals for 13, finance for 22, retail for 28, distribution for 9, logistics for 7, new materials for 7, food for 5, government for 19, construction for 19, real estate for 3, and education for 36. The staff consists of the following individuals: Sixty-three businesses have fewer than one hundred workers, eighty have between one hundred and three hundred, sixty-four have between three hundred and five hundred, sixty-two have between five hundred and two thousand, nineteen have between two thousand and four thousand, and twenty-eight have more than three thousand. The subjects have been included in the main businesses that need to be researched because, according to the fundamental analysis, there are no substantial variations in the hierarchical firms.

3.5 Data Analysis Method

Data analysis and hypothesis testing in the area of information technology often make use of structural equation models (SEM). As this is an exploratory research, we employ a SEM method based on Partial Least Squares to analyse the data. In order to assess the soundness of the theoretical model and delve into the connection between the latent tool.

3.6 XGBoost

To address a problem, ensemble learning takes numerous models and averages their results. The goal of ensemble learning is to recover performance above that of a single approach by employing many models as basis classifiers and then combining them. Ensemble learning outcomes are more accurate and resilient because of the variety of approaches and the mutual correction of mistakes. Ensemble learning algorithms have risen to prominence as a topic of study in the area of machine learning due to the benefits they provide in terms of model improvement and precision.

The two most common types of ensemble learning are bagging and boosting. Integrating base classifiers helps lessen the likelihood of overfitting while also increasing precision. Chen invented the extreme gradient boosting technique (XGBoost), an ensemble learning algorithm based on supervised gradient boosting. The purpose of this approach is to generate a K regression tree with the highest possible generalisation ability and the highest degree of accuracy in predicting the value of the tree group. XGBoost is a lifting method that produces many inefficient learners through residual fitting. A robust learner is obtained by accumulating the produced weak classifiers. In order to speed up the model's convergence during training, XGBoost increases the loss function's order to second-order Taylor and incorporates second-derivative information. Overfitting is avoided and model complexity is reduced by include a regularisation factor in the loss function, which is something XGBoost does. The following is a detailed explanation of how the XGBoost algorithm was developed. Create the data set $D = (x_i, y_i)$, where n is the total sum of samples and d is the total sum of features. Sample i's identifying label is denoted by x_i. We started with a classification and regression tree as our primary model (CART). XGBoost's ensemble model makes predictions by combining K independent models into a single accumulative appearance.

$$\hat{y}_i = \sum_{k=1}^{K} f_k(x_i) \tag{1}$$

where k is the tree's sum and $f_k()$ is the tree's expression.

The model's deviance may be reflected in the loss function. The model's difficulty was added to the optimisation target as a regularisation factor to boost its generalisation capabilities. The objective function of the XGBoost method is composed of the regularisation term and the loss function of the model, as shown below:

$$
\begin{cases}
Obj^{(k)} = \sum_{i=1}^{n} l\left[y_i, \hat{y}_i^{(k-1)} + f_t(x_i)\right] + \sum_k \Omega(f_k) \\
\Omega(f) = \gamma T + \frac{1}{2}\lambda\|\omega\|^2
\end{cases}
\tag{2}
$$

where y i is the actual value, y I is the forecast value, f k (x i) is the consequence of tree k, l is a curved loss function term, and are the regularisation and sum, and w and T are the value and node. Set I j is the collection of data collected from leaf node j. When y = _i((k − 1)), the loss function is expanded using the Taylor formula. The Taylor expansion's first and second derivatives are denoted by g i and h i, respectively. The goal function after Taylor expansion, after eliminating the constant component, looks like this:

$$
Obj^{(k)} = \sum_{j=1}^{T} \left[\left(\sum_{i\in I_j} g_i\right)\omega_j + \frac{1}{2}\left(\sum_{i\in I_j} h_i + \lambda\right)\omega_j^2\right] + \gamma T
\tag{3}
$$

the weight of leaf node j, denoted by w j. Next, we plug in the values we just calculated for G i = (iI j)g i and H i = (iI j)h i into Eq. (3). Hence, we may reduce the goal function to:

$$
Obj^{(k)} = \sum_{j=1}^{T} \left[G_i\omega_j + \frac{1}{2}(H_i + \lambda)\omega_j^2\right] + \gamma T
\tag{4}
$$

The value of the leaf node w j in Eq. (4) is undetermined. So, we compute the objective function Obj((k)) for the first copied of w j, and then solve for the optimum value w j* of the leaf node j as:

$$
\omega_j^* = \frac{G_i}{H_i + \lambda}
\tag{5}
$$

By reinserting w j* into the impartial function, we can calculate the least value of Obj((k)) as:

$$
Obj^{(k)} = -\frac{1}{2}\sum_{j=1}^{T} \frac{G_j^2}{H_i + \lambda} + \gamma T
\tag{6}
$$

XGBoost employs a greedy approach to divide features while constructing CART.

The greedy method starts at the root node and visits all the other nodes, collecting their characteristics along the way. At the split node, the highest-scoring node is chosen. After reaching the the tree and beginning to construct the remainder of the next tree, further splitting is no longer performed. At the end, the XGBoost model is compiled from all the produced trees.

There are three primary ways in which XGBoost enhances the Decision Tree (GBDT). To begin, XGBoost employs a second- with both the first and second orders as enhanced residuals, while regular GBDT relies on the residual resulting from the first-order Taylor expansion. This means the XGBoost model may be used in a wider variety of contexts. Second, XGBoost regulates the model's complexity by including a regularisation component in the goal function. The model's variance and overfitting may be lowered with the help of this regularisation term. Lastly, XGBoost makes use of the random forest column sampling technique to significantly lessen the likelihood of overfitting occurring. XGBoost is shown to have high-quality learning performance and fast training times.

3.6.1   KPCA-LDA-XGB

Fisher discriminant investigation. Using the original data's category information, LDA can extract useful features with high precision. LDA is more effective for feature extraction since it seeks projection modification to increase difference between classes and similarity within classes.

KPCA stands for kernel principal component investigation and is a non-linear technique for extracting features. KPCA's kernel function is what really does the non-linear mapping. Non-linear data dimensionality reduction is accomplished with maximum integrity of the original data preserved. The non-linear features of the data are better captured by KPCA.

Taking into account the issue of green plum pH prediction, we used KPCA and LDA to minimise the input of XGBoost in order to increase our model's feature extraction capabilities. Nonlinear mapping P: R! F was originally employed in KPCA to transform data from the original input space to the high-dimensional feature space F. The evidence in feature space F was then analysed using principal components. To minimise the dimensionality of the data while keeping the nonlinear info among variables, the kernel principal with a reasonably significant cumulative contribution rate was then chosen to construct a new data set. Since it was not known what shape P took, we had to create the kernel function k(x i, x j) = P(x j)P(x j). Several different flavours of kernel functions exist. Linear kernel functions, polynomial kernel functions, the radial basis function kernel, and the kernel are only some of the most popular types of kernel functions. With LDA, the original data were transformed into a linear representation of the training set samples. The results of applying a map were different depending on which straight line (w) was chosen. To make sure that projection points for comparable samples are as near together as feasible and projection points for heterogeneous samples are as far apart as possible, the LDA technique must locate this straight line.

To train the XGBoost model, we first extracted the feature vector from the raw spectral data using KPCA and LDA. Step 1: Feed Step 2: Calculate the Residual by Adding the Predicted Value set to 0; Step 3: Set the Input Features for the Segmentation to a Default Value. Achieve a change in the objective function both before and after segmentation for possible places of segmentation; (4) check whether the current base learner's depth has reached the maximum split depth. If the maximum depth is not reached, we must determine where to divide the graph into left and right leaf nodes and allocate samples there (3). Stop splitting if the maximum split depth is achieved, compute the weight of each leaf node, and finish laying the groundwork for the current base beginner; (5) check whether the current model's training has met the termination condition. If it doesn't, go back to step 2. (2). If so, XGBoost may be created by combining all of the trained base learners. Stop educating the model.

4   Discussions and Implications

These findings have theoretical contributions and practical consequences for the MCCDS, as discussed above.

4.1 Theoretical contributions

(1) The findings explain how BAAs fortify the MCCDS.

From the viewpoint of corporate cloud computing customers, the study develops a theoretical model, and then utilises empirical analysis to reveal the mechanism, which consists of BAAs, DRC, and MCCDS. The findings provide light on how businesses may use business analytics to create more efficient MCCDS, how IT affordances data security, and how affordances research can be expanded to include other developing IT areas.

(2) The findings provide light on the process by which MCCDS low-level affordances are transformed into high-level affordances.

As ecological psychology's affordances theory has been applied to the information technology (IT) domains, a wealth of research has been published investigating IT affordances at the institutional level. Yet, it is still unclear how MCCDS low-level affordances are transformed into high-level affordances. In contrast to DAC, BAAs are considered basic affordances. This research builds the theoretical model using BAAs and DAC based on the notion of IT affordances. The findings also show how BAAs enhance MCCDS through DAC and DRC under the joint influence of DDC and IBI, which improves the MCCDS's conversion process from low-level to affordances.

4.2 Practical Implications

The findings also have significant ramifications for the MCCDS in the real world.

(1) The findings aid the company in making intelligent, data-driven improvements to the MCCDS.

There is a dearth of literature on how to steer the organisation rationally strengthen the MCCDS using analytics, despite the fact that current literatures suggest that rational decision making based on business analytics is superior on experience. The research uses IT affordances to make up for the organization's insufficient skills, resources, and enthusiasm for enhancing the MCCDS via the use of business analytics. The outcomes aid in both DRC enhancement and rational MCCDS improvement through business analytics, as well as search, explanation, evaluation, and identification of data security needs inside the firm.

(2) The findings direct the company to focus on the function of DDC and IBI.

Business analytics play a crucial role, as seen by the decision-making process's growth from intuition to analysis, and they provide the impetus for businesses to fortify their MCCDS. Yet, there are limitations on how business analytics may be used, and this presents a conundrum for many organisations. The paper provides an empirical evaluation of the research perfect, which incorporates DDC and IBI. Based on the findings, it is clear that DDC and IBI play a significant role in the MCCDS. DDC makes business analytics easier to implement. When an organisation has a solid culture, its aims are more likely to be supported by its employees. Organizations are more likely to employ business analytics to judiciously fortify the MCCDS when their leaders trust the data's provenance, development process, and creators. Also, highly integrated IT and business departments work together more effectively, which facilitates the development of data-driven business rules and the use of business analytics inside an organisation to precisely define data security needs and logically enhance MCCDS.

4.3 Practical Implications

Performance Evaluation Metric

Each model and deep network was assessed using standard classification metrics such as accuracy, precision, recall, and F-measure. The proportion of right detections is measured by accuracy, the percentage of relevant examples is measured by precision, and the percentage of relevant instances is measured by recall. For assessing a method's overall performance, the F-measure is helpful since it considers both accuracy and recall (Table 2).

$$Accuracy = \frac{TP + TN}{TP + FP + FN + TN} \tag{7}$$

$$Recall = \frac{TP}{TP + FN} \tag{8}$$

$$Precision = \frac{TP}{TP + FP} \tag{9}$$

$$F - measure = 2.\frac{Precision.Recall}{Precision + Recall} \tag{10}$$

Table 2. Analysis of various ML techniques

Model	Accuracy	Recall	Precision	F-Measure
LDA	0.8834	0.8009	0.8175	0.8091
ELM	0.8862	0.7492	0.8657	0.8032
XGB	0.8605	0.8661	0.8035	0.7284
KPCA-LDA-XGB	**0.9442**	**0.9050**	**0.9203**	**0.9126**

The suggested model scored 94% on the accuracy test, whereas XGB scored 86% and ELM and LDA both scored 88%. ELM had a 74% recall, XGB had an 86% memory, LDA had an 80% recall, and the suggested model had a 90% recall when compared to all other models. When compared to the current models, the suggested model obtained 92% accuracy and 91% F-measure, whereas the previous models only achieved 80% to 86% precision and 72% to 80% F-measure, respectively. The experimental results shown in Figs. 1 and 2 provide conclusive evidence that the proposed model outperformed state-of-the-art ML classifiers.

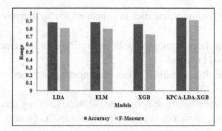

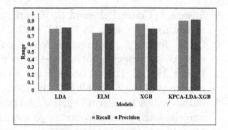

Fig. 1. Validation Analysis of Proposed Model

Fig. 2. Comparative Results of Various Techniques

5   Conclusions and Limitations

Based on empirical investigation, this paper explains how BAAs impact efficient MCCDS by means of KPCA-LDA-XGB. The KPCA-LDA-XGBoost model was presented as an extension of the XGBoost model for enhancing the predictive power of business analytics. KPCA and LDA were used to enhance the XGB model's feature extraction process. There are, however, significant caveats to this work that suggest additional avenues for investigation. To begin, future studies may include a statistically significant sample of businesses with varying data security needs in order to exclude CMV and get a more general result. Second, the influence of DDC and IBI was only examined in this study, while other aspects on MCCDS, such as data analysis capability, data quality, and so on, may be investigated in future studies.

References

1. Hussain, A.A., Al-Turjman, F., Sah, M.: Semantic web and business intelligence in big-data and cloud computing era. In: Ben Ahmed, M., Rakıp Karaş, İ, Santos, D., Sergeyeva, O., Boudhir, A.A. (eds.) SCA 2020. LNNS, vol. 183, pp. 1418–1432. Springer, Cham (2021). https://doi.org/10.1007/978-3-030-66840-2_107
2. Ionescu, L., Andronie, M.: Big data management and cloud computing: financial implications in the digital world. In: SHS Web of Conferences, vol. 92, p. 05010. EDP Sciences (2021)
3. Niu, Y., Ying, L., Yang, J., Bao, M., Sivaparthipan, C.B.: Organizational business intelligence and decision making using big data analytics. Inf. Process. Manage. **58**(6), 102725 (2021)
4. Kim, T.: Improved predictive unmanned aerial vehicle maintenance using business analytics and cloud services (2021)
5. Dawood, B.A., Al-Turjman, F., Nawaz, M.H.: Cloud computing and business intelligence in IoT-enabled smart and healthy cities. In: AI-Powered IoT for COVID-19, pp. 1–38. CRC Press (2020)
6. Potančok, M., Pour, J., Ip, W.: Factors influencing business analytics solutions and views on business problems. Data **6**(8), 82 (2021)
7. Tavera Romero, C.A., Ortiz, J.H., Khalaf, O.I., Ríos Prado, A.: Business intelligence: business evolution after industry 4.0. Sustainability **13**(18), 10026 (2021)
8. Xue, M., Xiu, G., Saravanan, V., Montenegro-Marin, C.E.: Cloud computing with AI for banking and e-commerce applications. Electron. Libr. **39**(4), 539–552 (2021)

9. Silva, A.J., Cortez, P., Pereira, C., Pilastri, A.: Business analytics in industry 4.0: a systematic review. Expert systems **38**(7), e12741 (2021)

10. Qi, X., Joghee, S., Mohammed, A.S.: E-commerce combined with enterprise management using cloud computing for business sector (2021)

11. Park, J., Han, K., Lee, B.: Green cloud? An empirical analysis of cloud computing and energy efficiency. Manage. Sci. **69**(3), 1639–1664 (2023)

12. Pazhayattil, A.B., Konyu-Fogel, G.: An empirical study to accelerate machine learning and artificial intelligence adoption in pharmaceutical manufacturing organizations. J. Generic Med. 17411343221151109 (2023)

13. Tu, W., He, J.: Can digital transformation facilitate firms' M&A: empirical discovery based on machine learning. Emerg. Mark. Financ. Trade **59**(1), 113–128 (2023)

14. Costa-Climent, R., Haftor, D.M., Staniewski, M.W.: Using machine learning to create and capture value in the business models of small and medium-sized enterprises. Int. J. Inf. Manage. 102637 (2023)

15. Chalapathi, M.M.V., Kumar, M.R., Sharma, N., Shitharth, S.: Ensemble learning by high-dimensional acoustic features for emotion recognition from speech audio signal. Secur. Commun. Netw. **2022**, 10 (2022). https://doi.org/10.1155/2022/8777026

16. Oubrahim, I., Sefiani, N., Happonen, A.: The influence of digital transformation and supply chain integration on overall sustainable supply chain performance: an empirical analysis from manufacturing companies in Morocco. Energies **16**(2), 1004 (2023)

17. Rudra Kumar, M., Pathak, R., Gunjan, V.K.: Machine learning-based project resource allocation fitment analysis system (ML-PRAFS). In: Kumar, A., Zurada, J.M., Gunjan, V.K., Balasubramanian, R. (eds.) Computational Intelligence in Machine Learning. LNEE, vol. 834, pp. 1–14. Springer, Singapore (2022). https://doi.org/10.1007/978-981-16-8484-5_1

18. Ramana, K., et al.: Early prediction of lung cancers using deep saliency capsule and pre-trained deep learning frameworks. Front. Oncol. **12**, 886739 (2022). https://doi.org/10.3389/fonc.2022.886739

Prediction of Stock Market in Small-Scale Business Using Deep Learning Techniques

D. Jayanarayana Reddy[1], B. Somanaidu[2($\boxtimes$)], G. Srivathsa[2], and K. Sreenu[2]

[1] Department of CSE, G. Pullaiah College of Engineering and Technology, Kurnool, India
[2] G. Pullaiah College of Engineering and Technology, Kurnool, India
somanaiduraju79@gmail.com

Abstract. The goal of utilising machine learning to forecast the stock is to create more reliable and precise models for doing so. Predictions in the stock market have been made using a wide variety of ensemble regressors and classifiers, each employing a unique combination of methods. While building ensemble classifiers and regressors, however, three risky situations spring to mind. The first issue is with the classification or regression method used as the foundation. The second factor is the number of regressors ensembled, and the third is the combining procedures utilised to build numerous components. As a result, there is a dearth of high-quality research that thoroughly investigates these issues. Existing approaches provide inadequate classification results due to the computational difficulty related with gathering features; as a result, it is vital to build a technique leveraging deep learning ideas for categorising data. The stock market is classified by a powerful and efficient classification model called Deep mahout network, which is based on the dolphin swarm algorithm (DSA). Using a Deep maxout network has the benefit of efficiently learning the data's inherent properties. With each new iteration, the fitness metric informs a change to the weight factor in the deep learning model, leading to improved performance through reduced error. From January 2012 through December 2018, we analysed stock-data from the Stock Exchange (NYSE), the Conversation (BSE-SENSEX), the Ghana Stock Exchange (GSE), and the (JSE) and associated their execution speeds, accuracy, and error measures. The results of the investigation demonstrate that the suggested method provides superior prediction accuracies.

Keywords: Stock-market prediction · Dolphin swarm algorithm · Deep maxout network · Johannesburg Stock Exchange · Multiple regressors

1 Introduction

Stock value forecasting is notoriously difficult [1] owing to the characteristic randomness of stock prices over the long run. Recent technical studies demonstrate that most stock prices are reflected in historical records; hence, the drive patterns are crucial to anticipate values efficiently [2]. The outdated holds that it is difficult to forecast stock values and that stocks behave arbitrarily. Moreover, political events, general financial circumstances, the commodity price index, investor expectations,, the psychology of investors, etc. [3] all

S. Kadry and R. Prasath (Eds.): MIKE 2023, LNAI 13924, pp. 365–376, 2023.
https://doi.org/10.1007/978-3-031-44084-7_34

have an impact on the groupings and movements of the market. Market capitalization is used to determine the worth of market indexes. Statistical information may be extracted from stock prices using a variety of technical factors [4]. Stock market indexes, which are calculated from the values of heavily traded equities, are frequently used as a proxy for a country's economic health. The size of a market, for instance, has been shown to have a beneficial effect on that country's financial growth [5]. Investors take on a high degree of risk due to the lack of clarity surrounding the causes and effects of stock price fluctuations.

Despite the misconception that the stock market is a close price. As most conventional time series prediction algorithms are built on steadfast patterns, is inherently difficult. In addition, there are several factors to think about while making stock price predictions. It is feasible to predict the market's movements over the medium to long term [6]. (ML) is the most effective tool since it uses many different algorithms to learn from experience and become better at solving a specific case study. Just using the most recent data for training would be insufficient, since it is evident that the stock market prediction process is related to both current information and past data. Making economic projections is a natural use for RNN because of its ability to utilise the network to recall recent occurrences and develop linkages between all the nodes.

This study proposes the DSA-based Deep maxout network as a method for time series categorization.

A network was developed with a useful classification framework for sorting stock market data using the proposed DSA-based (DMN). The suggested classifier is able to learn features automatically, allowing for reliable classification outcomes based on fitness values.

The remaining sections of the paper are as shadows: In Sect. 2, we label the relevant literature, and in Sect. 3, we provide a concise description of the suggested model. Section 4 provides a comparison of the proposed model to existing validation methods, and Sect. 5 draws conclusions.

2   Related Works

In order to synthesise and convey in-depth information about the situation and the allure of new target markets, Tsilingeridis et al. [11] present a multidisciplinary framework dubbed MULTIFOR. The primary goal of MULTIFOR is to aid stakeholders (including businesses, SMEs, academics, government officials at all levels, and others) in making informed decisions and developing effective strategies. As it incorporates marketing basics (for time-series framework and its Web service provide a novel solution. The scope of MULTIFOR in European markets has been studied, and it has been verified and verified through a use circumstance on (E&E) business, a robust industry with significant global influence. By using PESTEL analysis for preprocessing time-series data, and by selecting suitable indicators, MULTIFOR was able to boost the performance of LSTM networks.

Using deep embedded clustering, Kanchanamala et al. [12] split the data, with the master node employing the suggested Jaya Anti Coronavirus Optimization (JACO) method to fine-tune the settings. In this case, the technical indications are treated as processing enhancement characteristics. The next step is to augment the master node. Finally, the suggested JACO is used for training, and (Deep LSTM) is used for prediction at the master node. Mean absolute error of 0.113, mean of of 0.309 are all achieved by the suggested LSTM.

When it comes to tackling prediction glitches in the realm of machine learning, are two of the most effective ensemble approaches available. Gradient boosting and XGBoost are two algorithms that have been utilised extensively by top data experts in rivalries and have contributed to the recent success of tree-based models. In addition, deep learning (DL), a recent advancement in ML, may be considered a deep nonlinear topology in its own framework; it is quite good at extracting relevant information from financial time series [8]. In contrast to a standard (RNNs) have achieved remarkable success in the financial industry [9, 10].

Prediction framework using sentiment analysis is presented by BL and BR [13]. So, we also take into account stock market information and news sentiment analysis. Features based on technical indicators are retrieved from the stock market data. These indicators include the moving average (MA). At the same time, processes such as (1) pre-processing—during which the news data undergo keyword extraction and sentiment category—(2) keyword extraction—during which WordNet performed—(3) feature extraction—during which Proposed holoentropy based features are extracted— process the data to ascertain the sentiments. When it comes to step four (classification), a deep neural network is utilised to get an opinionated result. Lastly, the stock is predicted using an optimised deep belief network (DBN) that takes into account both the characteristics of the stock data and the sentiment findings from the news data. Here, the new SIWOA is used to fine-tune DBN's weights.

An (ECNN), denoising auto-encoder (DAE) models, and a model are combined in the innovative hybrid model SA-DLSTM suggested by Zhao et al. [14] to forecast stock market and simulation trading (LSTM). Initially, ECNN was utilised to extract the sentiment representation from Internet user comments that were used to supplement stock market data. Second, DAE may be used to extract the most important elements of stock market data, leading to more accurate forecasts. Third, build more trustworthy and realistic sentiment indices by considering the temporal nature of stock market emotion. In the end, LSTM is used to anticipate the stock market based on important aspects of stock data and sentiment indices. The experimental findings demonstrate that SA-DLSTM outperforms its competitors in terms of prediction accuracy. SA-DLSTM, nevertheless, shows promising results in terms of both return and risk. It can aid investors in making sound choices.

3 Proposed System

3.1 Research Data

We obtained four stock-datasets from four distinct countries (Ghana, South Africa, the United States, and India) to conduct our analysis [15]. As can be understood in Table 1, the sum of features (chosen independent variables) varied among the datasets.

Table 1. Details of dataset

Data Source	Data Size	No. of features	Retro
GSE [Ghana]	1100	9	Jan 2012 to Dec 2017
NYSE [United State]	1760	15	Jan 03, 2012 to Dec 2018
JSE	1749	7	Jan 2012 to Dec 2018
BSE [India]	984	12	Jan 2015 to Dec 2018

Market indexes were obtained from the GSE, JSE, NYSE and BSE from January 2012 through December 2018 for the purpose of evaluating ensemble techniques employing datasets from across the globe. As a result, Previous research showing that some ensemble approaches underperform on datasets from certain regions of the globe is supported by our own experiments. Daily stock data are included in the datasets (year high, year price, closing offer). In order to generalise the results of this research, we used five (5) widely used technical indicators: the (SMA), the (OBV) (OBV). The readings were based on five primary indicators. We aimed to use regression and classification to foretell the 30-day closing price and the movement. Prior to any further analysis, the downloaded datasets underwent two basic processes: I data cleaning, and (ii) data alteration.

3.2 Data Cleaning

Due to its inherent complexity and randomness, stock market data is perpetually prone to noise that might prevent a machine learning system from accurately analysing underlying patterns. Equation (1) uses a wavelet transform (WT) to remove noise and inconsistencies in the dataset. The data set X_ was converted using WT in the following way: coefficients (a, b) with standard deviations greater than zero were discarded (STD). To get rid of the noise in our fresh data, we used an inverse transformation on the updated coefficients. The WT was chosen because it has strong real-time frequency characteristics and multiresolution, in addition to being able to adopt and expand upon the localization-principle of the alter approach.

$$X_\omega(a, b) = \frac{1}{\sqrt{a}} \int_{-\infty}^{\infty} x(t)\varphi\left(\frac{t - b}{a}\right) dt \tag{1}$$

3.3 Data Transformation

When the contribution data is scaled to the similar range, machine learning algorithms perform more accurately and with lower error metrics. For this reason, we use the min-max normalisation procedures stated in Eq. (2), which ensures that all features will be on the same scale [0, 1].

$$b' = \frac{b - b_{min}}{b_{max} - b_{min}} \tag{2}$$

where b is the initial value of the data, b' is the normalised value of b, b(max) and b(min) are the highest and lowest values in the input data, and b(std) is data.

3.4 Classification Using Proposed Model

Let us reflect the dataset as E with h sum of input data stated by Eq. (3):

$$E = \{Z_1, Z_2, \ldots, Z_u, \ldots, Z_h\} \tag{3}$$

Dataset E is implied here by Z, whereas Z u represents the uth index of input data. Stock market classification is the final step of the suggested strategy, and it involves developing a deep learning classifier to categorise the data based on its attributes. As compared to other networks, Deep maxout network converges more quickly. Similar to the ReLU and the leaky ReLU, the Maxout Unit is a generalisation of these two function that, when combined with dropout, yields the inputs' maximum value. It takes 5 min to complete the training phase, but just 5–10 s to complete the testing phase. It demonstrates superior data-fitting ability and achieves higher accuracy in the testing phase. The suggested optimisation procedure is used to fine-tune for use in performing the classification strategy on the given set of features.

3.4.1 Structure of Deep Maxout Network

While classifying the stock market, the network classifier uses M as an input value. Both a dropout and a maxout layer are used to improve classification accuracy. This classifier's activation function strengthens the suggested model's reliability. Input, embedding, dropout, convolution, maxout are only few of the layers that make up the network architecture. In this network, the maxout node is denoted by Eq. (4):

$$Q(M) = {}_{z\in[1,\eta]}{}^{max}R_{yz} \tag{4}$$

where M represents input, represents weight, and represents a bias issue, and R yz = M _yz + _yz. The symbol for the feature map is. The input layer receives the M-dimensional input (of size [1698]), processes it, and sends the result to the implanting layer, which generates an output of size [1 698 50]. The convolution layer follows the dropout layer, with the third convolution layer yielding an output of [1 692 50]. A max-pooling layer is then employed after the convolution layer to provide an output of size [1 50]. The dropout layer comes after the thick layer and has a size of [1 50]. The input from the preceding dropout layer is fed into the maxout unit, which computes a result with a measurement of [150], which is then supplied as a layer, resulting in an output of [7]. The dense layer's output is then sent to the activation purpose, which determines the categorization as [7]-sized vector V.

3.4.2 Training Using DSA

The suggested DSA method, which will be detailed in the next subsection, is used to implement the training strategy of the network.

Coding the Answer: Using an encoding element, we can establish the vector arrangement to be L = [l], where is the weighting factor. In order to determine the weight parameter, an optimisation procedure is used.

The role of fitness: As shown in Eq. (5), the optimal weight may be found by using a wellness metric to convey and assess the value.

$$F = \frac{1}{k} \sum_{\omega=1}^{K} [O_\omega - V_\omega]^2 \qquad (5)$$

where F stands for fitness, K for the number of samples, O for the anticipated output, and V for the intended classification result.

3.4.3 Proposed Optimization (DSA)

Each dolphin in the DSA optimisation process stands in for a particle in the PSO, each of which represents a plausible solution to the optimisation issue. In this study, dolphins are denoted by the formula (j = 1,2,...,D) denotes the corresponding component. Namely, the optimal separate solution (represented by the letter L) and the neighbourhood solution are two key concepts in the DSA (shown as K). In addition, there are two crucial for each Dol i (i = 1,2,..,N), where L i refers to the optimal solution that Dol i discovers in a particular amount of time and K i refers to the ideal solution that Dol i receives from others.

The distance between two points, L i and K i, denoted by DLK i, the distance between two points, Dol i and K i, denoted by DK i, and the distance among two points, Dol i and DD j, denoted by DD j, are the three types of distances that must be described in DSA (i,j). The three separations are written as shadows:

$$DD_{i,j} = \|Dol_i - Dol_j\|, i, j = 1, 2, \ldots, N, i \neq j. \qquad (6)$$

$$DK_i = \|Dol_i - K_i\|, i = 1, 2, \ldots, N \qquad (7)$$

$$DKL_i = \|L_i - K_i\|, i = 1, 2, \ldots, N \qquad (8)$$

1) SEARCH STAGE

Dolphins emit their sonar in all M directions around them when they are actively hunting. The sound is defined in this study as V i = [M), where v j (j = 1,2,...,D) indicates the component of each measurement, named M represents the sum of sounds, and v D means the. of T1. For each time t between 0 and T1, there exists a new solution X Here is how X ijt is defined.

$$X_{ijt} = Dol_i + V_j t \qquad (9)$$

For X_{ijt}, its fitness worth E_{ijt} is defined as shadows:

$$E_{ijt} = Fitness(X_{ijt}) \tag{10}$$

If

$$E_{iab} = min_{j=1,2,...,M;t=1,2,...,T_1} E_{ijt}$$

$$min_{j=1,2,...,M;t=1,2,...,T_1} Fitness(X_{ijt}) \tag{11}$$

Then L_i of Dol_i is defined as

$$L_i = X_{iab} \tag{12}$$

If

$$Fitness(L_i) < Fitness(K_i) \tag{13}$$

Then L_i substitutes K_i; then, K_i fixes not differ. After all the Dol_i ($i = 1, 2, \ldots, N$) update, L_i and K_i, dolphins get hooked on the call stage.

2) RECEPTION PHASE

In DSA, the call stage occurs before the reception stage. First, a comprehensive illustration of the stage of reception. The information transfer between dolphins is represented by a NN- TS ij represents the sound to travel from dolphin I to dolphin j. To show that the noises spread on any component N), it is sufficient to show that all components TS ij in the TS drop when dolphins approach the receiving stage.

$$TS_{i,j} = 0 \tag{14}$$

This demonstrates that Dol i will have originated from Dol j. In addition, the'maximum transmission time' (T 2) will replace the current acquisition time represented by TS (i,j). This method will allow you to hear the associated noise. In addition, contrasting K_i and K_j, if

$$Fitness(K_i) > Fitness(K_j) \tag{15}$$

Either K i will be swapped out for K j, or K i will remain constant. After then, DSA moves into the predatory phase.

3) CALL PHASE

For K_i, K_j, and $TS_{i,j}$ if

$$Fitness(K_i) > Fitness(K_j) \tag{16}$$

$$TS_{i,j} > \left[\frac{DD_{i,j}}{A.speed} \right] \tag{17}$$

where A is the velocity multiplier. Then, the following equation is used to revise TS (i,j):

$$TS_{i,j} = \left[\frac{DD_{i,j}}{A.speed} \right] \tag{18}$$

After all the $TS_{i,j}$ is efficient, DSA enters the welcome stage.

4) **PREDATION STAGE**

At a range of R 2, the dolphins are now actively hunting for food. Furthermore, R 2 determines how far away from the optimal neighbourhood solution the prey's new position is. Maximum search distance (hence denoted by R 1) is calculated as follows.:

$$R_1 = T_1 \times speed \tag{19}$$

Then, $Dol_i(i = 1, 2, \ldots, N)$ is hypothetical to an instance for portraying the control of R2 and update the dolphin's location.

(a) For $Dol_i(i = 1, 2, \ldots, N)$, if

$$DK_i \le R_1 \tag{20}$$

Next, R_2 is calculated on the foundation of Eq. (21).

$$R_2 = \left(1 - \frac{2}{e}\right)Dk_j, e > 2 \tag{21}$$

where e means the radius discount coefficient.

After obtaining R_2, Doli's new position $newDol_i$ is got:

$$newDol_i = K_i + \frac{Dol_i - K_i}{DK_i}R_2 \tag{22}$$

(b) For $Dol_i(i = 1, 2, \ldots, N)$, if

$$DK_i > R_1 \tag{23}$$

And

$$DK_i \ge DKL_i \tag{24}$$

Next, R_2 is intended on the foundation of RECKONING (25).

$$R_2 = \left(1 - \frac{\frac{DK_i}{Fitness(K_i)} + \frac{DK_i - DKL_i}{Fitness(L_i)}}{e.DK_i \frac{1}{Fitness(K_i)}}\right)DK_i, e > 2 \tag{25}$$

After obtaining R_2, Doli's new position $newDol_i$ can be obtained:

$$newDol_i = K_i + \frac{Random}{\|Random\|}R_2 \tag{26}$$

(c) For $Dol_i(i = 1, 2, .., N)$, if it contents Eq. (23) and

$$DK_i < DKL_i \tag{27}$$

Next, R_2 is intended on the foundation of Eq. (23).

$$R_2 = \left(1 - \frac{\frac{DK_i}{Fitness(K_i)} + \frac{DKL_i - DK_i}{Fitness(L_i)}}{e.DK_i \frac{1}{Fitness(K_i)}}\right)DK_i, e > 2 \tag{28}$$

After obtaining R_2, Doli's new position $newDol_i$ is got by Eq. (26). After Dol_i moves to the position $newDol_i$, comparing $newDol_i$ fitness, if

$$Fitness(newDol_i) > Fitness(K_j) \tag{29}$$

Then K_i will be replaced by $newDol_i$, or K_i does not vary.

If the termination condition of the iteration has been met, DSA enters the end stage; then, it enters the search phase.

4 Results and Discussion

The PYTHON tool, Gaps 10 OS, Intel CPU, and 4 GB RAM power the designed model's implementation.

4.1 Evaluation Measures

Mean Absolute Percentage Error

Often used to evaluate the efficacy of prediction algorithms is the Mean Absolute Percentage Error (MAPE). In accuracy for forecasting systems, and it typically shows accuracy as a percentage. Its formula is shown by Eq. (30).

$$MAPE = \frac{1}{n} \sum_{t=1}^{n} \left| \frac{A_t - F_t}{A_t} \right| \times 100 \tag{30}$$

where At represents the current value and F t represents the expected one. The formula calls for dividing the absolute value of the gap between them by A t. For each predicted value, we add its divide it by the total number of observations. Multiplying the error by 100 provides the final error percentage.

Mean Absolute Error

The MAE is a statistic used to quantify the degree to which two numbers diverge from one another. The mean absolute error (MAE) is calculated by averaging the disparity between forecasted and observed numbers. In the field of machine learning, MAE is a shared metric for gauging the accuracy of a regression analysis's predictions. You may see the formula in Eq. (31).

$$MAE = \frac{1}{n} \sum_{t=1}^{n} |A_t - F_t| \tag{31}$$

where the actual value, A t, is subtracted from the predicted value, F t. For each predicted value, the formula divides the value of the discrepancy by the total number of samples and adds the resulting numbers.

Relative Root Mean Square Error

In regression analysis, the RMSE measures the spread of wrong predictions. The spread of the gap between actual data and a prediction model is represented by the prediction error or residual. This statistic displays the degree to which information clusters around the optimal model. Root-mean-squared error measures how far off expectations are from actual data. Similar to RMSE, relative (RRMSE) divides the absolute squared error by the absolute squared error of the forecaster perfect to achieve a standard deviation. Equation displays the formula (32).

$$RRMSE = \sqrt{\frac{1}{n} \sum_{t=1}^{n} \left(\frac{A_t - F_t}{A_t}\right)^2} \tag{32}$$

where A_t is the experiential value, F_t is the forecast value and n is the sum of samples.

Mean Squared Error

A low MSE is a good sign of a high-quality predictor (MSE) accounts for the bias and variance of the prediction model and is the second moment of the error. You can find the solution on the Eq. (33)

$$MSE = \frac{1}{n} \sum_{t=1}^{n} (A_t - F_t)^2 \tag{33}$$

where A_t is the experiential value, F_t is the forecast value and n is the sum of examples.

Table 2. Diversified financials 30 days ahead

Methods	MAPE	MAE	rRMSE	MSE
AE	2.83	48.39	0.0587	12,924.43
DBN	3.21	54.37	0.0467	8803.66
CNN	3.18	54.06	0.0465	8799.45
RNN	2.33	37.63	0.0374	5369.06
LSTM	7.48	26.69	0.0994	54,940.25
DSA-DMN	0.77	10.03	0.0121	376.82

In above Table 2 represent that the Diversified financials 30 days ahead in first model AE reached the MAPE of 2.83 and MAE of 48.390.058712,924.43. DBN model reached the MAPE of 3.21 and MAE of 54.37 and finally MSE of 8803.66. CNN model reached the MAPE of 3.18 and MAE of 54.06 and finally MSE of 8799.45. RNN model reached the MAPE of 2.33 and MAE of 37.63 and finally MSE of 5369.06. LSTM model reached the MAPE of 7.48 and MAE of 26.69 and finally MSE of 54,940.25. DSA-DMN model reached the MAPE of 0.77 and MAE of 10.03 and rRMSE of 0.0121 and finally MSE of 376.82 respectively.

In above Table 3 represent that the Average performance for diversified financials. In this analysis, AE reached the MAPE of 2.07 and MAE of 35.82 and also rRMSE value of 0.0396 and finally MSE value of 7984.30 in another DBN model reached the MAPE value of 1.91 and MAE value of 32.00 and rRMSE value of 0.0288 and finally the MSE value of 3973.92. CNN reached the MAPE of 1.70 and MAE value of 29.91 and rRMSE value of 0.0318 and finally the MSE value of 5662.68. RNN reached the MAPE of 3.86 and MAE value of 69.05 and rRMSE value of 0.0530 and finally the

Table 3. Average performance for diversified financials.

Methods	MAPE	MAE	rRMSE	MSE
AE	2.07	35.82	0.0397	7984.32
DBN	1.91	32.00	0.0288	3973.92
CNN	1.70	29.91	0.0318	5662.68
RNN	3.86	69.05	0.0530	22,409.96
LSTM	1.91	31.96	0.0288	3962.97
DSA-DMN	0.60	6.70	0.0093	148.77

MSE value of 22,409.96 LSTM reached the MAPE of 1.91 and MAE value of 31.96 and RMSE value of 0.0288 and finally the MSE value of 3962.97. DSA-DMN0 60 6.70 0.0093 and finally the MSE value of 148.77 repectively.

5 Conclusion

Stock-price changes are influenced by a wide variety of circumstances, leading many to label the stock market as a stochastic and difficult real-world setting. Many computational models based on soft-computing get around the difficulties of stock market analysis. Using stock from four nations, this work aimed to optimise DL for stock-market prediction. This research appears to be the first application of an improved DL model to the task of stock market forecasting. Given that the effectiveness of classifiers founded on these methods for predicting the stock market has not been thoroughly investigated. As a result, methods like genetic algorithms (GA) and (PCA) can be modified in the future to evaluate the impact of feature-selection. We also utilise the market indices dataset, which is utilised to anticipate the performance of stock market indices by removing failing companies from the top-line directories and replacing them with outperforming stocks to maintain market constancy. Predicting stock prices precisely using ensemble methods is another area of study.

References

1. Parmar, I., et al.: Stock market prediction using machine learning. In: 2018 First International Conference on Secure Cyber Computing and Communication (ICSCCC), pp. 574–576. IEEE (2018)
2. Khan, W., Ghazanfar, M.A., Azam, M.A., Karami, A., Alyoubi, K.H., Alfakeeh, A.S.: Stock market prediction using machine learning classifiers and social media, news. J. Ambient Intell. Humanized Comput. **13**, 1–24 (2020). https://doi.org/10.1007/s12652-020-01839-w
3. Jiang, W.: Applications of deep learning in stock market prediction: recent progress. Expert Syst. Appl. **184**, 115537 (2021)
4. Gandhmal, D.P., Kumar, K.: Systematic analysis and review of stock market prediction techniques. Comput. Sci. Rev. **34**, 100190 (2019)
5. Thakkar, A., Chaudhari, K.: Fusion in stock market prediction: a decade survey on the necessity, recent developments, and potential future directions. Inf. Fusion **65**, 95–107 (2021)

6. Hoseinzade, E., Haratizadeh, S.: CNNpred: CNN-based stock market prediction using a diverse set of variables. Expert Syst. Appl. **129**, 273–285 (2019)
7. Li, X., Wu, P., Wang, W.: Incorporating stock prices and news sentiments for stock market prediction: a case of Hong Kong. Inf. Process. Manage. **57**(5), 102212 (2020)
8. Umer, M., Awais, M., Muzammul, M.: Stock market prediction using machine learning (ML) algorithms. ADCAIJ: Adv. Distrib. Comput. Artif. Intell. J. **8**(4), 97–116 (2019)
9. Strader, T.J., Rozycki, J.J., Root, T.H., Huang, Y.H.J.: Machine learning stock market prediction studies: review and research directions. J. Int. Technol. Inf. Manage. **28**(4), 63–83 (2020)
10. Shah, D., Isah, H., Zulkernine, F.: Stock market analysis: a review and taxonomy of prediction techniques. Int. J. Financ. Stud. **7**(2), 26 (2019)
11. Tsilingeridis, O., Moustaka, V., Vakali, A.: Design and development of a forecasting tool for the identification of new target markets by open time-series data and deep learning methods. Appl. Soft Comput. **132**, 109843 (2023)
12. Kanchanamala, P., Karnati, R., Bhaskar Reddy, P.V.: Hybrid optimization enabled deep learning and spark architecture using big data analytics for stock market forecasting. Concurrency Comput. Pract. Exp. **35**, e7618 (2023)
13. Bl, S., Br, S.: Combined deep learning classifiers for stock market prediction: integrating stock price and news sentiments. Kybernetes **52**(3), 748–773 (2023)
14. Chalapathi, M.M., Kumar, M.R., Sharma, N., Shitharth, S.: Ensemble learning by high-dimensional acoustic features for emotion recognition from speech audio signal. Secur. Commun. Netw. **2022**, 10 (2022). Article ID 8777026, https://doi.org/10.1155/2022/8777026
15. Rudra Kumar, M., Pathak, R., Gunjan, V.K.: Diagnosis and medicine prediction for COVID-19 using machine learning approach. In: Kumar, A., Zurada, J.M., Gunjan, V.K., Balasubramanian, R. (eds.) Computational Intelligence in Machine Learning. LNEE, vol. 834, pp. 123–133. Springer, Singapore (2022). https://doi.org/10.1007/978-981-16-8484-5_10

Predictive Intelligence Based Semiconductor Substrate Fault Detection Model with User Interface

Abhinav Sharma[(✉)], Amrit Mohapatra, Soumya Sahoo, Aditya Prasad Panda, and Ayush Mandal

Department of Computer Science and Engineering, C.V Raman Global University, Bhubaneswar, Odisha, India
abhinavsharma.as13@gmail.com

Abstract. The semiconductor industry is growing day by day and currently, all different countries are trying to invest in semiconductor development as these chips are very much useful in almost every type of business-like automobile, IT, electrical industries, healthcare, transportation, and many more industries. So, a system should be there to check the proper making of semiconductors such that there is less wastage of materials. These wafers are circular in shape and have sensors installed in them. These wafers are subjected to pass through various chemical and radiation processes which cause defects in them and moreover, these wafer's size is also increased over time to fit in extra memory. Our objective is to make an ML model which will help industries to identify these defective wafers without physically examining them with the help of previously recorded data about defective wafers.

Keywords: Semiconductor · Wafer · Substrate · Fault detection · Production line · Tuning · Machine Learning

1 Introduction

An extremely thin slice of silicon known as a wafer is used in countless electronics and other components. They perform admirably as semiconductors. Slices of silicon or silicon substrates are other names for silicon wafers. These are constructed of silicon, which is a common material in modern computers and other cutting-edge electrical gadgets. In addition to these applications, the production of extremely efficient solar cells has seen an increase in the use of wafers. The silicon wafer serves as a substrate for microelectronic devices and is typically included in the base of the electronic component to be a dependable part of integrated circuits.

Single silicon crystals that are extremely pure are used to make silicon wafers. To generate the appropriate silicon wafers, a variety of methods are often used. The Kochanski Method, so named in honor of the Polish scientist Jan Kochanski, is the most well-known and trustworthy method for producing silicon wafers.

S. Kadry and R. Prasath (Eds.): MIKE 2023, LNAI 13924, pp. 377–386, 2023.
https://doi.org/10.1007/978-3-031-44084-7_35

Microchips, which are essential parts of many electronic gadgets such as computers, smartphones, and televisions, are made from silicon wafers. The silicon wafers are produced with high purity, and their flat, polished surface makes them perfect for use in the production of semiconductors.

The goal of this project is to develop a model that can correctly identify which wafer is defective, allowing only that portion of the production line to be shut down, the defective wafer to be replaced, and then that portion of the production line to resume normal operation. so as not to obstruct the entire production process. Based on this, the customer will configure an alert on their end to go off anytime we declare a specific wafer to be bad.

2   Literature Review

In [1], process monitoring and profile analysis are essential for spotting a variety of aberrant events in semiconductor production, which entails incredibly intricate, interrelated, and time-consuming wafer fabrication processes for yield improvement and quality control. Identification of these problems precisely and promptly has become crucial. Several electrical tests known as "Parametric Tests" are performed on wafers once the back-end processing is finished.

In [2], wafer detection is a significant problem in the semiconductor business, according, which can have an impact on a company's manufacturing process and yield. Three categories—training, testing, and validation—were created from the dataset that was used. 707 photos were obtained with defect features, such as bumps, burn marks, and foreign objects, for training and testing purposes.

For the validation step, an additional set of photos was employed. Model validation is the process of assessing a trained model using a testing data set. To identify the best classifier, the captured images were embedded using the InceptionV3 transferring learning method. A model for image identification called InceptionV3 has been discovered to perform more than 78.1% better than the ImageNet dataset.

In [3], this paper suggests a brand-new data-driven methodology for SVID data fault detection and diagnosis in semiconductor production. In a data-driven approach, decisions are made based on data rather than on intuition. A data-driven strategy gives quantifiable benefits. That's because a data-driven strategy relies more on hard data than it does on intuition.

In [4], using NN-based object detention architecture, this research proposed a deep learning-based defeat classification model (DLD) to identify and categorize wafer surfer defeats from metal layers. The Fester R-NN model is used by the proposed DLD system to first input Review-SEM images for flaw and scale bar detection. With the discovered objects (defect and scale bar), sizing computation is further conducted for Classes "sp" and "ed".

In [5], The Optimal March Test, a testing method for locating and identifying flaws in DRAMs, was introduced. The authors sought to reduce test time while enhancing fault localization and detection. They explained how the system used particular test patterns and marching sequences to find errors and pinpoint where they were located in the memory array. The Optimal March Test outperformed existing methods, according to experimental results. The research, taken as a whole, introduced a revolutionary algorithm that improved the speed and precision of memory testing for DRAMs.

In [6], the binomial test, a statistical technique for locating process issues in the semiconductor production process, is discussed in the paper. It outlines the procedures for carrying out the test, offers suggestions for interpreting and making decisions based on the test results, and presents a case study to illustrate its use. The benefit of the binomial test as a straightforward and useful technique for raising process quality in semiconductor production is emphasized by the authors.

In [7], describes a wafer categorization approach for semiconductor production that uses support vector machines (SVMs). It discusses the fundamentals of SVMs, lists the procedures for preparing data for preprocessing, and gives experimental findings that show how effective the SVM-based classification strategy is. The authors stress the potential advantages of incorporating this technique into the semiconductor industry.

In [8], a technique for forecasting wafer yield in semiconductor manufacturing using derived spatial factors is shown. The authors introduce the idea of geographical factors obtained from defect data and discuss the significance of yield prediction. They outline the process for obtaining geographical information and show how their method performs better than conventional models. The practical implications and prospective uses of the generated spatial variables in semiconductor production are also covered in the article.

In [9], a flaw detection technique utilizing the random forest similarity distance is shown. The authors go over the formula for calculating similarity distance and analyze how well it works for fault detection. They also emphasize the method's robustness and computational effectiveness. The research highlights its applicability in industrial settings and its potential for fault finding.

In [10], the research provides a voting ensemble classifier for detecting manufacturing defect patterns on semiconductor wafers. It combines various base classifiers and represents data using statistical attributes. In terms of accuracy, precision, recall, and F1-score, the suggested strategy performs better than individual classifiers and alternative ensemble strategies. Up until a certain point, performance is enhanced by adding more base classifiers. The study emphasizes the value of ensemble approaches for identifying semiconductor defect patterns.

In [11], a light-weighted convolutional neural network (CNN) model for wafer structural defect detection in the semiconductor manufacturing process is introduced by the study. The model is made to be effective and capable of operating in real time. It outperforms baseline models and detects faults on wafer pictures with high accuracy. The proposed model has the potential to be used in the field, which would increase reliability and quality control.

In [12], the research offers a thorough analysis of deep learning-based object identification techniques. It focuses on two-stage and one-stage detectors and includes conventional and deep learning-based techniques. We discuss well-known models such as Faster R-CNN, SSD, YOLO, and RetinaNet. The report also discusses benchmark datasets and evaluation metrics. The obstacles and future directions in object detection research are covered in the conclusion. Overall, it is a useful resource that provides an overview of developments, difficulties, and potential paths for deep learning-based object detection.

In [13], the dataset includes wafer map information for the SM-811K and WM-811K. In the semiconductor manufacturing sector, wafer maps are frequently used to visualize and analyze faults or patterns on semiconductor wafers. The dataset might be used for study or analysis in the field of semiconductor manufacturing or other relevant fields.

3   Proposed Methodology

The main aim of our proposed work is to identify whether a given substrate/wafer is faulty or not in a wafer production line or industry. We have divided the whole work into smaller pieces such that the team can work on the small modules of the whole paper. The small problems are like how to read the data, how to make sure the data is according to the data sharing agreement, how to do data pre-processing, how to do data clustering, which models to be used, how to identify the best model among the models used, hyper parameter tuning, how to make an end to end application. The main parameter which will be used is the clustering approach which will help to create a customized machine-learning model. Data gathering for training is first taken in several batch files that will be first checked for all the conditions that are mentioned in the data sharing agreement. In pre-processing first of all we will check if categorical values are present, we will first convert these categorical values to numeric values, the second step will be to balance the imbalanced data, third step will be to normalize the given data drop the column whose standard deviation is 0. While making the data models for training purposes, to increase the accuracy of the data models we are doing data clustering, the whole dataset will be divided into separate clusters, each cluster will have different cluster numbers, and we will train individual models for individual clusters and at the end we will have a comparison about the models which models will have the best performance for individual clusters. After the data clustering, we will apply 4–5 different algorithms on different clusters to compare the models and find the best model. This comparison will be done on the basics of different parameters that will be found in hyper parameter tuning.

A benchmark model uses a set of sensor data as a characteristic to predict the state (working or not working) of a semiconductor wafer. Benchmark models are compared to more advanced classification models (SVM, decision tree, and AdaBoost ensemble learners) based on MCC and accuracy metrics (Fig. 1).

3.1   System Architecture

The system for predicting wafer fault detection is described in this section. The first component of the system is data sources, which include sensor data that is collected through various sensors at the production line. These data can be collected from various sources, such as the industry's IT applications. The second component of the system is data preprocessing, which involves cleaning and transforming the data to make it suitable for analysis. This includes data cleaning, data normalization, feature selection, and feature engineering. The third component of the system is machine learning models, which are used to make predictions about faulty wafer numbers. Various machine learning algorithms, such as decision trees, random forests, and support vector machines, can be used to build prediction models (Fig. 2).

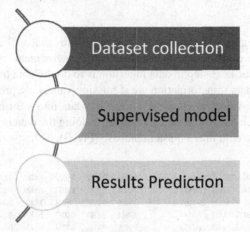

Fig. 1. Methodology

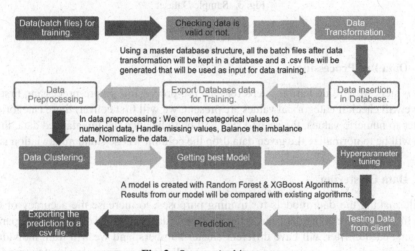

Fig. 2. System Architecture

The fourth component of the system is model training, where the machine learning models are trained on the preprocessed data. This involves selecting the appropriate model, tuning its hyperparameters, and optimizing its performance. The fifth component of the system is model validation, where the trained models are evaluated on a validation dataset to assess their accuracy and performance. This step is important to ensure that the models are robust and can generalize well to new data. The sixth component of the system is prediction and recommendation, where the trained models are used to make predictions about student academic performance and provide personalized recommendations to improve their outcomes. The seventh component of the system is the user interface, which provides an interactive interface for users, such as workers, owners, and administrators, to access the predictions generated by faulty wafers, so the numbers generated by the system can be used for precautionary purposes.

3.2   Dataset

Data gathering for training is taken in several batch files that will be checked for all the conditions that are mentioned in the data-sharing agreement. After checking the validity of the data, the next step in data ingestion is to do the data transformation. The conditions used for data transformation are if missing values are present convert them to null. If categorical values are present convert the data into a format understandable by the database. When the whole data is ready after doing the data ingestion steps, then we will put the whole data into a master database (Fig. 3).

Wafer-800	3002.22	2462.06	2202.122	1833.377 ...	0.5021	0.0071	0.0024	1.4182	0.0545	0.0184	0.0055
Wafer-500	3047.28	2186.06	2235.056	1302.661 ...	0.4998	0.0131	0.0033	2.6229	0.0222	0.0182	0.006
Wafer-501	3076.81	2158.75	2208.233	1517.015 ...	0.5016	0.0152	0.004	3.0319	0.0465	0.0299	0.009
Wafer-502	2951.62	2511.92	2253.511	1397.506 ...	0.4953	0.0105	0.0037	2.1266	-0.0012	0.0252	0.0081
Wafer-503	2930.42	2505.17	2235.056	1302.661 ...	0.4958	0.0111	0.0033	2.2296	-0.0012	0.0252	0.0081
Wafer-504	2997.28	2357.99	2141.067	1236.521 ...	0.4962	0.0086	0.0024	1.7297	-0.0012	0.0252	0.0081

Fig. 3.   Sample Dataset

3.3   Data Pre-Processing

In this step, we will do various steps for data pre-processing. In pre-processing first of all we will check if categorical values are present, we will first convert these categorical values to numeric values, the second step will be to balance the imbalanced data, third step will be to normalize the given data drop the column whose standard deviation is 0.

3.4   Data Clustering

While making the data models for training purposes, to increase the accuracy of the data models we are doing data clustering, the whole dataset will be divided into separate clusters, each cluster will have different cluster numbers, and we will train individual models for individual clusters and at the end we will have a comparison about the models which models will have the best performance for individual clusters. For dividing the dataset k-means is used (Fig. 4).

For model training we are using the clustering approach, we will cluster the whole data and then use bagging and boosting algorithms on the whole data and save those different clusters in a folder either with .sav or. Pickle file. For example – Model1.pickle/.sav after saving the models we will compare the models to figure out which is the best model with the help of the Area Under Curve (AUC score). We have used Random Forests and XG-boost Models which are better than others for binary classification models (Fig. 5).

For the clusters first, we have dropped the labels and clusters name such that we have divided our data into cluster features and cluster labels. Then we split our cluster as x_train, x_test, y_train, and y_test for model training. We have defined a class named Model_finder_inside which we have a method called get_best_model here we are comparing the performance of the XG-boost algorithm and random forest algorithm for individual clusters with different parameters under the AUC score. Inside get_best_model

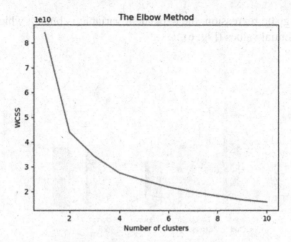

Fig. 4. Elbow Curve for Clustering

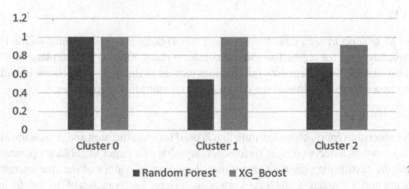

Fig. 5. Model Selection from the clustering approach

we are doing hyperparameter tuning for XG-Boost by taking parameters like learning rate [0.5,0.1,0.01,0.001], max depth [3,5,10,20], n_estimator [10,50,100,200]. For random forest, we have used n_estimator [10,50,100,130] and max depth [1, 2, 4]. Then we will calculate the roc AUC score for both algorithms and compare them to return the best algorithm. Also, we will be getting one exception here if the nod of unique parameters in test_y is 1 then the AUC score fails to provide the result and, in that case, we use the accuracy score. A benchmark model uses a set of sensor data as a characteristic to predict the state (working or not working) of a semiconductor wafer. Benchmark models are compared to more models of Machine Learning Algorithms.

a) Logistic Regression

A categorical dependent variable and one or more independent variables which may be categorical or continuous are analyzed using the statistical method known as logistic regression. In logistic regression, the dependent variable is binary, which means there are only two potential values (Fig. 6).

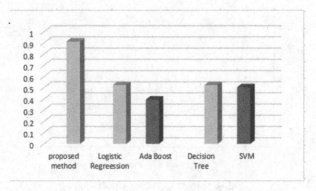

Fig. 6. Visualization of different algorithms

b) SVM

SVM, or support vector machines, is used. It is a potent machine-learning method that is frequently used for regression and classification tasks. Finding a hyperplane in SVM that best divides the data points into distinct classes is the goal. A decision boundary known as a hyperplane divides the data points in a multidimensional space.

c) Decision Tree

An algorithm for machine learning that is used for classification and regression analysis is known as a decision tree. To forecast the value of a target variable, it operates by recursively partitioning the data into subsets based on the values of the characteristics. There are nodes, branches, and leaves in a decision tree. The branches of the dataset indicate the various features while the root node symbolizes the complete dataset. Each leaf node represents a class name or a numeric value, whereas each interior node corresponds to a characteristic. The decision tree algorithm builds the tree by deciding which feature at each node offers the greatest information gain or impurity reduction. By dividing the entropy or Gini impurity, information gain quantifies the decrease.

d) XGBoost

The term "extreme gradient boosting" is XGBoost. For classification, regression, and ranking tasks, it is a well-liked machine learning method. Gradient boosting is an ensemble learning technique that combines several weak models into a stronger model. XGBoost is an advanced implementation of this technique. In XGBoost, the algorithm successively creates decision trees, repairing the errors of the preceding tree with each new tree. To increase forecast accuracy, the method starts with a basic prediction and iteratively builds new decision trees. The algorithm tests the effectiveness of the current model after each iteration and then builds a new tree to cut down on residual errors.

e) Random Forest

A common machine learning technique called Random Forest is utilized for classification, regression, and anomaly detection tasks. It uses an ensemble learning technique to merge different decision trees to build a more reliable and precise model. On various subsets of the training data and various subsets of the characteristics, the Random Forest algorithm constructs several decision trees. A randomly chosen subset of the training data and a randomly chosen subset of the characteristics are used to train each decision tree in the forest. This randomization aids in improving generalization and lowering overfitting (Tables 1 and 2).

Table 1. Accuracy comparison between base model and proposed model.

	RandomForest	XG-Boost	Model Chosen
Cluster0	1.0	1.0	RandomForest
Cluster1	0.5454	1.0	XG-Boost
Cluster2	0.718	0.912	XG-Boost

Table 2. AUC Score comparison between different algorithms.

Models	Accuracy	Auc_roc
Logistic Regression	0.481	0.525
SVM	0.462	0.504
DecisionTree	0.519	0.524
AdaBoost	0.385	0.394

4 Conclusion and Future Work

The proposed methodology shown in this paper suggests a way that can be adopted for finding faulty wafers in a production line. The proposed methodology gives a way better solution than finding the faulty wafers manually by using various machine learning algorithms like Random forests, XG-Boost, etc. Currently, the work is in a development state that why many algorithms are not used. In the coming time, we would like to expand our work to make it more user-friendly, use more algorithms to give better comparison studies, deploy our whole work on a cloud-based environment for more user interactions and we would like to optimize our code for faster operations.

We would like to add more models to our work in the future and also, and we would like to complete the prediction modules a web-based application with a simple user interface will be added and the whole model will be deployed to a cloud-based environment.

Author Contributions. Abhinav Sharma, Amrit Mohapatra, SoumyaSahoo, Aditya Prasad Panda and Ayush Mandal contributed equally. Abhinav Sharma, Amrit Mohapatra, Aditya Prasad Panda, Ayush Mandal and SoumyaSahoo conceived and planned the experiments. Abhinav Sharma, Amrit Mohapatra, Aditya Prasad Panda, Ayush Mandal carried out the experiments. Abhinav Sharma, Amrit Mohapatra, SoumyaSahoo, Aditya Prasad Panda planned and carried out the simulations. Ayush Mandal contributed to sample preparation. All authors provided critical feedback and helped shape the research, analysis and manuscript.

References

1. Chandrasekar, K., et al.: Exploiting expendable process margins in DRAMs for run-time performance optimization. In: Proceedings of the Design, Automation & Test in Europe Conference & Exhibition, Dresden, Germany, pp. 1–6, 24–28 March 2014
2. Huang, P.S., et al.: Warpage, stresses and KOZ of 3D TSV DRAM package during manufacturing processes. In: Proceedings of the 14th International Conference on Electronic Materials and Packaging (EMAP), Lantau Island, Hong Kong, China, pp. 1–5, 13–16 December 2012
3. Hamdioui, S., Taouil, M., Haron, N.Z.: Testing open defects in memristor-based memories. IEEE Trans. Comput. **64**, 247–259 (2013)
4. Guldi, R., Watts, J., Paparao, S., Catlett, D., Montgomery, J., Saeki, T.: Analysis and modeling of systematic and defect related yield issues during early development of new technology. In: Proceedings of the Advanced Semiconductor Manufacturing Conference and Workshop, Boston, MA, USA, vol. 4, pp. 7–12, 23–25 September 1998
5. Shen, L., Cockburn, B.: An optimal march test for locating faults in DRAMs. In: Proceedings of the 1993 IEEE International Workshop on Memory Testing, San Jose, CA, USA, pp. 61–66, 9–10 August 1993
6. Kaempf, U., Ulrich, K.: The binomial test: a simple tool to identify process problems. IEEE Trans. Semicond. Manuf. **8**, 160–166 (1995)
7. Baly, R., Hajj, H.: Wafer classification using support vector machines. IEEE Trans. Semicond. Manuf. **25**, 373–383 (2012)
8. Dong, H., Chen, N., Wang, K.: Wafer yield prediction using derived spatial variables. Qual. Reliab. Eng. Int. **33**, 2327–2342 (2017)
9. Puggini, L., Doyle, J., McLoone, S.: Fault detection using random forest similarity distance. IFAC PapersOnLine **48**, 583–588 (2015)
10. Saqlain, M., Jargalsaikhan, B., Lee, J.Y.: A voting ensemble classifier for wafer map defect patterns identification in semiconductor manufacturing. IEEE Trans. Semicond. Manuf. **32**, 171–182 (2019)
11. Chen, X., et al.: A light-weighted CNN model for wafer structural defect detection. IEEE Access **8**, 24006–24018 (2020)
12. Zhao, Z.-Q., Zheng, P., Xu, S.-T., Wu, X.: Object detection with deep learning: a review. IEEE Trans. Neural Netw. Learn. Syst. **30**, 3212–3232 (2019)
13. Kaggle SM-811K Wafer Map. WM-811K Wafer Map. https://www.kaggle.com/qingyi/wm811k-wafer-map. Accessed 14 Feb 2019

Improving Sustainability with Deep Learning Models for Inland Water Quality Monitoring Using Satellite Imagery

Lokesh Kumar and Yasir Afaq[✉]

Lovely Professional University, Phagwara, Punjab, India
khyasir2@gmail.com

Abstract. Inland water sources like lakes, rivers, and streams are important for the environment and human well-being. Monitoring these water sources is essential to ensure that they remain healthy and productive. This paper presents a study of deep learning-based inland water image classification using neural networks through satellite. The objective of the study is to develop VGG-16 neural network architecture that can be used to accurately distinguish normal images from water images. To assess the performance of the proposed network, several performance metrics are employed. The performance of the neural network is compared to existing methods to ascertain the efficacy of the proposed network. The results of the study show that the proposed neural network architecture is capable of accurately distinguishing normal images from water images, thus demonstrating its potential for successful implementation in real-world applications.

Keywords: Inland water · VGG-16 · AlexNet · Water Quality · Deep learning · Sustainability

1 Introduction

Monitoring inland water resources is an important aspect of sustainable water management. In recent years, advances in satellite technology have enabled the monitoring of inland water resources such as rivers, lakes, and wetlands from space. This has opened up new opportunities for the use of deep learning techniques to better understand and manage these resources. Deep learning is a type of machine learning which utilizes artificial neural networks to learn from large datasets. It can be used to detect and classify objects in satellite imagery, identify changes in land cover and surface water characteristics, and monitor the health of inland water bodies. This paper will discuss the potential of deep learning for monitoring inland water resources, focusing on the application of deep learning to satellite imagery.

The use of deep learning for monitoring inland water resources has been gaining traction in recent years as a way to better understand and manage these resources. The ability to detect and classify objects in high-resolution satellite imagery has allowed researchers to gain valuable insight into the health of inland water bodies, such as changes in water levels, water quality, and surface features. This information can be used to better inform and guide decision-making related to water resources management. Satellite imagery has become an invaluable tool for monitoring inland water resources. High-resolution satellite imagery allows for detailed analysis of water bodies, such as changes in water levels, water quality, and surface features.

Many of the researchers examined alternative methods to classify and detect the water body image and some of these are described more below.

[1] The random under sampling boosted (RUSBoost) technique is used to develop a high-resolution machine learning (ML) approach for identifying inland water content using CYGNSS data. [3] applied a deep learning-based approach to satellite imagery for detecting inland water sources in the Yellow River Basin in China. The authors used a CNN to classify water bodies according to their size. [4] used deep learning techniques to detect water bodies in satellite imagery of a coastal area [5]. The study found that deep learning was more accurate than traditional methods, and it was able to detect more than 90% of water bodies in the imagery [6]. The authors suggest that deep learning techniques could be used to monitor inland water sources in a more efficient way than traditional methods.

Contribution of the Study
The primary goal of this research is, deep learning based inland water image classification using a neural network through satellite. The study explores the possibility of using neural network to classify the normal image as well as water image.

Objectives of the Study
The following are the objectives that need to be accomplished in order to do this study.

- To model a neural network architecture, that can be used to classify the normal image and water image through satellite.
- To analyze the performance of neural network by using performance metrics.

2   Background

In this section, Background of our proposed model VGG 16 architectures as well as other architectures like (AlexNet, and GoogLeNet).

2.1 VGG16 Architecture

VGG16 proved to be a defining moment in humanity's attempt to make computers "see" the world. For decades, a lot of work has been invested into enhancing this capacity under the field of Computer Vision (CV) [7]. The major development known as VGG16 paved the way for several other developments in this field. Andrew Zisserman and Karen Simonyan of the University of Oxford created the Convolutional Neural Network (CNN) model. A manual contest called the ImageNet Wide Scale Visual Recognition Challenge (ILSVRC) evaluated image categorization (and object identification) techniques on a significant scale.

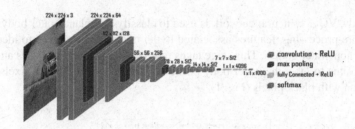

2.2 AlexNet Architecture

The network's initial two convolutional layers are connected to overlapping layers of max-pooling to get the most features feasible [8]. The outputs of the convolutional layer and fully connected layer are all coupled to the ReLu non-linear activation function.

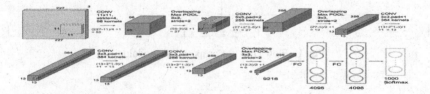

2.3 GoogLeNet Architecture

The GoogLeNet Architecture is comprised of 27 pooling levels and has a total of 22 stacked layers. In all, there are nine inception modules that are arranged in a linear fashion. The global average pooling layer is linked to the terminals of the inception modules. The graphic that follows shows the whole of the GoogLeNet architecture at a reduced scale.

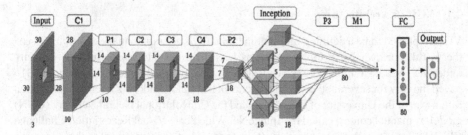

3 Methodology

In this study, VGG 16 neural network is used to classify the inland water body images. Through pre-processing, features associated to the images are retrieved to identify the water and non-water bodies. This stage takes raw satellite imagery and uses an internal threshold value to label the water body portion of the image with white pixels and the land portion with black pixels (Fig. 1).

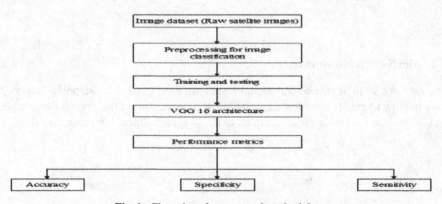

Fig. 1. Flow chart for proposed methodology

3.1 Dataset Description

The data we collected at this website https://www.kaggle.com/datasets/franciscoescobar/satellite-images-of-water-bodies From this dataset, we have to select water body images. The Kaggle satellite images of water bodies data set is used for training and testing of the listed models. The data set consists of the images of both normal and water body images. These images are enhanced and classified before going into training phase.

3.2 Pre-Processing Steps for Image Classification

The objective is to demonstrate how the accuracy changes when some well-known pre-processing methods are applied to certain basic convolutional networks. The following lists a few pre-processing methods.

Read Image: To read the image, we constructed a method to load picture-containing folders into arrays after storing the path to our image dataset in a variable.

Resize Image: In resizing image, we will write two methods to show the photos in this phase, one to display one image and the other to display two images, in order to see the change. Following that, we develop a method called processing that only accepts the photos as an input.

Remove Noise: To Remove the noise a Gaussian function is used to blur a picture, producing a gaussian blur. It is a typical graphics application effect that is often used to decrease visual noise. In order to improve picture structures at various sizes, computer vision algorithms also use gaussian smoothing as a pre-processing step.

Segmentation: The segmentation done by dividing the picture into its background and foreground using segmentation, and we will then use further noise reduction to further enhance our segmentation.

Morphology: The processing of pictures based on forms encompasses a wide range of image processing processes. The output image produced by morphological processes is the same size as the input image after it has had a structural element applied to it.

Train_test Split: After all the preprocessing steps are done, the data set is split into training and testing sets based on the user's split ratio. Later, this split train data will be used to train the models, and the test data will be used to test the models.

Train the Network: The suggested VGG-16 model and Alex, Google Nets are trained using the train data. The suggested model's performance is evaluated and compared using these two extra Alex and Google nets. The following metrics may be used to assess the model's performance.

Performance Metrics

The effectiveness of a technique is assessed in view of the confusion matrix's accuracy, sensitivity, precision, and F1-score

$$Accuracy = \frac{TP + TN}{TP + TN + FP + FN} \tag{1}$$

$$Sensitivity = \frac{TP}{TP + FN} \tag{2}$$

$$Pr\,ecision = \frac{TP}{TP + FP} \tag{3}$$

$$F1 - score = 2 * \frac{Pr\,ecisison * Re\,call}{Pr\,ecision + Re\,call} \tag{4}$$

$$Specificity = \frac{TN}{TN + FP} \tag{5}$$

4   Results

A confusion matrix is a type of table that is used to describe the performance of a classification model on a set of test data for which the true values are known. It is a table of correct predictions and incorrect predictions broken down by each class, allowing you to see where the model is making mistakes. The rows of the matrix represent the predicted classes, while the columns represent the actual classes. The cells of the matrix contain the number of correct predictions and incorrect predictions for each class. The below shown figures are the confusion matrix of the proposed and existing models.

The Fig. 2 is the confusion matrix of the proposed VGG-16 architecture. In figure normal images are represented by zeros and water body images are represented by ones. The proposed method identified 64 times images as normal images and there is a misclassification of 2 images in normal images. 152 times, it is correctly predicted as water body images and there are 2 misclassifications.

Figure 3 presents the image of confusion matrix of the AlexNet architecture is seen above. Normal images are represented by zeros in the figure, whereas water body images are represented by ones. AlexNet identified 60 images as normal images, with 3 images misclassified as normal images. It is properly predicted as water body images 151 times, with 6 misclassifications.

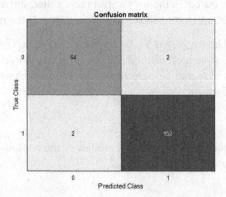

Fig. 2. Confusion matrix of the proposed VGG-16

Figure 4 presents the confusion matrix for the GoogLeNet is seen in the above graphic. Normal images are represented in the figure by zeros, whereas images of water bodies are represented by ones. GoogLeNet wrongly identified 11 images as normal images while identifying 62 times as normal images. There were 4 incorrect classifications out of 143 predictions that the images were of bodies of water.

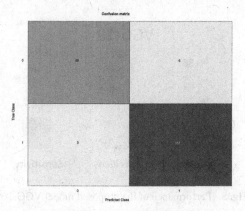

Fig. 3. Confusion matrix of the AlexNet architecture

It is clear from the Fig. 5 graph that the performance of the proposed model VGG-16 is measured and compared using the two additional models AlexNet and GooLeNet. Therefore, the recommended model provided greater performance, with an accuracy score of 96.36, in comparison to AlexNet's score of 95.91 and GoogLeNet's score of 93.18, respectively.

Below accompanying Fig. 6 graph makes it evident that the two additional models AlexNet and GooLeNet are used to evaluate and compare the performance of the proposed model VGG-16. As a result, the suggested model performed better than AlexNet and GooLeNet, with a specificity score of 93.05 and 91.86, respectively, for the proposed model.

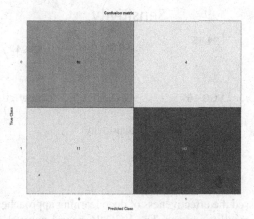

Fig. 4. Confusion matrix for the GoogLeNet

The following Fig. 7 graph demonstrates that the performance of the proposed model VGG-16 is assessed and compared using remaining two models, AlexNet and GoogLeNet. With a sensitivity score of 94.45, the suggested model demonstrated superior performance in contrast to AlexNet (90.91) and GoogLeNet (89.94).

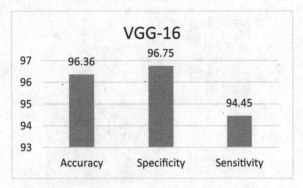

Fig. 5. Performance of the proposed model VGG-16

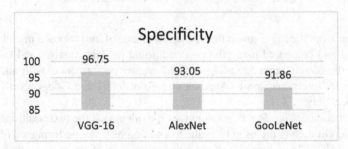

Fig. 6. Specificity

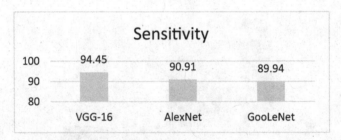

Fig. 7. Sensitivity

5   Conclusion

This study has assessed the effectiveness of deep learning approaches to monitor inland water bodies using satellite images. The VGG-16 neural network is used to classify the inland water body images. The evaluation of the model was done by comparing the results with two additional models: AlexNet and GoogLeNet. The proposed model showed higher accuracy, sensitivity and specificity scores than the other two models. The proposed model VGG-16 shows improved performance in comparison to the benchmark models AlexNet and GoogLeNet, with an accuracy score of 96.36, specificity score of 93.05 and 91.86, respectively. This model can be used to extract useful information from satellite images and identify the water body portion of the image with more accuracy. This research provides an important step towards better understanding of the inland water resources and their distribution.

References

1. Ghasemigoudarzi, P., Huang, W., De Silva, O., Yan, Q., Power, D.: A machine learning method for inland water detection using CYGNSS data. IEEE Geosci. Remote Sens. Lett. **19**, 1–15 (2022). https://doi.org/10.1109/LGRS.2020.3020223
2. Shen, C.: A transdisciplinary review of deep learning research and its relevance for water resources scientists. Water Resour. Res. **54**(11), 8558–8593 (2018). https://doi.org/10.1029/2018WR022643
3. Manocha, A., Afaq, Y., Bhatia, M.: Mapping of water bodies from sentinel-2 images using deep learning-based feature fusion approach. Neural Comput. Appl. **35**, 9167–9179 (2023). https://doi.org/10.1007/s00521-022-08177-2
4. Afaq, Y., Manocha, A.: Fog-inspired water resource analysis in urban areas from satellite images. Eco. Inform. **64**, 101385 (2021)
5. Afaq, Y., Manocha, A.: Analysis on change detection techniques for remote sensing applications: a review. Eco. Inform. **63**, 101310 (2021)
6. Qian, J., et al.: Water quality monitoring and assessment based on cruise monitoring, remote sensing, and deep learning: a case study of Qingcaosha Reservoir
7. Li, L., Yan, Z., Shen, Q., Cheng, G., Gao, L., Zhang, B.: Water body extraction from very high spatial resolution remote sensing data based on fully convolutional networks. Remote Sens. **11**(10) (2019). https://doi.org/10.3390/rs11101162
8. Hassan, G., Shaheen, M.E., Taie, S.A.: Prediction framework for water quality parameters monitoring via remote sensing. In: Proceedings of the - 2020 1st International Conference of Smart Systems and Emerging Technologies SMART-TECH 2020, pp. 59–64 (2020). https://doi.org/10.1109/SMART-TECH49988.2020.00029

Machine Learning Based Prediction of Student's Performance Based on Psychological and Behavioral Data

Ankit Kumar Saha(✉), Abhishek Kumar Sharma, Soumya Sahoo,
Shaikh Ejaz Hussain, and Nikhil Kumar Sahoo

Department of Computer Science and Engineering, C. V. Raman Global University,
Bhubaneswar, Odisha, India
ankitcvrgu@gmail.com

Abstract. In recent years, there has been an increase in interest in the field of education to use machine learning approaches to improve student achievement. Our study emphasizes the significance of including behavioral and psychological data in addition to standard academic data for a thorough understanding of student performance. We can discover important new information about the non-cognitive elements that have a big impact on learning outcomes by combining these many data sources. By enabling proactive interventions and customized instructional strategies that meet the specific needs of individual pupils, this multidimensional approach has the potential to revolutionize current educational practices. This study highlights the capability of machine learning methods to forecast students' academic achievement based on psychological and behavioral information. In order to promote academic performance and improve educational outcomes for all children, educators and policymakers can use data analytics to make knowledgeable decisions and put evidence-based solutions into practice.

Keywords: Psychological data · GFG · DFS · ROC Curve · Decision Tree · XG Boost

1 Introduction

Both scholars and educators have shown an interest in the prediction of pupils' academic achievement. The performance of a student can be affected by a variety of circumstances, yet psychological and behavioural elements are frequently disregarded.

Based on these frequently disregarded elements, machine learning algorithms present a potential strategy for forecasting pupils' academic achievement. Machine learning algorithms may find patterns and links in vast quantities of data, including a student's behavior, demographics, and academic background, and use those patterns and relationships to forecast a student's success in the future.

By enabling teachers to spot at-risk kids early and offer them targeted interventions to promote their achievement, this technique has the potential to enhance educational

S. Kadry and R. Prasath (Eds.): MIKE 2023, LNAI 13924, pp. 396–408, 2023.
https://doi.org/10.1007/978-3-031-44084-7_37

results. This approach can also offer insightful information on what influences academic achievement, which can help to guide practice and policy in education.

In this research, we'll investigate how machine learning algorithms may be used to forecast students' academic achievement using psychological and behavioral data. By gathering information on students' actions and psychological characteristics like motivation and self-efficacy, we can train machine learning algorithms to forecast academic success. By doing this, we seek to advance the corpus of knowledge about machine learning in education and enhance student learning outcomes.

The main motivation behind this research is we want to help the confident and qualified applicants about getting into their dream colleges' in abroad for graduate programs. Numerous internet resources offer statistics on admission, but they do not offer information on specific profile data, leaving applicants with no choice except to guess and wait to hear whether they were accepted or rejected. There is a need to implement a more effective and efficient automated way to handle the admission process because the number of candidates has greatly increased in recent years. Since there are more students in India willing go abroad, it is necessary to use more effective technologies that accurately handle the admission procedure from both viewpoints. It's estimated that 1.8 million Indians will be spending US $85 billion on education overseas by 2024.In the first three months of 2022, 133,135 students left India for academic pursuits, an increase from 2020 when 259, 655 students studied abroad. In 2021, there were 4, 44, 553 Indians – an overall increase of 41% in just one year. Interestingly, Gen Z's are motivated by 'self-dependence' and 'living life on their own terms.

Our main contribution in this paper is for feature engineering, we used Deep Feature Synthesis and Genetic Feature Generation. We have concluded after a careful review that numerous papers have employed these strategies, although not all at once. So, in an effort to improve our model's predictive powers, we attempted incorporating these two strategies for extracting relevant features.

Rest of the paper is followed by Literature Review, proposed methodology, system architecture, datasets, data pre-processing, data visualization, Comparison of DFS and GFG and machine learning algorithms and future work.

2 Literature Review

In [1], the authors used various machine learning algorithms, including decision tree, random forest, k-nearest neighbor, and support vector machine, to predict students' academic performance based on demographic, psychological, and behavioral data. The study found that the random forest algorithm had the best performance in predicting students' academic performance.

In [2] used machine learning techniques such as logistic regression, decision tree, random forest, and neural networks to predict students' academic performance based on demographic, psychological, and behavioral data. The study found that the neural network algorithm had the best performance in predicting students' academic performance.

In [3] used machine learning techniques such as decision tree, random forest, k -nearest neighbor, and logistic regression to predict students' academic performance based on demographic, psychological, and behavioral data. The study found that the random forest algorithm had the best performance in predicting students' academic performance.

In [4], the authors used machine learning techniques such as decision tree, random forest, k-nearest neighbor, and logistic regression to predict students' academic performance based on demographic, psychological, and behavioral data. The study found that the random forest algorithm had the best performance in predicting students' academic performance.

Overall, these studies demonstrate that machine learning techniques can be effective in predicting students' academic performance based on psychological and behavioral data.

In [5], the author investigates the use of ML algorithms to predict student academic performance based on psychological and behavioral data. The authors use a dataset of 147 students and compare the performance of different ML algorithms. They found that the Decision Tree algorithm performed the best.

In [6], the author provides a comprehensive review of studies that have used ML techniques to predict student academic performance. The authors review 37 studies and identify the most used features, such as attendance, grades, and demographics. They also compare the performance of different ML algorithms.

In [7], the author investigates the use of personality traits to predict student academic performance. The authors use a dataset of 320 students and found that personality traits are important predictors of academic performance. They also compare the performance of different ML algorithms and found that the Random Forest algorithm performed the best.

In [8], this paper investigates the use of personality traits to predict student academic performance. The authors use a dataset of 300 students and found that personality traits are important predictors of academic performance. They also compare the performance of different ML algorithms and found that the Support Vector Machine algorithm performed the best.

In [9], This paper investigates the use of learning behavior data to predict student academic performance. The authors use a dataset of 66 students and found that learning behavior data can be used to predict academic performance. They also compare the performance of different ML algorithms and found that the Artificial Neural Network algorithm performed the best.

In [10], this paper investigates This paper investigates the use of students' interactions with an online learning environment to predict their academic performance. The authors use a dataset of 515 students and compare the performance of different ML algorithms. They found that the Random Forest algorithm performed the best.

3 Proposed Methodology

This methodology's objective is to examine and forecast students' academic performance based on their psychological and behavioural data by using machine learning techniques. This can provide significant insights for educators to adjust interventions and support systems to improve students' academic results.

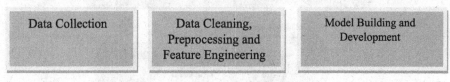

Fig. 1. Methodology

3.1 System Architecture

The system for predicting student performance is described in this section. Numerous demographic, psychological, behavioral and academic performance data are used as input in proposed approach for forecasting students' success. Block diagrams for the suggested approaches are shown in Figs. 1 and 2. The system architecture of a students' performance prediction system typically involves the following components:

 The first component of the system is data sources, which include demographic, psychological, behavioral, and academic performance data. These data can be collected from various sources, such as student information systems, learning management systems, surveys, and other sources. The second component of the system is data preprocessing, which involves cleaning and transforming the data to make it suitable for analysis. This includes data cleaning, data normalization, feature selection, and feature engineering [11–14]. The third component of the system is machine learning models, which are used to make predictions about student academic performance. Various machine learning algorithms, such as decision trees, random forests, and support vector machines, can be used to build prediction models. The fourth component of the system is model training, where the machine learning models are trained on the preprocessed data. This involves selecting the appropriate model, tuning its hyper-parameters, and optimizing its performance. The fifth component of the system is model validation, where the trained models are evaluated on a validation dataset to assess their accuracy and performance. This step is important to ensure that the models are robust and can generalize well to new data. The sixth component of the system is prediction and recommendation, where the trained models are used to make predictions about student academic performance and provide personalized recommendations to improve their outcomes. The seventh component of the system is the user interface, which provides an interactive interface for users, such as teachers, counselors, and administrators, to access the predictions and recommendations generated by the system [15, 16].

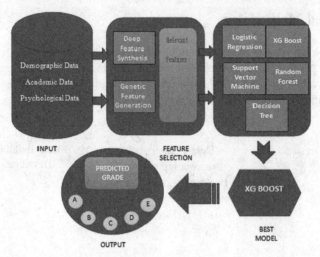

Fig. 2. System Architecture

3.2 Dataset

Our data was gathered from several online archives, including the open-source and Online internet resources [25]. For the prediction we evaluated numerous psychological aspects for psychological data and other dimensions for a smooth and effective predication. The dataset which we have used is named as ("Student Performance Dataset").

Features Used for Psychological Analysis

- family size
- Parental Status
- Parents education
- Parents job
- Travel time from home to school
- Family educational support
- extracurricular activities
- Romantic relationship
- Family Relationship

Other Features used

- Age
- address (Urban or Rural)
- Internet access
- Grades (G1, G2, G3)

The dataset represents the academic achievement of secondary school students in two Portuguese schools. The data used includes various factors such as student grades, demographic information, social background, and school-related features. The data was collected through school reports and questionnaires. Two datasets are provided, focusing on the performance of students in two different subjects: Mathematics (mat) and Portuguese language (por). The datasets were analyzed using binary/five-level classification and regression tasks. It is important to note that the target attribute, G3 (final year grade), is strongly correlated with attributes G2 (2nd period grade) and G1 (1st period grade). Predicting G3 accurately is more challenging without considering G2 and G1, but such predictions are highly valuable.

3.3 Data Pre-Processing

For each of the six subjects we picked, we created a unique dataset. Datasets are verified for null values, duplicate values, and incorrect values. Because each dataset has the same set of qualities and similar types of data, they are combined vertically to make the dataset more effective and extend it. This is a supposedly critical aspect of data preparation. The model cannot handle non-numeric features; hence a number of non-numeric and categorical attributes must be encoded. The ideal assumption would be one-hot-encoding, which was our first choice because categories are retrieved as features, amplifying the influence of the feature, but because the dataset only contains a few categories per column (maximum of five), this is not the case.

3.4 Data Visualization

To study and observe the behavior of data, attributes, relationships between attributes, and target variables are graphically visualized, allowing the pattern of data to be closely examined and the dependency and weightage of attributes to be explored, allowing reliable features to be extracted and a dependable model with robust features to be developed [18–24].

The following are the key plots that are visualized: Individual attributes are displayed against the frequency of their occurrence [to observe the data]. Student Grade Prediction and Characteristics in Relation to the Goal Variable ('Final Grade') [to study dependency and weighting] (Figs. 3 and 4).

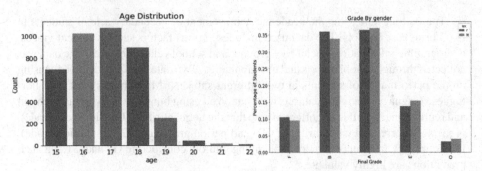

Fig. 3. Age Distribution   **Fig. 4.** Grade by Gender

3.5 Comparison of DFS and GFG

Deep Feature Synthesis (DFS) and Genetic Feature Generation (GFG) are two different approaches used in machine learning for feature engineering. Both methods aim to extract meaningful features from raw data, but they differ in their techniques and applicability.

Using DFS, we can automatically create new features by combining the existing variables in the dataset. For example, we can create a feature that captures the total amount spent by a customer in the past six months, another feature that captures the number of different product types purchased by a customer, and a third feature that captures the average amount spent per transaction.

Using GFG, we can generate new features by applying genetic algorithms to the existing features. For example, we can create a population of feature combinations, where each combination consists of a subset of the existing features. We can then evaluate the fitness of each feature combination based on its ability to predict the target variable (i.e., repeat purchases).

The choice of which method to use depends on the specific problem at hand and the expertise of the data scientist. DFS may be more suitable as it can quickly generate a large number of relevant features that capture complex relationships between the variables. GFG, on the other hand, may be more suitable in situations where there is a need for domain-specific features, and the data has complex relationships that are difficult to model using traditional feature engineering methods. However, GFG can be computationally expensive, especially when dealing with large datasets.

3.6 Machine Learning Algorithms

a) Logistic Regression

　　Logistic regression is a statistical method used to analyze the relationship between a categorical dependent variable and one or more independent variables, which can be categorical or continuous. The dependent variable in logistic regression is binary, meaning it takes only two possible values (e.g., yes/no, true/false, 0/1).

b) SVM

SVM stands for Support Vector Machines. It is a powerful machine learning algorithm that is widely used for classification and regression tasks. In SVM, the objective is to find a hyperplane that best separates the data points into different classes. A hyperplane is a decision boundary that separates the data points in a multi-dimensional space.

c) Decision Tree

A decision tree is a machine learning algorithm that is used for classification and regression analysis. It works by recursively partitioning the data into subsets based on the values of the features, in order to predict the value of a target variable. A decision tree consists of nodes, branches, and leaves. The root node represents the entire dataset, and the branches represent the different features in the dataset. Each internal node corresponds to a feature, and each leaf node represents a class label or a numerical value. The decision tree algorithm constructs the tree by selecting the feature that provides the most information gain or reduction in impurity at each node. Information gain measures the reduction in entropy or Gini impurity achieved by splitting the data at a particular node. The aim is to maximize the information gain or reduction in impurity at each node, so as to create a tree that accurately classifies the data.

d) XGBoost

XGBoost stands for Extreme Gradient Boosting. It is a popular machine learning algorithm used for classification, regression, and ranking tasks. XGBoost is an advanced implementation of gradient boosting, which is an ensemble learning method that combines multiple weak models into a stronger model. In XGBoost, the algorithm builds decision trees sequentially, with each tree correcting the mistakes of the previous tree. The algorithm starts with an initial prediction and then iteratively adds new decision trees to improve the prediction accuracy. At each iteration, the algorithm evaluates the performance of the current model and then creates a new tree to reduce the residual errors.

e) Random Forest

Random Forest is a popular machine learning algorithm used for classification, regression, and anomaly detection tasks. It is an ensemble learning method that combines multiple decision trees to create a more robust and accurate model. In Random Forest, the algorithm builds multiple decision trees on different subsets of the training data and different subsets of the features. Each decision tree in the forest is trained on a randomly selected subset of the training data and a randomly selected subset of the features. This randomness helps to reduce overfitting and improve generalization.

3.7 ROC Curve

See Figs. 5 and 6.

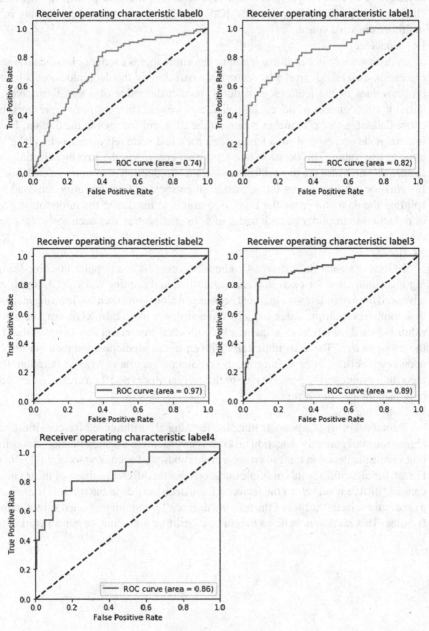

Fig. 5. ROC Curve

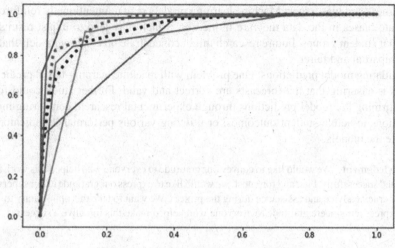

Fig. 6. AUC of XGBOOST for all labels

With its high and nearly constant accuracy for each input, it is evident from the above table that XG Boost is the best model for prediction (Table 1).

Table 1. Results of Different Algorithms

Classifiers	Logistic Regression	Random Forest	SVM	Decision Tree	XG Boost
Accuracy	86.7	94.1	94.5	97.2	98.6
Recall	85.5	80.0	81.9	93.6	97.2
Precision	88.6	90.0	92.0	92.6	96.2
F-Score	86.9	83.3	85.8	93.0	96.7

4 Conclusion and Future Work

The machine learning-based prediction model for students' academic success has a number of potential extensions and improvements. Here are some suggestions:

Integrating new data sources: Today, most machine learning-based prediction models for academic achievement rely on demographic, psychological, and behavioral data. However, other data sources, such as social network data, academic programme data, and health data, may provide useful information for enhancing forecast accuracy.

Investigating various machine learning algorithms: Although numerous machines learning algorithms, including random forests, decision trees, and neural networks, have been utilized in academic performance prediction models, other methods might increase prediction accuracy. For example, new breakthroughs in deep learning could be studied to increase the accuracy of prediction models.

· Fairness and bias issues: Machine learning models may unintentionally reinforce or aggravate biases in the data they are trained on, which may lead to unjust results for particular student groups. Future research might concentrate on creating models that are more impartial and fairer.

Validating model predictions: One problem with machine learning-based prediction models is ensuring that the forecasts are correct and valid. Further study could focus on confirming the model predictions through experimental research, comparing model predictions to actual student outcomes, or utilizing various performance indicators to evaluate the models.

Acknowledgment. We would like to convey our gratitude to everyone who helped this work to be completed successfully. First and foremost, we would like to express our gratitude to our supervisor for their crucial advice and assistance during the project. We want to take this opportunity to once again express our sincere gratitude to everyone who helped make this initiative a success.

References

1. Siddique, A., Jan, A., Majeed, F., Qahmash, A.I., Quadri, N.N., Wahab, M.O.A.: Predicting academic performance using an efficient model based on fusion of classifiers. Appl. Sci. **11**, 11845 (2021). https://doi.org/10.3390/app112411845
2. Hussain, S., Khan, M.Q.: Student-performulator: predicting students' academic performance at secondary and intermediate level using machine learning. Ann. Data Sci. **10** (2021). https://doi.org/10.1007/s40745-021-00341-0
3. Yağcı, M.: Educational data mining: prediction of students' academic performance using machine learning algorithms. Smart Learn. Environ. **9**(1), 1–19 (2022). https://doi.org/10.1186/s40561-022-00192-z
4. Sharma, N., Sharma, M., Garg, U.: Predicting academic performance of students using machine learning models. In: 2023 International Conference on Artificial Intelligence and Smart Communication (AISC), Greater Noida, India, pp. 1058–1063 (2023). https://doi.org/10.1109/AISC56616.2023.10085214
5. Altabrawee, H., Ali, O.A.J., Ajmi, S.Q.: Predicting students' performance using machine learning techniques. J. Univ. Babylon Pure Appl. Sci. **27**, 194–205 (2019). https://doi.org/10.29196/jubpas.v27i1.2108
6. Baashar, Y., Alkawsi, G., Ali, N., Alhussian, H., Bahbouh, H.T.: Predicting student's performance using machine learning methods: a systematic literature review. In: 2021 International Conference on Computer & Information Sciences (ICCOINS), Kuching, Malaysia, pp. 357–362 (2021). https://doi.org/10.1109/ICCOINS49721.2021.9497185
7. Al-Alawi, L., Al Shaqsi, J., Tarhini, A., et al.: Using machine learning to predict factors affecting academic performance: the case of college students on academic probation. Educ. Inf. Technol. (2023). https://doi.org/10.1007/s10639-023-11700-0
8. Koutina, M., Kermanidis, K.L.: Predicting postgraduate students' performance using machine learning techniques. In: Iliadis, L., Maglogiannis, I., Papadopoulos, H. (eds.) Artificial Intelligence Applications and Innovations. EANN AIAI 2011 2011. IFIP AICT, vol. 364, pp. 159–168. Springer, Berlin, Heidelberg (2011). https://doi.org/10.1007/978-3-642-23960-1_20
9. Yakubu, M.N., Abubakar, A.M.: Applying machine learning approach to predict students' performance in higher educational institutions. Kybernetes **51**(2), 916–934 (2022). https://doi.org/10.1108/K-12-2020-0865

10. Alamri, H.L., Almuslim, R.S., Alotibi, S.M., Alkadi, K.D., Ullah Khan, I., Aslam, N.: A Comparison of Machine Learning Techniques for the Prediction of the Student's Academic Performance (2020). https://doi.org/10.1007/978-3-030-32150-5_107

11. Banerjee, D., Kukreja, V., Hariharan, S., Sharma, V.: Fast and accurate multi-classification of kiwi fruit disease in leaves using deep learning approach. In: 2023 International Conference on Innovative Data Communication Technologies and Application (ICIDCA), Uttarakhand, India, 2023, pp. 131–137 (2023). https://doi.org/10.1109/ICIDCA56705.2023.10099755

12. Raj, A., Sharma, V., Shanu, A.K.: Comparative analysis of security and privacy technique for federated learning in IOT based devices. In: 2022 3rd International Conference on Computation, Automation and Knowledge Management (ICCAKM), pp. 1–5. IEEE, November 2022

13. Manikandan, N., Ruby, D., Murali, S., Sharma, V.: Performance analysis of DGA-driven botnets using artificial neural networks. In: 2022 10th International Conference on Reliability, Infocom Technologies and Optimization (Trends and Future Directions) (ICRITO), pp. 1–6. IEEE, October 2022

14. Ali, M.A., Balamurugan, B., Sharma, V.: IoT and blockchain based intelligence security system for human detection using an improved ACO and heap algorithm. In: 2022 2nd International Conference on Advance Computing and Innovative Technologies in Engineering (ICACITE), pp. 1792–1795. IEEE, April 2022

15. Juyal, V., Saggar, R., Pandey, N.: An optimized trusted-cluster–based routing in disruption-tolerant network using experiential learning model. Int. J. Commun. Syst. **33**(1), e4196 (2020)

16. Juyal, V., Pandey, N., Saggar, R.: A heuristic light weight security algorithm for resource constrained DTN routing. In: 2016 IEEE International Conference on Computational Intelligence and Computing Research (ICCIC), pp. 1–4. IEEE, December 2016

17. Raghuwanshi, S., Singh, M., Rath, S., Mishra, S.: Prominent cancer risk detection using ensemble learning. In: Mallick, P.K., Bhoi, A.K., Barsocchi, P., de Albuquerque, V.H.C. (eds.) Cognitive Informatics and Soft Computing. LNNS, vol. 375, pp. 677–689. Springer, Singapore (2022). https://doi.org/10.1007/978-981-16-8763-1_56

18. Mukherjee, D., Raj, I., Mishra, S.: Song recommendation using mood detection with Xception model. In: Mallick, P.K., Bhoi, A.K., Barsocchi, P., de Albuquerque, V.H.C. (eds.) Cognitive Informatics and Soft Computing. LNNS, vol. 375, pp. 491–501. Springer, Singapore (2022). https://doi.org/10.1007/978-981-16-8763-1_40

19. Sinha, K., Miranda, A.O., Mishra, S.: Real-time sign language translator. In: Mallick, P.K., Bhoi, A.K., Barsocchi, P., de Albuquerque, V.H.C. (eds.) Cognitive Informatics and Soft Computing. LNNS, vol. 375, pp. 477–489. Springer, Singapore (2022). https://doi.org/10.1007/978-981-16-8763-1_39

20. Mishra, Y., Mishra, S., Mallick, P.K.: A regression approach towards climate forecasting analysis in India. In: Mallick, P.K., Bhoi, A.K., Barsocchi, P., de Albuquerque, V.H.C. (eds.) Cognitive Informatics and Soft Computing. LNNS, vol. 375, pp. 457–465. Springer, Singapore (2022). https://doi.org/10.1007/978-981-16-8763-1_37

21. Patnaik, M., Mishra, S.: Indoor positioning system assisted big data analytics in smart healthcare. In: Mishra, S., González-Briones, A., Bhoi, A.K., Mallick, P.K., Corchado, J.M. (eds.) Connected e-Health. Studies in Computational Intelligence, vol. 1021, pp. 393 415. Springer, Cham (2022). https://doi.org/10.1007/978-3-030-97929-4_18

22. Periwal, S., Swain, T., Mishra, S.: Integrated machine learning models for enhanced security of healthcare data. In: Mishra, S., Tripathy, H.K., Mallick, P., Shaalan, K. (eds.) Augmented Intelligence in Healthcare: A Pragmatic and Integrated Analysis. Studies in Computational Intelligence, vol. 1024, pp. 355–369. Springer, Singapore (2022). https://doi.org/10.1007/978-981-19-1076-0_18

23. De, A., Mishra, S.: Augmented intelligence in mental health care: Sentiment analysis and emotion detection with health care perspective. Augment. Intell. Healthc. Pragmat. Integr. Anal. 205–235 (2022)
24. Dutta, P., Mishra, S.: A comprehensive review analysis of Alzheimer's disorder using machine learning approach. Augment. Intell. Healthc. Pragmat. Integr. Anal. 63–76 (2022)
25. http://archive.ics.uci.edu/ml/datasets/Student+Performance

Task Scheduling Based Optimized Based Algorithm for Minimization of Energy Consumption in Cloud Computing Environment

M. Sri Raghavendra[1]($\boxtimes$), S. Sai Sahithi Reddy[2], P. Nikhitha[2], P. Sai Priya[2], and N. Madhura Swapna[2]

[1] Department of CSE, G. Pullaiah College of Engineering and Technology, Kurnool, India
sr.meeniga@gmail.com
[2] G. Pullaiah College of Engineering and Technology, Kurnool, India

Abstract. Allocating virtual machineries in the cloud optimally for workloads is difficult. In the cloud, finding the best way to schedule tasks is an NP-hard issue due to the huge sizes of the tasks involved. The optimal strategy includes allocating work to a data centre full of virtual machines in such a way as to minimise energy consumption, make time, and cost. For that purpose, this paper introduces a hybrid optimisation approach for scheduling jobs. As the butterfly optimisation algorithm (BOA) only takes into account the scent perception criteria, it easily becomes stuck at a local maximum. The suggested butterfly optimisation algorithm (HFBOA) is more in accordance with the real foraging behaviours of butterflies in countryside since it incorporates an additional operator, namely, a colour perception rule, as opposed to the original BOA. In addition, the HFBOA uses logistic mapping to implement an up-to-date technique for controlling parameters, hence improving global optimisation. The simulation findings show that the cloud data canter's energy consumption, maketime, and cost may all be reduced by improving the scheduling of tasks. In conclusion, the article suggests that using metaheuristic algorithms to agenda tasks on the cloud is a viable option. The simulation answers show that the suggested technique performs very well when used to the solution of difficult, real-world engineering constraints.

Keywords: Optimal allocation · Cloud computing · Virtual machines · Butterfly optimization algorithm · Energy usage · Foraging characteristics

1 Introduction

Many businesses now have the option of migrating, computing, and hosting their apps on the cloud, thanks to the accessibility and convenience of the cloud computing paradigm [1]. In addition, each of these offerings may be modified to better suit the requirements of the client. Early adoption of big data was hampered by the inability of standard hardware to handle the influx of diverse workloads from many sources [2, 3]. As a result, many IT firms are looking to make the transition to a cloud environment in order to better support their customers' increasingly complex and varied workloads. Because

S. Kadry and R. Prasath (Eds.): MIKE 2023, LNAI 13924, pp. 409–423, 2023.
https://doi.org/10.1007/978-3-031-44084-7_38

of its many benefits, including those listed above [4], the cloud computing approach is becoming more popular. With these benefits in mind, businesses are eager to move their on-premises infrastructures into the cloud, where they can take use of virtualized designs in data centres to provide more customer freedom and reliability. Every user in every industry is increasingly likely to use cloud computing applications due to their convenience, accessibility, and .the fact that cloud paradigm users are geographically dispersed. As a result, we need an efficient scheduler to delivery cloud users with virtual resources to deal with the influx of requests we receive from multiple users at once [6]. A real and lively task scheduler is required [7] to grip these types of needs and to provision capitals to numerous cloud users in a situation where users are making requests for services of the cloud environment simultaneously from customers using a wide variety of heterogeneous and diversified resources. The cloud paradigm necessitates a task scheduler that is both efficient and adaptable. And it has to function according to the load being input into the cloud management interface. The loss of confidence in the cloud service provider and the resulting decline in business might be attributed to an inefficient scheduling mechanism in the cloud paradigm [8, 9].

Consequently, under the cloud paradigm, an efficient scheduler should be used, one that dynamically schedules jobs in response to workloads uploaded to the cloud control panel. Consequently, both the cloud provider and the cloud customer may profit from using an efficient scheduler in the cloud paradigm. In this research, we outlined a scheduling method, in which trust is calculated using service level agreement (SLA) metrics including availability, success rate, and turnaround efficiency. These variables have an indirect impact on makespan and energy usage [11, 12], as well as the quality of service. Several metaheuristic and nature-inspired procedures, such as PSO, GA, ACO, and many more, have been used by previous writers to develop task schedulers [13]. No authors have yet into account, despite the fact that many authors have addressed parameters with names like makespan, SLA violation, and energy consumption [14]. In this manuscript, we minimised improved availability, success rate to boost trust value. This was accomplished by carefully evaluating priorities for both errands and VMs whale optimisation algorithm. It was discovered via this study that may be reduced by optimising the trust parameters, and evil versa.

In order to enhance the optimisation accuracy of the unique BOA, a new hybrid-flash butterfly optimisation method (HFBOA) is presented for tackling the restricted manufacturing glitches. The BOA's subpar optimum accuracy results from its exclusive focus on the scent perception rule in forage and mating. Studies in the relevant eco-logical context have shown the importance of butterfly vision in the foraging process (collecting pollen). The HFBOA is more realistic in its representation of butterfly forag-ing behaviour because it incorporates both olfactory and visual cues for global and local search. Moreover, logistic mapping is employed to update the control parameters, which improves the HFBOA's capacity to do global optimisation. The no free lunch (NFL) hypothesis suggests using a novel optimisation approach when facing a particular class of problems.

Here is how the remaining paper is formatted: The relevant literature is reviewed in Part 2, and the suggested model is briefly described in Sect. 3. In Sect. 4, we provide the experimental analysis and its verification. Section 5 delivers a final summary and analysis.

2 Related Works

A Multi impartial trust aware scheduler was developed by Mangalampalli et al. [15], which prioritizes jobs and virtual machines (VMs) and allocates them to the most suitable virtual resources, all while reducing make span and energy consumption. Our scheduler is modeled after the whale optimization method. Cloudsim was used throughout the whole of the simulation. This simulation makes use of both synthetic workload based on previous work logs and real-time work logs obtained from HPC2N and NASA. The suggested technique was compared to the state-of-the-art metaheuristic methods currently in use, such as ACO, GA, and PSO. Makespan, energy use, overall runtime, and turnaround efficiency) have all been shown to increase significantly in the simulation findings.

A new method for secure and efficient task scheduling in the cloud is projected in [16] by Badri et al. To maximize throughput and reduce makespan, a new CNN optimized modified butterfly optimization (CNN-MBO) approach is suggested for preparation the jobs. Second, a safe RSA method variant is used to encrypt the data before it is sent. Lastly, the suggested method is evaluated by running simulations in a cloudlet simulator and analyzing the resulting data. In addition, other task scheduling-based strategies are examined and compared to the suggested method in terms of numerous presentation measures, including resource usage, reaction time, and energy consumption. The testing findings showed that for a job with a size of 100, the suggested method obtained minimums in energy 180 kWh, reaction time of the 20 s, utilization of 98%.

To properly assign tasks to suitable VMs, Mangalampalli et al. [17] offer an effective task scheduling technique that takes both task and VM priority into account. The concept of firefly optimization is used to represent this scheduling technique. For this method's workload analysis, we analyzed both synthetic datasets with varying deliveries and the real-time work logs of NASA. We used the Cloudsim simulation environment to test this method, and we compare it to the ACO, PSO, and GA as well as our own suggested technique. Our suggested method significantly improved upon the baseline methods in terms of reducing the makespan, availability, efficiency, as evidenced by the simulation results.

In this study, we present the electric earthworm optimization algorithm (EEOA) for IoT requests in a cloud-fog architecture, which is a multi-objective job scheduling approach inspired by nature. This technique was developed by fusing the earthworm optimization algorithm (EOA) with the (EFO) to increase the utility of EFO in the search for the optimal solution. The effectiveness of the projected scheduling method was evaluated using large-scale examples of real-world workloads like CEA-CURIE and HPC2N to measure execution time, cost, makespan, and energy consumption. The simulation findings show that compared to the state-of-the-art methods, our suggested method significantly increases performance in all of the benchmarked cases we tested it on. Extensive simulations show that the proposed method yields improved scheduling outcomes compared to the state-of-the-art scheduling methods.

In order to explain and solve the scheduling issue, Chandrasekhar et al. [19] develop and implement the Hybrid Weighted method, an enhanced version of the previously extant Ant Colony Optimization Algorithm. In terms of cost, this method is evaluated and compared to others in the literature. It is evident that the proposed HWACO is superior to more conventional algorithms.

A two-stage metaheuristic algorithm, CSSA-DE, was suggested by Khaleel [20]. To begin, we use a clustering technique to organize all of the computers into functional groups. The leader of the mega cluster is chosen based on the node that achieves the greatest Performance-to-Power Ratio (PPR) during training with varying degrees of resource usage (MCH). Next, we combined the (DE) procedure with the sparrow search algorithm (SSA) to increase the already impressive search efficiency of pairing tasks with virtual machines. The number of underutilized and over utilized VMs may be used during integration to optimize resource allocation. CSSA-DE outperforms state-of-the-art procedures in a number of scenarios, and its performance is competitive overall.

3   Proposed System

As can be seen in Fig. 5, the suggested framework is broken down into four distinct parts. Allocating $T = T1, T2, T3,...,Tm$ tasks across $V = V1, V2, V3,...,Vn$ virtual machines in a cloud computing environment may be optimised for cost. The voltage range assumed for $V(i)$ in this study is [V Min i], [V Max i]. There might be as many as st k subtasks in each Task Ti. Next, we'll go through the specifics of the primary modules:

3.1   Energy Usage Computation Module

This section uses a DVFS system to calculate the energy consumption of m cloud-based virtual machines. The DVFS method calculates the total energy consumption of each resource while operating at its own unique voltage and frequency. Using Eq., we can determine how much power a virtual machine needs to do a certain subtask (1).

$$EU = a \times V^2 \times W_{freq} \tag{1}$$

where V is the operating voltage of the subtask Ti. is a constant multiplied by both the flip frequency and the load capacitance. May take on any value between zero and one.

The term "working frequency" or "Wfreq" is used to describe the operational frequency. Using Eq, one may estimate how much energy will be needed to do a certain job (2).

$$EU_{Task_i} = \sum_{i=1}^{k} a \times V_i^2 \times W_{freq_i} \times CT_i \tag{2}$$

where, gives completion period for Task Ti on a assumed virtual machine.

3.2 Task Estimation Module

The suggested framework entails creating a scheduling model for tasks using the same assessment elements used in multi-objective scheduling methods. Most multi objective scheduling techniques evaluate tasks based on their cost, energy consumption, and time to completion, waiting time, flow time, and dependability.

In this study, we employed the task estimate module to concentrate on the three most popular metrics: energy consumption, cost, and makespan. The energy required for each job is calculated by the task estimate module using Eq. (2). Makespan is the amount of time it takes to finish the very last thing. It is the primary yardstick by which multi-objective scheduling methods are judged. It allows you to do the job faster and within budget. When making a timetable for virtual machines, a smaller makespan value is indicative of a better plan. It is calculated as the longest time that any schedule allows a virtual machine to do all tasks. To keep things simple, we define makespan in this work as the amount of time it takes a virtual machine to finish its last job (expressed as an integer) (3).

$$MS = Min_{Schedule_j}(Max_{Task_i}) \tag{3}$$

The expense of the time and energy spent on a computer is what is meant by "cost." To calculate prices, we utilise Eq. (4).

$$Cost = \sum_{i=1}^{n}(ET_i \times RC_i) + \sum_{i=1}^{n}(CT_i \times TT_i) \tag{4}$$

where ET j is the time it took to complete activity TT i. The value of RC j indicates how much time and resources in computation cost. "CT" i and "TT" i stand for "task transfer time" and "task completion time," respectively, for task T i.

3.3 Task Scheduling Module

By using NSGA-II, this component facilitates the optimum scheduling of $T = T1; T2; T3;; Tm$ workloads across $V = V1; V2; V3;; Vn$ machines in a cloud data centre. The task scheduler creates a relationship between T and V that helps achieve the goal more effectively. In this paper, we optimise the energy consumption, makespan, and cost given by Eqs. (3) and (4).

This section is dedicated to scheduling work on virtual machines with the goal of minimising energy consumption, making the most efficient use of available resources, and keeping costs low. Estimates of energy consumption, makespan, and cost, coupled with requests for user-initiated tasks, are read by the module. Furthermore, the module uses a metaheuristic strategy to establish an optimal schedule of work on the provided virtual machines, which in turn optimises the evaluation criteria. The job scheduling issue may be formulated as an optimisation problem in mathematics, as stated in (5).

$$Best_{Schedule} = minimize(EU, MS, Cost) \tag{5}$$

Finding the optimal option among the feasible ones is the primary goal of optimisation. Hence, all conceivable solutions in the search space need to be compared using a set of evaluation criteria known as objective functions. It is possible for a cloud computing environment to have many goal functions, such as the reduction of energy consumption, the minimization of maketime, and the minimization of cost, all of which must be optimised concurrently.

This section presents the scheduling of tasks as a multi-objective optimisation problem with the goals of minimising total system energy consumption, total system makespan, and total system cost. This module uses optimisation techniques to the multi-objective task scheduling issue in order to determine the best course of action.

The following is a mathematical description of the task scheduling issue.

Features with Objectivity:

1. Minimize (energy usage) $= \sum EU_{ji}$, Where, EU_{ij} gives the energy usage for executing task T_i on virtual machine V_j.
2. Minimize(makespan) $= Min_{Schedule_j}\left(Max_{Task_i}\right)$
3. Minimize $(cost) = \sum_{i=1}^{n}(ET_i \times RC_i) + \sum_{i=1}^{n}(CT_i \times TT_i)$

Design constraints:

- Each task T must be allocated to a single virtual machine, and all tasks must be completed by their respective due dates.
- The sum total of a virtual machine's resource needs across all running jobs must not exceed the virtual machine's maximum capability.

Tasks and VMs are not interdependent in any way. Given that fitness functions are used to evaluate the quality of solutions, we suggested in this work to employ three goal functions for directly and concurrently decreasing energy consumption, makespan, and cost as fitness functions.

3.4　Task Allocating Module

The task-allocation module then uses the optimal values of the evolution issues to machines according to the determined non-inferior timetable. Domain experts may hand-pick tasks for the non-inferior task plan, or the schedule can be generated automatically depending on cloud providers' needs.

3.5　Model of the Basic BOA

The BOA [21], [is inspired by the butterfly's natural behaviours, such as its search for food and its courtship rituals. The standard BOA outlines three stages: the scent phase, the phase. The aroma is contributed by:

$$f_i = cI^a \tag{6}$$

where f i is the scent's perceived intensity, c is the sense involved, I is the stimulus's strength, and an is a power exponent depending on how much of the fragrance is taken in.In the prime BOA search phase, the sensual modality c is defined as:

$$c_{t+1} = c_t + \frac{0.025}{c_t}.T_{max} \tag{7}$$

parameter c starts at 0.01, and T max is the maximum sum of evolutionary repetitions. Parameter c's allowed range of values is given by Eq. (7) as (0, 1, 0.3).To toggle among global and local searches, a value of p between 0 and 1 is needed. The butterfly's flight pattern throughout its worldwide quest is described as

$$x_i^{t+1} = x_i^t + (r^2 \times gbest - x_i) \times f_i \tag{8}$$

where r is a random value between 0 and 1 and x it is vector xi during the t-iteration. Currently, the best solution across all stages is denoted by gbest. While doing a local search, we have:

$$x_i^{t+1} = x_i^t + \left(r^2 \times x_i^k - x_j^t\right) \times f_i \tag{9}$$

where x jt is the jth butterfly in the solution space and x ik is the kth butterfly. If x ik and x jt are from the same iteration, then the butterfly transforms into walk. If not, the solution will get more complex as a result of this sort of random motion.

3.6 Hybrid-Flash Butterfly Optimization Algorithmx

To achieve local optimisation in HFBOA, we drew inspiration from the firefly algorithm's (FA) [24–26] search techniques and used butterfly vision. Detailed explanations of HFBOA's inception, optimisation, global search, parameter configuration are given.

Initialization Phase:

The butterfly population's starting locations are determined by the chance function, with the following general formula for the first butterfly's position in the population:

$$X_{i,j} = X_{lb,j} + rand \times \left(X_{ub,j} - X_{lb,j}\right) \tag{10}$$

where i[1,2,3,,n],j[1,2,3,,Dim], and x (i,j) is the i-th solution for the j-th dimension. Upper and lower limits for the issue are denoted by x (ub,j) and x (lb,j), whereas rand is a unchanging random integer in [0, 1]. The population positions in swarm intelligence algorithms are often initialised using this method.

Optimization Phase:

Chiefly, F_i^t signifies the cologne of the i-th it can be intended as:

$$F_i^{t+1} = c.\left(F_i^t\right)^a \tag{11}$$

During the HFBOA's search phase, you may assign a random integer between 0 and 1 to the parameter c, where c represents the sensory modality. As parameter c is in the interval [0, 1], we employ the chaotic technique to revise its value via a one-dimensional disordered mapping, henceforth referred to as the logistic mapping. We consider the power exponent an of the suggested approach to be 0.1 since this is the value used in the original Proposal.

Global Search:

The parameter r is considered, and the value and is substituted in its place. So, the suggested methods exact model of the butterfly's global search motions may be expressed as follows:

$$X_i^{t+1} = X_i^t + \left(a^2 \times gbest - X_i\right) \times F_i^t \tag{12}$$

where X it stands for the i-th butterfly's solution vector X i at the t-th iteration and an is a random value in the range [0, 1]. (0, 1). At this level, gbest represents the optimal answer among all possible ones. Parameter a may be thought of as a scaling factor that modifies the gap between butterfly I and the optimal solution.

Local Search:

While looking for the best possible value, people should alternate between two stages of the HFBOA. During the HFBOA's local search phase, we include butterfly vision. Thus, the following describes the butterfly's search phase::

$$X_i^{t+1} = X_i^t + \beta \times \left(X_i^k - X_j^t \right) + a.\epsilon \tag{13}$$

where X jt and X ik are agents selected at random from the set of all possible agents. In addition, e is an arbitrary number in the range [-0.5,0.5]. an is an arbitrary number between 0 and 1.

One such expression for the attractiveness b is:

$$\beta = \beta_0.e^{-R_{ij}} \tag{14}$$

where the attraction at R = 0 is denoted by 0. Parameter b is typically initialised at 1, or _0 = 1. The distances among two points X i and X j, denoted by R ij, is determined using the 2-norm. The resulting formula of R_{ij} is:

$$R_{ij} = \|X_i - X_j\|_2 \tag{15}$$

Switch Parameter sp:

The typical global search is changed to an intense local search using the switch option sp. It produces a number between 0 and 1 at random on each iteration and uses it together with the switch probability sp to determine whether to do a global search or a local search. By default, the global search stage is skipped when sp is 0. Instead, setting sp = 1 merely activates the global search phase.

Chaotic Map and Parameter a:

In nonlinear systems, chaos tends to occur often. One kind of traditional one-dimensional mapping is the logistic map, which is defined as:

$$z_{n+1} = \mu.z_n.(1 - z_n) \tag{16}$$

where (0, 4]) represents the chaotic factor. The chaotic system is defined by its sensitivity to the beginning conditions. As a result, we examine the uniqueness of the logistic mapping, where chaotic may be shown in Fig. 2. The logistic mapping are (0, 1) and 0.6839 when = 4 and z(0) = 0.35, respectively.

We know that the iterative value will not generate a fixed point of the logistic mapping method if the starting value is beyond the range [0.25, 0.5, 0.75]. So, in the following tests, with m = 4, and c0 = 0.35, the parameter c of the projected HFBOA is updated using this chaotic mapping.

Each iteration, the rand strategy of (0, 1) is replaced with the chaotic approach of updating the parameter a. During optimisation, the parameter a starting value is 0.2 and the maximum number of iterations is 500. The initial value of an is 0.22 since the difference between parameters c and an is exactly 0.35.

4 Results and Discussion

Here, we detail the hardware and software infrastructure used to carry out the extensive range of tests required for this study. Both real and fabricated data sets for review are outlined. In the sections that follow, we will show and analyse experimental findings in terms of three distinct assessment criteria: energy consumption, makespan, and cost.

4.1 Experimental Setup

Evaluating scheduling algorithms and other world Cloud Computing context is a very difficult undertaking. Due to the pay-as-you-go nature of the actual cloud computing environment, it would be financially prohibitive to perform a series of consistent experiments there. Researchers in the area often choose to test the efficacy of their scheduling approaches in a simulated environment based on benchmark data sets. In this study, we used CloudSim, the industry standard in cloud simulation software (version 5.0).

The most used simulation programme for assessing optimisation strategies is CloudSim. Elements of the cloud system, such as data centres, tasks, and virtual machines, are simulated. In addition, it can simulate a wide range of workloads and supports a variety of scheduling algorithms and energy consumption models. To do this, we mimic a cloud service provider's infrastructure as a service (IaaS) using a single data centre. There are a total of eight physical hosts available in the proposed data centre, split evenly between two types. Each physical computer hosts three distinct virtual machines.

In Table 1 we can see how our mock cloud data centre is set up. In Table 2, we see Amazon EC2's suggested categorization of virtual machine settings. Physical host configurations are shown in Table 3. We ran tests using a Windows 10 64-bit PC with an Intel (R) Core (TM) i5–4210 CPU running at 1.70 GHz, 8GBs of Memory, and a 1TB hard disc.

Table 1. Shape of cloud simulated information centre.

Parameter	Value
Main memory	2 GB
HDD	1 TB
Bandwidth	10 gbps
Scheduler	Time Sharing
Virtual machine manager	Xen

4.2 Synthetic Data Set

In this series of tests, we employed a synthetic data set to compare and contrast the suggested method to the state-of-the-art alternatives. Pseudorandom generating numbers are used to create the synthetic data collection. 512, 1024, 2048, 128 and 256 randomly

Table 2. Setting of the fake virtual machines.

VM type	Type 1 (small)	Type 2 (medium)	Type 3 (large)
Pe	4	7	20
MIPS	500	1000	1500
Capacity	2000	7000	30000

Table 3. Setting of the fake physical hosts.

Physical host	H-I	H-II
Processor	Intel core 2 Extreme X6800	Intel core 2 Extreme 3960X
PE	2	6
MIPS	27079	177730

generated heterogeneous jobs were included in this data collection. We used a standard layout based on the HCSP data set, which included grid sizes of (51216), (102432), (204864), 1284096, and 2568192. Here, 16 VMs will handle 512 tasks, 1024 tasks, 2048 tasks, 128 VMs, and 256 VMs will handle 8192 tasks. Fully consistent, moderately consistent, and inconsistent clouds were all things we thought about. We also took into account high and low levels of task and cloud heterogeneity.

4.3 Performance of Proposed Scheme in Terms of Makespan

It is primarily aimed at reducing the amount of time needed, i.e. the whole duration of the work schedule, where all work has been completed or the time taken for it from the beginning to the end. The following equation demonstrates the makeup formula.

$$Makespan = \sum_{i=1}^{n} F_{t_i} \tag{17}$$

where, F is the finishing time of the task $t(n)$.

In this section, the performance of projected is compared with existing techniques in terms of makespan (time in seconds), which is presented in Table 4.

When the tasks are 2000, FFA has 917s, Whale has 921s, GWO has 836s, Butterfly has 816s and proposed model has 800s. From this analysis, it is clearly proves that the proposed model achieved better performance (Fig. 1), (Table 5).

After determining the makespan, we computed the scheduler's energy consumption, which is an essential step in the design process from the points of view of both the Cloud user and the service provider. The cloud service provider saves money on electricity costs, the cloud user saves money on cloud services, and the environment benefits all at the same time by reducing energy use. Hence, we determined Energy Consumption by distributing 1000–6000 jobs throughout 50 simulation runs. The accompanying data

Table 4. Makespan comparison with a different algorithm

Number of tasks	Makespan (time in seconds)				
	FFA	Whale	GWO	Butterfly	HFBOA
1000	434	428	421	415	395
2000	917	921	836	816	800
3000	1302	1293	1264	1251	1229
4000	1688	1686	1635	1619	1602
5000	1979	2000	1994	1891	1875

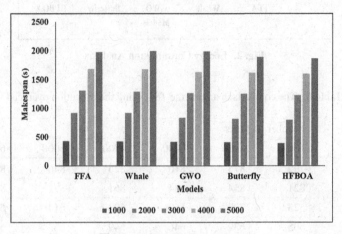

Fig. 1. Makespan Comparison

Table 5. The comparison among the TSA using the energy consumption

Algorithms	Number of Tasks					
	1000	2000	3000	4000	5000	6000
FFA	2453	2723	3473	4142	4663	5081
Whale	2273	2657	3845	4361	4534	4921
GWO	2265	2558	3518	3965	4307	4709
Butterfly	2159	2471	3350	3659	3878	4548
HFBOA	**2055**	**2206**	**3032**	**3435**	**3726**	**3976**

definitively demonstrates that the suggested model has the lowest energy consumption of all currently available models (Table 6).

There are 826 in FFA, 861 in Whale, 593 in GWO, 649 in butterfly, and 542 in HFBOA at node 4000. All of our simulations were conducted using synthetic data sets

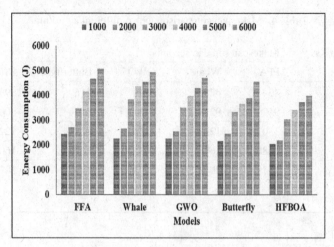

Fig. 2. Energy Consumption Analysis

Table 6. The comparison among the TSA using the execution overhead

Algorithms	Number of Tasks					
	1000	2000	3000	4000	5000	6000
FFA	792	875	810	826	846	884
Whale	821	854	844	861	871	892
GWO	523	708	567	593	614	627
Butterfly	605	639	646	649	663	689
HFBOA	**482**	**503**	**537**	**542**	**551**	**561**

and real-time activity records. Inferences about the extent to which our suggested model enhanced parameters and created schedules successfully are drawn from the data (Fig. 3).

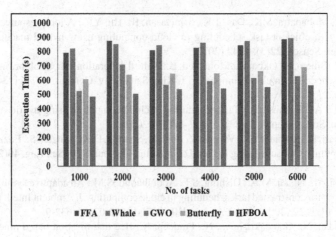

Fig. 3. Execution Overhead

5   Conclusion

This study introduces a novel metaheuristic framework, the hybrid-flash butterfly optimisation algorithm (HFBOA), for optimal task scheduling and dynamic allocation of virtual machines in the cloud. The HFBOA combines olfactory and visual signals to improve foraging efficiency on a global and local scale. Also, the HFBOA uses logistic mapping to include real-time updates of the control parameters for better global optimal performance. The HFBOA framework presents a set of energy-efficient, maketime- and cost-effective alternatives to the traditional task-scheduling methods. Experimental results that the HFBOA approach beats the other existing methods and cost using the HCSP as a benchmark dataset and a synthetic dataset. The GWO, butterfly, FFA, and Whale algorithms are some of the alternatives. Significant gains in terms of energy efficiency, maketime, and cost have been achieved in comparison to the benchmark data set. Going forward, the following tasks will be prioritised:

- Using Markov chain theory, we will demonstrate the proposed HFBOA's convergence and stable qualities.
- Although the HFBOA's central framework is somewhat complicated, we want to improve it further on the premise of guaranteeing the optimization's accuracy using Quantum theory.

References

1. Houssein, E.H., Gad, A.G., Wazery, Y.M., Suganthan, P.N.: Task scheduling in cloud computing based on meta-heuristics: review, taxonomy, open challenges, and future trends. Swarm Evol. Comput. **62**, 100841 (2021)
2. Bezdan, T., Zivkovic, M., Bacanin, N., Strumberger, I., Tuba, E., Tuba, M.: Multi-objective task scheduling in cloud computing environment by hybridized bat algorithm. J. Intell. Fuzzy Syst. **42**(1), 411–423 (2022)

3. Bal, P.K., Mohapatra, S.K., Das, T.K., Srinivasan, K., Hu, Y.C.: A joint resource allocation, security with efficient task scheduling in cloud computing using hybrid machine learning techniques. Sensors **22**(3), 1242 (2022)
4. Imene, L., Sihem, S., Okba, K., Mohamed, B.: A third generation genetic algorithm NSGAIII for task scheduling in cloud computing. J. King Saud Univ.-Comput. Inf. Sci. **34**(9), 7515–7529 (2022)
5. Khan, M.S.A., Santhosh, R.: Task scheduling in cloud computing using hybrid optimization algorithm. Soft. Comput. **26**(23), 13069–13079 (2022)
6. Rajakumari, K., Kumar, M.V., Verma, G., Balu, S., Sharma, D.K., Sengan, S.: Fuzzy based ant colony optimization scheduling in cloud computing. Comput. Syst. Sci. Eng. **40**(2), 581–592 (2022)
7. Abdullahi, M., Ngadi, M.A., Dishing, S.I., Abdulhamid, S.M.: An adaptive symbiotic organisms search for constrained task scheduling in cloud computing. J. Ambient Intell. Humanized Comput. **14**, 1–12 (2021). https://doi.org/10.1007/s12652-021-03632-9
8. Manikandan, N., Gobalakrishnan, N., Pradeep, K.: Bee optimization based random double adaptive whale optimization model for task scheduling in cloud computing environment. Comput. Commun. **187**, 35–44 (2022)
9. Zhang, A.N., Chu, S.C., Song, P.C., Wang, H., Pan, J.S.: Task scheduling in cloud computing environment using advanced phasmatodea population evolution algorithms. Electronics **11**(9), 1451 (2022)
10. Otair, M., Alhmoud, A., Jia, H., Altalhi, M., Hussein, A.M., Abualigah, L.: Optimized task scheduling in cloud computing using improved multi-verse optimizer. Clust. Comput. **25**(6), 4221–4232 (2022)
11. Pirozmand, P., Javadpour, A., Nazarian, H., Pinto, P., Mirkamali, S., Ja'fari, F.: GSAGA: a hybrid algorithm for task scheduling in cloud infrastructure. J. Supercomput. **78**(15), 17423–17449 (2022)
12. Gupta, S., et al.: Efficient prioritization and processor selection schemes for heft algorithm: a makespan optimizer for task scheduling in cloud environment. Electronics **11**(16), 2557 (2022)
13. Amer, D.A., Attiya, G., Zeidan, I., Nasr, A.A.: Elite learning Harris hawks optimizer for multi-objective task scheduling in cloud computing. J. Supercomput. **78**(2), 2793–2818 (2021). https://doi.org/10.1007/s11227-021-03977-0
14. Mangalampalli, S., Swain, S.K., Mangalampalli, V.K.: Multi objective task scheduling in cloud computing using cat swarm optimization algorithm. Arab. J. Sci. Eng. **47**(2), 1821–1830 (2022)
15. Mangalampalli, S., Karri, G.R., Kose, U.: Multi objective trust aware task scheduling algorithm in cloud computing using whale optimization. J. King Saud Univ.-Comput. Inf. Sci. **35**, 791–809 (2023)
16. Badri, S., et al.: An efficient and secure model using adaptive optimal deep learning for task scheduling in cloud computing. Electronics **12**(6), 1441 (2023)
17. Mangalampalli, S., Karri, G.R., Elngar, A.A.: An Efficient trust-aware task scheduling algorithm in cloud computing using firefly optimization. Sensors **23**(3), 1384 (2023)
18. Kumar, M.S., Karri, G.R.: EEOA: cost and energy efficient task scheduling in a cloud-fog framework. Sensors **23**(5), 2445 (2023)
19. Chandrashekar, C., Krishnadoss, P., Kedalu Poornachary, V., Ananthakrishnan, B., Rangasamy, K.: HWACOA scheduler: hybrid weighted ant colony optimization algorithm for task scheduling in cloud computing. Appl. Sci. **13**(6), 3433 (2023)
20. Khaleel, M.I.: Efficient job scheduling paradigm based on hybrid sparrow search algorithm and differential evolution optimization for heterogeneous cloud computing platforms. Internet Things, **22**, 100697 (2023)

21. Kumar, V.K.A., et al.: Dynamic wavelength scheduling by multiobjectives in OBS Networks. J. Math. **2022**, 10 (2022). Article ID 3806018, https://doi.org/10.1155/2022/3806018

22. Ramana, K., et al.: A vision transformer approach for traffic congestion prediction in urban areas. IEEE Trans. Intell. Transp. Syst. **24**(4), 3922–3934 (2023). https://doi.org/10.1109/TITS.2022.3233801

23. Arora, S., Singh, S.: Butterfly optimization algorithm: a novel approach for global optimization. Soft Comput. **23**, 715–734 (2019)

24. Ramana, k., et al.: Leaf disease classification in smart agriculture using deep neural network architecture and IoT. J. Circuits Syst. Comput. **31**(15), 2240004 (2022). https://doi.org/10.1142/S0218126622400047

25. Dwaram, J.R., Madapuri, R.K.: Crop yield forecasting by long short-term memory network with Adam optimizer and Huber loss function in Andhra Pradesh, India. Concurrency Comput. Pract. Exp. **34**(27), e7310 (2022)

26. Yang, X.S.: Nature-Inspired Metaheuristic Algorithms. Luniver Press, Bristol, UK (2010)

Author Index

S. Kadry and R. Prasath (Eds.): MIKE 2023, LNAI 13924, pp. 425–426, 2023.
https://doi.org/10.1007/978-3-031-44084-7

Printed in the United States
by Baker & Taylor Publisher Services

Printed in the United States
by Baker & Taylor Publisher Services